Advances in Solid State Physics
Volume 47

Advances in Solid State Physics

Advances in Solid State Physics is a book series with a history of about 50 years. It contains the invited lectures presented at the Spring Meetings of the "Arbeitskreis Festkörperphysik" of the "Deutsche Physikalische Gesellschaft", held in March of each year. The invited talks are intended to reflect the most recent achievements of researchers working in the field both in Germany and worldwide. Thus the volume of the series represents a continuous documentation of most recent developments in what can be considered as one of the most important and active fields of modern physics. Since the majority of invited talks are usually given by young researchers at the start of their career, the articles can also be considered as indicating important future developments.
The speakers of the invited lectures and of the symposia are asked to contribute to the yearly volumes with the written version of their lecture at the forthcoming Spring Meeting of the Deutsche Physikalische Gesellschaft by the Series Editor. Colored figures are available in the online version for some of the articles.
Advances in Solid State Physics is addressed to all scientists at universities and in industry who wish to obtain an overview and to keep informed on the latest developments in solid state physics. The language of publication is English.

Series Editor

Prof. Dr. Rolf Haug

Abteilung Nanostrukturen
Institut für Festkörperphysik
Universität Hannover
Appelstr. 2
30167 Hannover
Germany
haug@nano.uni-hannover.de

Rolf Haug (Ed.)

Advances in Solid State Physics

47

With 175 Figures and 2 Tables

Springer

Prof. Dr. Rolf Haug (Ed.)
Abteilung Nanostrukturen
Institut für Festkörperphysik
Universität Hannover
Appelstr. 2
30167 Hannover
Germany
haug@nano.uni-hannover.de

Physics and Astronomy Classification Scheme (PACS): 60.00; 70.00; 80.00

ISSN print edition: 1438-4329
ISSN electronic edition: 1617-5034
ISBN-13 978-3-642-09369-2 e-ISBN-13 978-3-540-74325-5

Springer is a part of Springer Science+Business Media

springeronline.com

Softcover reprint of the hardcover 1st edition 2008

Final processing: DA-TEX · Gerd Blumenstein · www.da-tex.de

Cover concept using a background picture by Dr. Ralf Stannarius, Faculty of Physics and Earth Sciences, Institute of Experimental Physics I, University of Leipzig, Germany
Cover design: WMXDesign GmbH, Heidelberg

Printed on acid-free paper 56/3180/YL 5 4 3 2 1 0

Preface

The 2007 Spring Meeting of the Arbeitskreis Festkörperphysik was held in Regensburg between March 26 and March 30, 2007 in conjunction with the 71th Annual Meeting of the Deutsche Physikalische Gesellschaft. The number of participants reached almost 5000 and there were 4120 scientific contributions. Comparing these numbers with the numbers of the meeting 2006 in Dresden shows that this Spring Meeting has a steady growth and seems to be attractive for a steadily rising number of scientists. This year's meeting was certainly one of the largest physics meetings in Europe.

The present volume 47 of the Advances in Solid State Physics contains the written version of a large number of the invited talks and gives a nice overview of the present status of solid state physics. Low-dimensional systems are surely dominating the field. Therefore, quite a number of articles is related to quantum dots and quantum wires. The importance of magnetic materials and the present day interest into magnetism is reflected by the large number of contributions to the part dealing with ferromagnetic films and particles. One of the most exciting achievements of the last couple of years is the successful electrical contacting and investigation of a single layer of graphene. This exciting physics is covered in Part IV of this book. Terahertz physics is another rapidly moving field which is presented here in five contributions. Achievements in solid state physics are only rarely possible without a thorough knowledge of material physics. Therefore, the last two parts of this book cover material aspects.

Hannover, April 2007 *Rolf J. Haug*

Contents

Part I Quantum Dots and Wires

Part VI Defects, Dislocations and Strain

Part VII Materials

Part I

Quantum Dots and Wires

Emission Characteristics, Photon Statistics and Coherence Properties of high-β Semiconductor Micropillar Lasers

S. M. Ulrich[1], S. Ates[1], P. Michler[1], C. Gies[2], J. Wiersig[2], F. Jahnke[2], S. Reitzenstein[3], C. Hofmann[3], A. Löffler[3], and A. Forchel[3]

[1] Institut für Halbleiteroptik und Funktionelle Grenzflächen, Universität Stuttgart,
Allmandring 3, 70569 Stuttgart, Germany
s.ulrich@physik.uni-stuttgart.de

[2] Institut für Theoretische Physik, Universität Bremen,
Otto-Hahn-Allee, 28359 Bremen, Germany

[3] Technische Physik, Universität Würzburg,
Am Hubland, 97074 Würzburg, Germany

Abstract. We report on complementary experiment-theory investigations regarding the photon emission statistics and coherence properties of quantum dot-based semiconductor micropillar lasers with high β factors, i.e., a large coupling of spontaneous emission into the lasing mode. In terms of power-dependent first- and second-order photon correlation measurements, our results consistently reveal a smooth transition between the regimes of spontaneous and mainly stimulated emission. The gradual onset of lasing is accompanied by strong photon number fluctuations and distinct changes of the field coherence length. In particular, the regime of stabilized coherent emission is found to establish at significantly increased excitation levels above the lasing onsets. As was verified by detailed semiconductor-theoretical calculations on the characteristics of these type of resonator devices, the smooth transition from thermal to coherent emission should indeed become increasingly harder to determine with $\beta \rightarrow 1$.

1 Introduction

Due to remarkable progress in semiconductor epitaxial growth and processing within the last decade, the realization of several novel types of tailored optical micro-resonator geometries [1–8] with a high photonic application potential (e.g., low-threshold microcavity lasers) has been achieved. As a common feature, those structures can provide very small mode volumes V_m in combination with high resonator quality factors Q. As number of eigenmodes in these resonators is strongly reduced, the β-factor of a mode, i.e., the fraction of spontaneous emission (SE) into that mode with respect to the total SE rate, can approach a value of 1. This has serious implications

R. Haug (Ed.): Advances in Solid State Physics,
Adv. in Solid State Phys. **47**, 3–15 (2008)

on the nature of light emission: With increasing β, the well-known step-like *threshold* as observable in the output of conventional $\beta \ll 1$ lasers gradually disappears up to the ultimate $\beta = 1$ regime of thresholdless lasing [9].

As was motivated on the basis of different semi-classical and quantum-mechanical theoretical approaches to the dynamics of lasing in those QED microresonators [10], an adequate interpretation of the transition from spontaneous into stimulated emission (lasing) and the nature of the emitted light should be given in terms of a *statistical analysis* of the photon emission process itself. In particular for high-β state-of-the-art devices, such analyses appear indispensable to gain fundamental insight into the complex dynamics around the onset of stimulated decay. This lasing transition should manifest itself as a significant change of the corresponding *photon field fluctuations*. From early investigations on micro-lasers with $\beta \approx 6 \cdot 10^{-3}$ [11], which were restricted to two-point counting statistics between coincident photons (i.e., zero delay $\tau = 0$) and temporally uncorrelated emission events ($\tau \gg 0$), indications for distinct changes in the field fluctuations from thermal to coherent behavior have indeed been reported.

In this article, we review recent results [12, 13] of comprehensive optical studies, focusing on the emission characteristics, photon statistics and coherence properties of *high*-β (0.04 and 0.12) semiconductor micropillar lasers at low temperature ($T = 4$ K).

Systematic studies were performed under the conditions of both continuous wave (cw) and pulsed optical excitation. Strong photon number fluctuations could be clearly identified from second-order $g^{(2)}(\tau)$ correlation measurements, where *photon bunching* was observed within a characteristically *smooth* transition into coherent light emission. Our experimental findings are supported by a microscopic semiconductor theory on coupled light-matter dynamics to describe the threshold behavior of QD micropillars under explicit consideration of semiconductor-typical effects different from atoms [14, 15].

In addition to our emission statistics analyses, first-order $g^{(1)}(\tau)$ correlation measurement series have been performed to directly trace the expected changes in field coherence within the lasing onset regime. In particular, the emission and coherence properties of micropillars with an elliptical cross-section have also been analyzed.

2 Sample Growth and Preparation

The micropillar structures under study were grown by the technique of molecular beam epitaxy (MBE). Prior to the main growth sequence, the (001)-GaAs substrate wafer was prepared by overgrowth of a 400 nm thick GaAs buffer layer.

As is schematically shown in Fig. 1, 27 and 23 distributed Bragg reflector (DBR) periods of alternating $\lambda/4$ thick AlAs (66 nm)/GaAs (76 nm) layer pairs define the bottom and top mirrors of a high-Q resonator structure. The

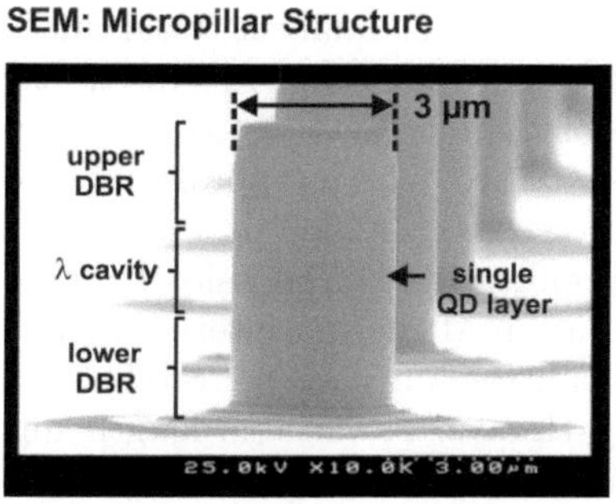

Fig. 1. Scanning electron microscopy (SEM) image of a 3 µm micropillar laser structure within an epitaxially grown array of identical resonators (figure by courtesy of Univ. of Würzburg). A single active layer of $In_{0.3}Ga_{0.7}As/GaAs$ QDs is embedded in a λ-cavity with 27 (23) top (bottom) DBR mirror periods

DBRs are separated by two 130 nm GaAs barriers, forming a 1λ-cavity centered around a single layer (4.5 nm) of self-assembled $In_{0.30}Ga_{0.70}As$ QDs as the active medium. The average lateral QD density in our sample structures was verified as $\sim 6 \times 10^9$ cm^{-2}. In a post-growth step of electron beam lithography in combination with Ar/Cl_2 plasma-induced reactive ion etching, high quality arrays of cylindrical micropillar structures with variable nominal diameters were processed. The scanning electron microscope (SEM) image of such a 3 μm pillar in Fig. 1 reveals the almost perfect structural shape of these pillars together with a smooth morphology of the cavity side walls. A detailed description of the growth and preparation process is given in [16]. All experimental data presented in the following has been obtained from two selected micropillars of 3 µm and 4 µm diameter.

3 Experimental Procedures

Optical investigations were performed on a confocal micro-photoluminescence (µ-PL) setup. Being mounted on the cold finger of a He-flow cryostat (T = 4 K), the sample was non-resonantly excited by either a continuous-wave (cw) Ti:Sapphire ring laser operated at 800 nm, or by a mode-locked Ti:Sa ($\lambda = 800 \pm 1$ nm) providing $\Delta t_{las} \approx 150$ fs (or 1.4 ps) wide pulses at $f_{exc} = 82\,(76)$ MHz. A 1 m double-monochromator equipped with a low-noise CCD camera (overall resolution of $\Delta E \approx 30\,\mu$eV) served to spectrally image and/or filter the emission signal.

In second-order $\tilde{g}^{(2)}(\tau)$ *Hanbury Brown & Twiss* (HBT) photon correlation measurements [17], the collimated and pre-filtered emission of an individual micropillar mode has been divided by a 50/50 non-polarizing beam splitter into two orthogonal optical pathways, each equipped with a sensitive avalanche photo diode (APD) for detection. The combined techniques of start-stop coincidence counting and multi-channel analysis yield measurement histograms which reflect the conditional probability of subsequent single-photon emission events as a function of their delay $\tau = t_{stop} - t_{start}$, thus giving direct insight into the emission statistics. The temporal response accuracy (IRF) of our HBT correlation setup is $\Delta t_{IRF} \approx 600$ ps.

First-order $g^{(1)}(\tau)$ field-correlation measurements on the emission field coherence have been performed by a high-resolution Michelson interferometer, consisting of a 50/50 non-polarizing beam splitter and two retro-reflectors mounted in a 90°-arrangement of optical pathways. A variable length difference between the interferometer arms can be introduced by a computer-controlled linear translation stage. In order to dynamically measure the interference fringes of a single micropillar mode, the spectrally filtered output signal of the interferometer is synchronously recorded by a single APD.

4 Experimental Results

4.1 Micropillar Emission Modes: Experiment-Theory Comparison

In order to initially identify the mode structure as well as to gain information on the quality factors of the cavity modes in view of following investigations, low-temperature (T = 4 K) μ-PL measurements have been performed on different diameter pillar structures. For these measurements, the inhomogeneously broadened emission of the QD ensemble serves as a cavity-internal light source. As is depicted in the topmost spectra of Fig. 2 for the selected 4 μm and 3 μm micropillars under study, one can observe a series of distinct narrow emission modes. The sequence of these eigenmodes has been verified by detailed theoretical calculations on the basis of an *extended transfer matrix method* [18]. For this, all relevant structural parameters (i.e., material refractive indices including dispersion, individual layer thicknesses, the pillar size, and its geometry) have been explicitly taken into account. As becomes evident from a direct comparison between our experimental spectra with the theoretical data shown in the center of Fig. 2, excellent conformity is found. In addition, the calculated transverse (cross-sectional) intensity patterns, characterized by their radial/angular quantum numbers (m, l) are depicted in the lower part of Fig. 2, where factors $1\times, 2\times, ...$ denote the corresponding mode degeneracies.

As an estimation of the cavity quality factors of these micropillars, values of $Q_{(4\,\mu m)} \approx E/\Delta E = 12300 \pm 500$ and $Q_{(3\,\mu m)} = 8600 \pm 300$ could be derived from the fundamental emission mode energies ($E_{(4\,\mu m)} \sim 1.3234$ eV; $E_{(3\,\mu m)} \sim 1.3346$ eV) and their linewidths ΔE under the conditions of low optical excitation power.

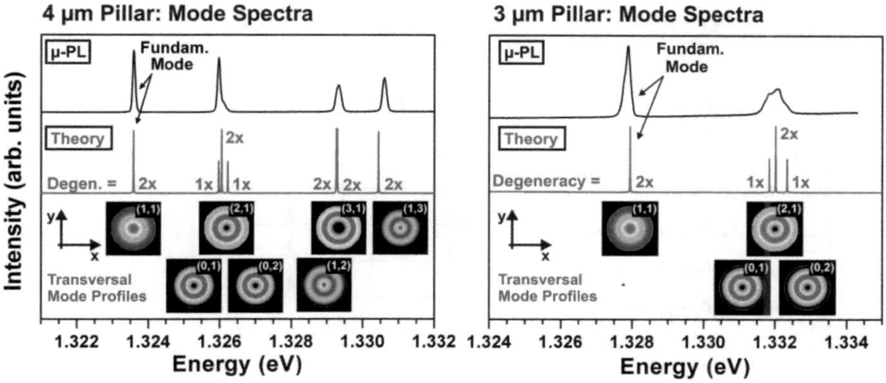

Fig. 2. Experimental μ-PL emission spectra (top trace) in direct comparison with theoretical mode calculations (bottom trace; 1×, 2× = mode degeneracy) for the 4 μm (left) and 3 μm (right) circular micropillar structures under study (marked by *arrows*). All measurements discussed in the current article were performed on the spectrally filtered fundamental modes of the micropillars. Inset figures: Calculated transverse emission profiles with (m, l) denoting the corresponding radial and angular quantum numbers of each mode, respectively

4.2 Emission Statistics and Field Coherence Properties of Micropillar Lasers

In order to study the emission characteristics of our micropillar structures with special focus on the lasing transition regime, power-dependent HBT autocorrelations were performed under continuous-wave and pulsed excitation. Representative results of these experiments on the spectrally filtered fundamental mode of the 3 μm and 4 μm pillar structures are shown in Fig. 3.

From the *pulsed* measurements on the 3 μm structure (Fig. 3b), the effect of photon bunching is clearly reflected by a $\tau \approx 0$ *center* correlation peak increase with respect to neighboring signals at $\tau = n \cdot \Delta t_{exc} = n/f_{exc}$ (which denote uncorrelated events with delays of multiple laser excitation cycles). This bunching is observable within a broadened excitation power range around the stimulated emission onset. As is evident from Fig. 3a, autocorrelation measurements under *cw* excitation have also revealed a strongly enhanced two-photon coincidence probability at short delays $|\tau| \leq \tau_c$ (τ_c: coherence time) within an intermediate power regime. For an adequate description of the *experimental* cw correlation traces $\tilde{g}^{(2)}(\tau)$ under explicit consideration of our temporal resolution $\Delta t_{IRF} \sim 600$ ps in HBT, a fit function defined as a convolution of the *idealized* form $g^{(2)}(\tau) = 1 + b_0 \cdot \exp(-2|\tau|/\tau_c)$ (b_0: bunching amplitude) with a Gaussian distribution of width $2\sigma = \Delta t_{IRF}$ has been derived:

$$\tilde{g}^{(2)}(\tau) = 1/\sqrt{2\pi\sigma^2} \int_{-\infty}^{\infty} d\tau' g^{(2)}(\tau - \tau') \exp\left(-\tau'^2/2\sigma^2\right) . \quad (1)$$

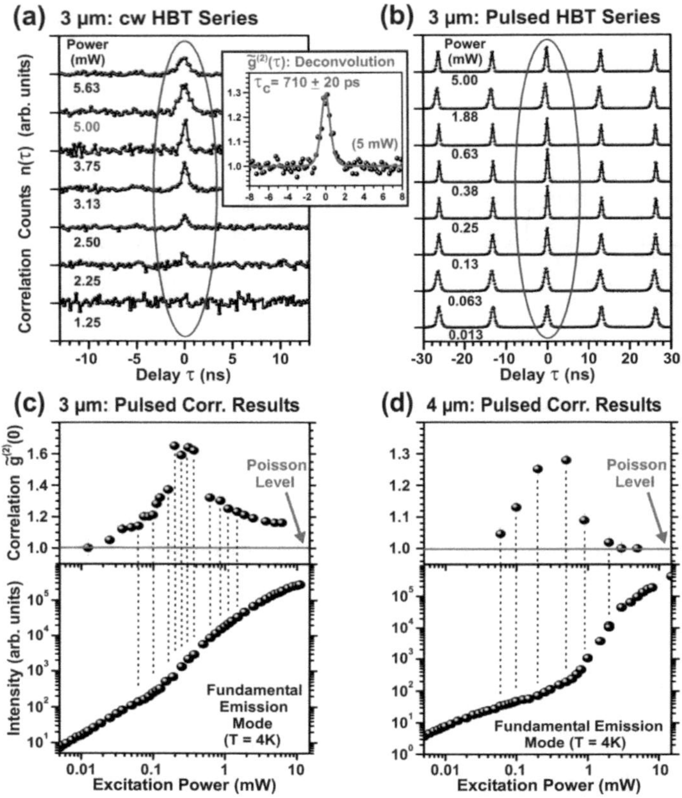

Fig. 3. (**a**) Power-dependent $\tilde{g}^{(2)}(\tau)$ correlations observed on the 3 μm pillar's fundamental mode under variable cw excitation, revealing the effect of bunching $\tilde{g}^{(2)}(0) > 1$ at intermediate powers. Inset: Coherence time fit of the 5 mW trace in (a) by convolution with $\Delta t_{IRF} \sim 600$ ps. (**b**) Correlation measurements of the same micropillar under pulsed optical pumping, also reflecting distinct bunching from the zero-delay correlation peak. (**c**) & (**d**) Power-dependent fundamental mode intensity traces (bottom) of the 3 and 4 μm pillars under pulsed excitation, in comparison with the normalized $g^{(2)}(\tau \approx 0)$ correlation values (top). A characteristically broad transition regime into lasing reflects in the smooth s-shape of the output intensity

Applying the above expression to the normalized cw $\tilde{g}^{(2)}(\tau)$ data (see the inset of Fig. 3a), values of $\tau_c = 710 \pm 20$ ps for $P_0 = 5.00$ mW and 480 ± 30 ps (4.375 mW; not shown) could be evaluated for the field coherence times, which are in excellent agreement with independent direct measurements of τ_c under variable power, as will be shown in the following.

For the 3 and 4 μm micropillars under pulsed excitation, the bottom parts of Figs. 3c & d display the integrated fundamental mode intensities in double-logarithmic scale. Both power series clearly reveal a *smooth* s-shaped intensity transition [19, 20] at intermediate excitation powers of ~ 200 μW (3 μm) and

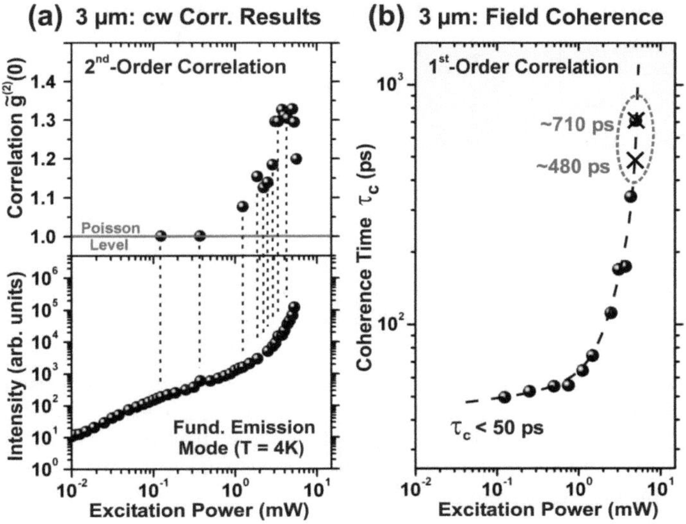

Fig. 4. (**a**) $\tilde{g}^{(2)}(0)$ results derived from power-dependent cw-HBT correlation series on the 3 μm pillar (see Fig. 3a), together with the fundamental mode's emission intensity. (**b**) Power-dependent field coherence time τ_c from first-order $g^{(1)}(\tau)$ correlation series (Michelson interferometry), which reveal a strong increase of τ_c around the lasing onset. Inset numbers (highlighted): Direct comparison between τ_c data (from $g^{(1)}(\tau)$) with corresponding de-convolution fit values derived from $\tilde{g}^{(2)}(\tau)$ (Fig. 3a), showing excellent agreement

$\sim 500\,\mu$W ($4\,\mu$m). In full accordance with the pulsed experiments, such an intensity trend was also found under *cw* pumping, as becomes evident from Fig. 4a (bottom). Due to experimental limits, the high-power branch of the s-shaped threshold region could not be fully traced under cw pumping.

In Figs. 3a & b and 4a (upper parts), indications of strong photon field fluctuations are consistently observed over the full onset regions (about ± 1 decade) of stimulated emission in all traces of $\tilde{g}^{(2)}(0)$. Well below the lasing onset, the measured $\tau = 0$ correlations remain close to 1 – in contrast to an expected *thermal* behaviour characterized by $g^{(2)}(0) > 1$ [10]. To clarify this point, we performed cw power-dependent $g^{(1)}(\tau)$ *first-order* field correlation measurements by Michelson interferometry on the same 3 μm pillar. As shown in Fig. 4b, the coherence time τ_c is found to strongly decrease from ~ 700 ps down to < 50 ps within the nonlinear transition onset regime. We therefore attribute the observed $\tilde{g}^{(2)}(0) \rightarrow 1$ decrease at low excitation powers to the temporal detection limits of the HBT setup (Δt_{IRF}), which no longer resolves the 'real' $g^{(2)}(\tau = 0)$ result but a *convolution* over extended time scales $\tau \gg \tau_c$ where $g^{(2)}(\tau > \tau_c) \approx 1$.

From a direct comparison between our first- and second-order correlation results especially at the regime of high cw power, indeed nice conformity

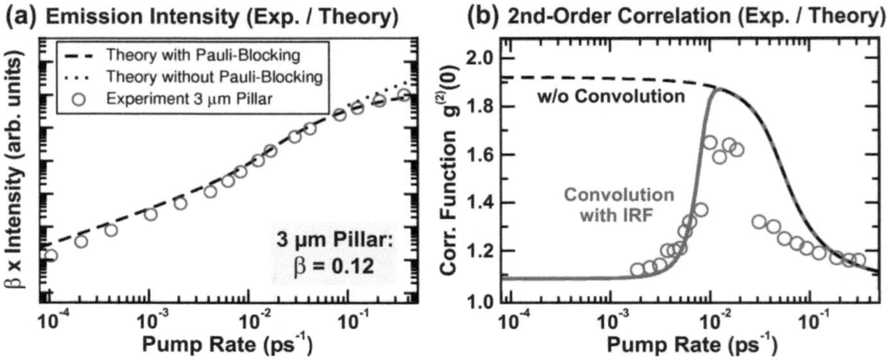

Fig. 5. (**a**) Calculated power-dependence of the fundamental mode in comparison with experimental data of the 3 μm pillar (see Fig. 3c), yielding a high $\beta = 0.12$ value for this device. (**b**) Corresponding calculated $g^{(2)}(\tau = 0)$ (*dashed line*) in dependence on pump rate. The solid line represents a convolution with the experimental IRF to reproduce the experimental $\tilde{g}^{(2)}(0)$ data

between the directly measured $g^{(1)}(\tau)$ values of τ_c and the coherence times extracted by de-convolution of $g^{(2)}(\tau)$ traces (see Fig. 3a, inset) is found.

Interesting to note from Figs. 3c and d, the Poisson level $\tilde{g}^{(2)}(0) = 1$ (which is to be expected for stabilized coherent emission) is only gradually approached in our experiments, where clear bunching behaviour is observable over extended power ranges up to levels of $> 10 \times$ (3 μ m) and $5 \times$ (4 μ m) increased excitation *above* the lasing transition onsets.

For an adequate interpretation of our experimental findings, extensive calculations have been performed in the framework of a novel semiconductor-theoretical model (for a detailed explanation, see [12] and references therein). Our model approach is based on the full semiconductor Hamiltonian for the interacting carrier-photon system, from which the dynamics of carrier population and photon number are derived by use of the Heisenberg picture of motion. From the resulting hierarchy of coupled dynamics, we determine the second-order photon-correlation function $g^{(2)}(\tau) = \left(\langle n^2\rangle - \langle n\rangle\right)/\langle n^2\rangle$ (with n: photon number operator) of the light field in the cavity, as well as the input-output characteristics of emission.

Figure 5a shows a comparison of these calculations for the pulsed measurements on the 3 μm pillar (Fig. 3), for which structural parameters of $\tau_{cav} = 13$ ps for the cavity photon lifetime ($\tau_{cav} \propto Q^{-1}$), a SE time constant into the laser mode of $\tau_{lase} = 250$ ps, and a total number of 42 QDs in resonance with the fundamental mode have been taken into account. Yielding a high SE coupling constant of $\beta \approx 0.12$, we find good agreement with the observed emission intensity trace. In addition, saturation effects become visible in the high excitation regime, indicative of a blocking within the rapidly filled pump levels. Similarly, a value of $\beta = 0.04$ has been verified to best reproduce the $4\,\mu$m pillar results (not shown).

Figure 5b depicts the calculated autocorrelation function (dashed line), which reveals the expected transition from thermal (spontaneous) to coherent (lasing) emission statistics. For a direct comparison with our $\tilde{g}^{(2)}(\tau)$ experiments (limited by Δt_{IRF}), we have adapted the measured coherence times τ_c to our results in Figs. 5a & b. Since the upper output intensity branch is not accessible for cw-pumping, this can only be done qualitatively. We determine $g^{(2)}(\tau)$ for each pump intensity from the calculated $g^{(2)}(0)$ via $g^{(2)}(\tau) = 1 + [g^{(2)}(0) - 1] \cdot \exp(-2|\tau|/\tau_c)$, and convolve with an apparatus function as discussed in the first section (1). The $\tau = 0$ value of this convolution is shown as a bold line in Fig. 5c, thus reproducing the *measured* $\tilde{g}^{(2)}(0)$ correlation 'peak' and its incomplete decay into full coherence, the latter being caused by saturation effects.

4.3 Coherence Properties of Elliptical Micropillar Lasers

As was already demonstrated in the preceding discussion, detailed studies of the first-order field coherence properties appear indispensable for a comprehensive description of the complex transition regime from spontaneous to mainly stimulated emission. In addition to our investigations on micropillars with a circular cross-section, also power-dependent $g^{(1)}(\tau)$ measurements for the case of elliptically-shaped resonators have recently been performed on micropillar structures with $\epsilon = 20\,\%$ nominal ellipticity [13].

From a theoretical perspective, even slight deviations of the resonator in-plane shape from an ideal rotational (circular) symmetry are expected to lift the initial eigenmode degeneracies. This would manifest in a characteristic spectral eigenmode splitting into non-degenerate doublets of orthogonal linear polarization along the corresponding principal symmetry axes [21–24].

As is demonstrated for the fundamental 1.3351 eV emission mode of a $4\,\mu$m structure under study (see the uppermost spectrum of Fig. 6b), polarization-*insensitive* μPL investigations have indeed revealed a distinct fine-structure splitting $\Delta E = 45 \pm 5\,\mu$eV. As shown in the lower μPL traces, each individual doublet component could be fully discriminated under the conditions of perpendicular linear polarization detection. A polar plot representation of both component's integrated intensity as a function of detection angle is shown in Fig. 6c, clearly verifying the expected orthogonality of the two polarization components. All further investigations of the current work have been performed on this spectrally selected fundamental (lowest energy) cavity mode, also designated in the inset spectra of Fig. 6a.

The mode's integrated output intensity as a function of optical excitation power is shown in a double-logarithmic representation in Fig. 6a: As for the circular micropillars discussed before, also here a characteristically broadened transition from spontaneous into non-linear stimulated emission (lasing) is traced around a comparable cw excitation power of $P \geq 1$ mW.

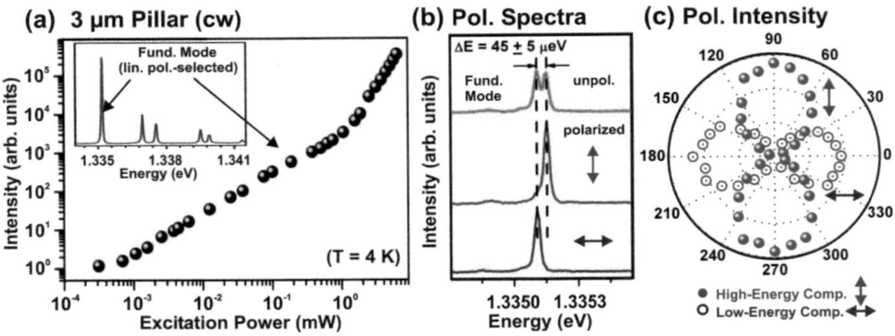

Fig. 6. (**a**) Input-output characteristics of the polarization-selected fundamental emission mode of an elliptical ($\epsilon = 20\,\%$) 4 μm pillar under non-resonant cw excitation ($T = 4$ K). Inset: Corresponding emission mode spectra at $P_0 = 5$ mW. (**b**) Polarization-dependent μPL spectra of the fundamental mode, resolving both non-degenerate components under orthogonal polarizations (marked by *arrows*). The observed mode splitting is $\Delta E \sim 45$ μeV. (**c**) Polar plot of the low and high energy mode components' intensities as a function of the polarization detection angle

With increasing excitation power, also a spectral narrowing of the fundamental mode is observed (not shown), which indicates that the cavity quality factor Q is limited by the QD absorption at low powers [21]. From the narrow fundamental mode emission of the 4 μm microcavity structure, a quality factor of $Q = \Delta E/E \approx 19000$ has been estimated on the basis of the respective low-power emission spectra around the transparency point.

For a detailed supplemental investigation of the smooth transition between spontaneous and stimulated emission, we addressed the fundamental mode emission coherence properties in terms of the *first-order* field correlation function [25] $g^{(1)}(\tau) = \langle E^*(t)E(t+\tau)\rangle / \langle E^2(t)\rangle$, which carries direct information about the coherence time τ_c. In the Michelson interferometer experiment, the photon field coherence properties can be evaluated from an analysis of the *visibility* (i.e., the contrast) of subsequent interference fringes, defined as $V(\tau) = (I_{max} - I_{min})/(I_{max} + I_{min}) = \left|g^{(1)}(\tau)\right|$. Therefore, the τ_c value for the lasing mode at a specific excitation power can be derived from the decay of its self-interference fringe contrast with increasing delay τ.

For the conditions of polarized mode detection of the 4 μm sample (see Fig. 6), the extracted visibility values as a function of delay time τ and in dependence on cw pump power variation ($P_0 = 0.01 \ldots 6.25$ mW) are composed in Fig. 7a. Beginning with the regime of lowest excitation at $P_0 = 0.01$ mW, the visibility profile of the lasing mode reveals a *Gaussian* shape $\sim exp\left[-(\pi/2)/(\tau/\tau_c)^2\right]$ due to the inhomogeneously broadened PL spectrum of QDs being coupled to the mode. In this regime of dominantly spontaneous emission within the pillar mode, a very short field coherence time of $\tau_c \approx 25$ ps, equivalent to a full width at half maximum of $\Delta\nu = (\sqrt{((2\ln 2)/\pi)}/\tau_c \sim 110$ μeV is derived from a fit to the visibility trace.

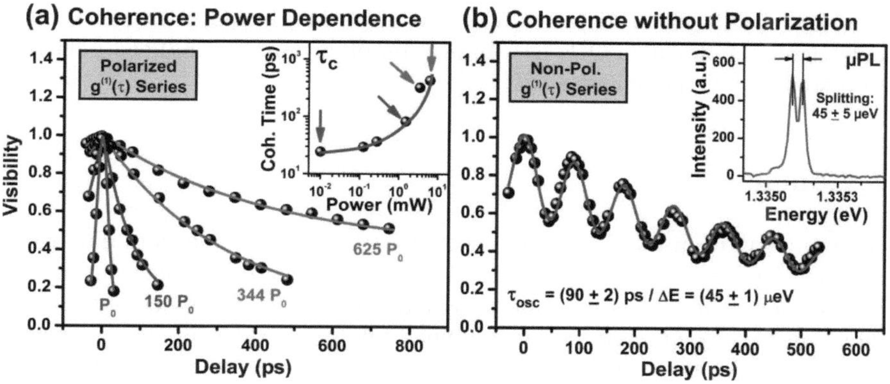

Fig. 7. (**a**) Power-dependent visibility curves of one polarization-selected lasing pillar mode component. With increasing pump powers ($P_0 = 0.01$ mW), the visibility profiles change from a Gaussian to a Lorentzian shape. Inset: Log-log representation of the extracted lasing mode τ_c power dependence. (**b**) Visibility curve for the same fundamental mode as in (a), without polarization selection ($P = 5$ mW). From a damped sinusoidal fit to the data (*solid line*), an oscillation period of 90 ± 2 ps, fully consistent with the mode fine structure $\Delta E \sim 45 \mu$eV (see inset figure) is observed

This value of $\Delta\nu$ reveals full consistence with the observed emission linewidth in μ-PL spectra at the same power. Under a systematic stepwise increase of excitation up to $P = 625 \times P_0$, a gradual change to an *exponential* shape $\sim exp[-|\tau|/\tau_c]$ is observable around and above the onset regime of stimulated emission. Moreover, a strong increase of τ_c up to ~ 430 ps was found in this power range, thus reflecting the expected change of the emission characteristics from thermal (spontaneous) to increasingly coherent light (lasing). The corresponding extracted τ_c values as a function of excitation power are depicted in the inset plot of Fig. 7a.

In addition to the former analyses, we also performed first-order interference measurements at a power of $P = 5\text{ mW} = 500 \times P_0$ on the same lasing mode *without* linear polarization selection. The result of the corresponding $g^{(1)}(\tau)$ series is depicted in Fig. 7b: In contrast to the monotonic $V(\tau)$ visibility decrease observable on a single polarization component (see above discussion of traces in Fig. 7a), damped sinusoidal *oscillations* are now found in the visibility curve. Indeed, this behavior is a direct consequence of the afore demonstrated emission mode splitting under lowered resonator symmetry. From the data, the oscillation period in visibility has been extracted as $\tau_{osc} \approx 90$ ps. Being equivalent to a fine-structure of $\Delta E \approx 45\,\mu$eV, this result shows excellent agreement with the corresponding high resolution μ-PL spectra depicted in the inset of Fig. 7b. From the fit, also a damping time constant of the curve is obtained as $\tau_c = 385 \pm 30$ ps. This value agrees well with the coherence time τ_c which was observed for the former *polarization-selected* fundamental mode component under identical excitation power conditions.

Worth noting, such beating behavior was also recently reported from $g^{(1)}(\tau)$ investigations on single QDs as a indication of asymmetry-induced polarization splitting in the emission [26].

5 Conclusion

In summary, we investigated the emission properties of quantum-dot based microcavity lasers with large spontaneous emission coupling factors. In this regime, the photon-statistical properties as well as the coherence of the light emission have been analyzed by combined photon-correlation and first-order coherence measurements. A smooth transition from spontaneous decay – with strong photon bunching and short field coherence – to stimulated emission with a Poissonian statistics and significantly increased coherence time is observed in detail. Our results are fully supported by semiconductor-theoretical calculations on the intensity and intensity fluctuations in such devices.

Acknowledgements

This work has been financially supported by the DFG research groups *Quantum Optics in Semiconductor Nanostructures* (FOR 485) and *Positioning of Single Nanostructures – Single Quantum Devices* (FOR 730).

References

[1] J.-M. Gérard: *Solid-State Cavity-Quantum Electrodynamics with Self-Assembled Quantum Dots*, in: *Single Quantum Dots - Fundamentals, Applications and New Concepts*, Topics in Applied Physics **90** (P. Michler (Ed.), Springer 2003)
[2] K. J. Vahala: Nature **424**, 839 (2003), and references therein
[3] M. Pelton et al.: Phys. Rev. Lett. **89**, 23, 233602 (2002)
[4] J. P. Reithmaier et al.: Nature **432**, 197 (2004)
[5] P. Michler et al.: Science **290**, 2282 (2000)
[6] E. Peter et al.: Phys. Rev. Lett. **95**, 067401 (2005)
[7] T. Yoshie et al.: Nature **432**, 200 (2004)
[8] A. Badolato et al.: Science **308**, 1158 (2005)
[9] F. De Martini and G. R. Jacobovitz: Phys. Rev. Lett. **60** (17), 1711 (1988)
[10] P. R. Rice and H. J. Carmichael: Phys. Rev. A **50** (5), 4318 (1994)
[11] R. Jin et al.: Phys. Rev. A **49** (5), 4038 (1994)
[12] S. M. Ulrich et al.: Phys. Rev. Lett. **98**, 043906 (2007)
[13] S. Ates et al.: Appl. Phys. Lett. **90**, 161111 (2007)
[14] M. Schwab et al.: Phys. Rev. B **74**, 045323 (2006)
[15] N. Baer et al.: Eur. Phys. J. B **50**, 411 (2006)
[16] A. Löffler et al.: Appl. Phys. Lett. **86**, 111105 (2005)
[17] R. Hanbury Brown and R. Q. Twiss: Nature **177**, 27 (1956)

[18] D. Burak and R. Binder: IEEE J. Quant. Electr. **33**, 1205 (1997)
[19] J. Hendrickson et al.: Phys. Rev. B **72**, 193303 (2005)
[20] S. Reitzenstein et al.: Appl. Phys. Lett. **89**, 051107 (2006)
[21] B. Gayral et al.: Appl. Phys. Lett. **72**, 1421 (1998)
[22] E. Moreau et al.: Appl. Phys. Lett. **79**, 2865 (2001)
[23] D. C. Unitt et al.: Phys. Rev. B. **72**, 033318 (2005)
[24] A. Daraei et al.: Appl. Phys. Lett. **88**, 051113 (2006)
[25] R. Loudon: *The Quantum Theory of Light*, (Oxford Univ. Press, N. Y., 2000)
[26] C. Santori et al.: New J. Phys. **6**, 89 (2004)

Optical Microtube Ring Cavities

Tobias Kipp

Institut für Angewandte Physik und Zentrum für Mikrostrukturforschung,
Universität Hamburg,
Jungiusstr. 11, 20355 Hamburg, Germany
tkipp@physnet.uni-hamburg.de

Abstract. By exploiting the self-rolling mechanism of strained layer systems we fabricate optical microtube ring cavities. In these structures either self-assembled InAs quantum dots or InGaAs quantum wells are embedded as optically active material. The optical properties of these microcavities are investigated by micro photoluminescence spectroscopy. We find spectra of sharp polarized cavity modes. The measured mode spacing is in very good agreement to theoretical calculations, modeling the microtube as a closed dielectric waveguide. We demonstrate confinement of light in direction of the tube axis induced by an axially varying geometry which is explained in an expanded waveguide model.

1 Introduction

Optical microtube ring cavities combine two fields in semiconductor physics that have been studied in recent years with increasing intensity: (i) the formation of three-dimensional objects out of two-dimensional epitaxial layers by use of strain relaxation and (ii) confinement of light in semiconductor structures with dimensions comparable to the wavelength of light.

Pseudomorphically strained semiconductor multilayer systems can roll-up to three-dimensional objects like lamellas or tubes if they are lifted off the substrate [1, 2]. The roll radius is determined by the amount of strain incorporated in the bilayer, which can be adjusted by the material composition and layer thickness [3, 4]. Standard semiconductor lithography processes can be used to precisely control the geometry of the objects, like – in the case of tubes – length, number of revolutions and position on the substrate [5–7]. Due to the possibility of using different material systems, controlling the shape of the objects and embedding electrically or optically active material inside the strained layer system, many possible applications of these objects have been proposed or already demonstrated. Concerning optical experiments and applications we want to exemplarily refer to [8–13].

The field of semiconductor microcavities was pushed, on the one hand, by concepts of new optoelectronic devices like low-threshold lasers or single-photon sources, and, on the other hand, by fundamental research on light-matter interaction. Three prominent examples for lithographically structured microcavities are microdisks [14], micropillars [15] and photonic-crystal microcavities [16]. For all of them lasing has been achieved and cavity quantum

R. Haug (Ed.): Advances in Solid State Physics,
Adv. in Solid State Phys. **47**, 17–28 (2008)

electrodynamic effects, like the Purcell effect or Rabi splitting, have been demonstrated [17–21].

In this paper we review some of our recent experiments on microtube ring cavities, which we fabricated exploiting the self-rolling mechanism of strained InGaAlAs layer systems. Compared to the microcavities mentioned above, the optical modes in these new resonators resemble mostly the modes in microdisks, however they exhibit some important differences due to the particular geometries of the resonators. The mode structure of our microtube ring resonators is probed by photoluminescence (PL) of an optically active material embedded in the microtube, i.e., either self-assembled InAs quantum dots (QDs) or InGaAs quantum wells (QWs). As an intrinsic property of a microtube ring resonator, the embedded optically active material is located close to the maximum optical field intensity. We observe sharp polarized modes in very good agreement to our theoretical models. Besides confinement of light in a ring on the circumference of a microtube, confinement along the tube axis is of importance for a complete three-dimensional resonator. We observe such a confinement and explain it in a simple theoretical model.

2 Sample Structure and Experiment

Starting point for the fabrication of our microtube cavities are molecular-beam epitaxy (MBE) grown samples (Fig. 1). The AlAs layer on top of a GaAs substrate serves as a sacrificial layer in the later processing. On top of this, the strained layer system is grown which will form the actual microtube. In the case of QD microtubes, this system consists of strain-containing 20 nm $In_{0.2}Ga_{0.8}As$ and 30 nm GaAs, centrally containing one layer of self-assembled InAs QDs. In the case of our QW microtubes, 14 nm $In_{0.15}Al_{0.21}Ga_{0.64}As$, 6 nm $In_{0.19}Ga_{0.81}As$, 41 nm $Al_{0.24}Ga_{0.76}As$, and 4 nm GaAs will roll-up in the later processing. Both In-containing layers are pseudomorphically strained

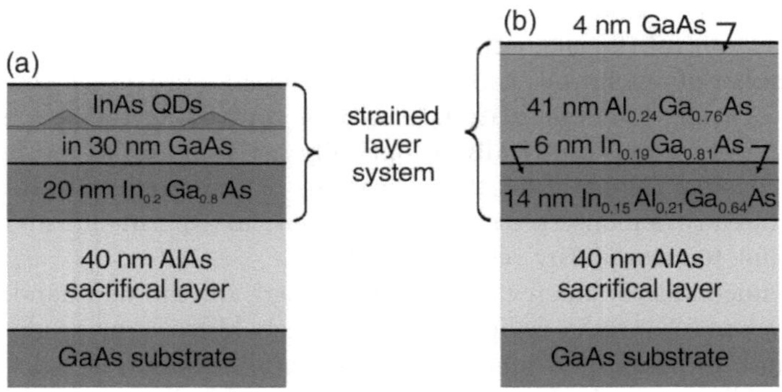

Fig. 1. Schematic sample structures for the preparation of (**a**) QD and (**b**) QW microtubes

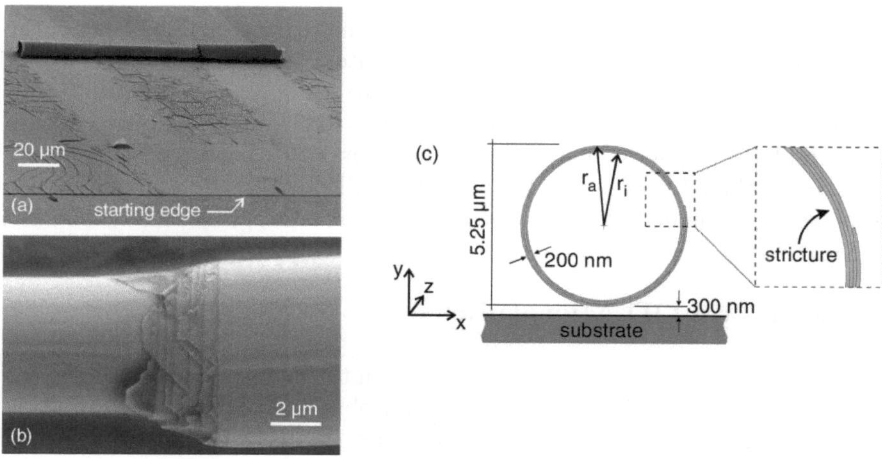

Fig. 2. (**a**) SEM picture of a microtube ring resonator. (**b**) Magnified topview on the microtube. (**c**) Scaled schematic cross section of the self-supporting part of the microtube

grown. The InGaAs layer forms a QW sandwiched between higher bandgap barriers.

Figure 2 shows scanning electron microscope (SEM) pictures of the QD microtube which is investigated in the following. The actual preparation process [6, 7] starts with the definition of a U-shaped strained mesa by etching into the strained InGaAs layer. In a next step, a starting edge [see Fig. 2a] is defined by etching through the AlAs layer. Here, the AlAs is now uncovered and, in the last step, the highly-selective HF solution starts to undercut the strained layer system. This leads to a bending of the strained mesa over its whole width resulting in the formation of a microtube. After a distinct distance defined by the U-shaped mesa (60 µm for the tube shown in Fig. 2), only the side pieces of the tube continue rolling [about 6 times in Fig. 2b]. This raises the center tube, leading to a self-supporting microtube "bridge", where in the middle part the tube is separated from the substrate (about 300 nm in Fig. 2). Figure 2c shows a scaled cross section of the center part of the tube. The outer diameter is about 5.25 µm whereas 3.8 revolutions lead to an overall tube wall thickness of only 200 nm. Since microtubes have the shape of rolled carpets, they exhibit discontinuities at the inside and outside surface. In the particular case of the tube in Fig. 2 these edges leads to a formation of a stricture in the tube wall with a thickness of only 150 nm.

The QW microtube which is dealt with later in the paper was similarly fabricated, but here, we further improved our preparation technique by etching deeply into the AlAs layer in the region between the legs of the U-shaped mesa. During the following selective etching step this region is protected by photoresist. This process leads to a larger and more controllable lifting of the center part of the microtube from the substrate.

Our microtubes were investigated by micro PL spectroscopy at low temperatures (T = 5–7 K). A He-Ne laser ($\lambda = 633$ nm) was focussed onto the sample by a microscope objective (×80 or ×50), having a spot diameter smaller than 1.5 μm. The PL light was collected by the same objective, dispersed by a monochromator and detected by a cooled CCD camera.

3 Results on QD Microtubes

Figure 3 shows PL spectra of the microtube in the energy range, in which the QDs emit. Figure 3a compares a spectrum obtained on the self-supporting part of the microtube to a spectrum obtained on its bearing. In the former case, we observe a regular sequence of sharp peaks superimposed on the broad QD luminescence. These sharp peaks are optical resonance arising from light which is guided around the tube axis inside the tube wall and which constructively interferes with itself. The spectrum taken on the microtube bearing upper curve in Fig. 3a shows much weaker and broader features with a smaller spacing. The larger outer radius of the microtube bearing leads to an increased cavity length of the resonator resulting in a smaller mode spacing. The broadness of the features can be explained by strong losses into the substrate, on which the bearing of the microtube lies. Figure 3b compares two spectra obtained on one and the same position on the center part of the microtube but for different polarization configurations. The upper curve corresponds to the transversal electric (TE) polarization which we define as having the electric field vector parallel to the tube axis. We prove the optical modes to be TE polarized. Their appearance also in the TM spectrum (which is much less pronounced than in the TE case) is due to a poor polarization selectivity of our first micro-PL setup.

We use two different theoretical models to explain the experimental results. In the first, so-called waveguide model, we regard the microtube wall as a dielectric waveguide and apply periodic boundary conditions. The height of the waveguide is given by the overall tube wall thickness $h = (r_a - r_i)$, see Fig. 2c. We calculate the modes of the planar waveguide and assign them to an effective refractive index n_{eff} [22]. To ensure phase matching of guided light after one round trip, we apply the periodic boundary condition $n_{\text{eff}} l = \lambda m$ (with the tube circumference $l = 2\pi(r_a - h/2)$, the vacuum wavelength of the propagating light λ and the azimuthal mode number $m \in N$). It is this model that prompts us to name modes with the electric field vector parallel to the tube axis TE polarized. We assume a radially-averaged energy-dependent refractive index of $n(E) = 3.46 + (E[\text{eV}] - 1.1)/2$ [23,24] for the tube wall. The positions of the lowest lying radial TE modes calculated within this model, using $h = 200$ nm and $2r_a = 5.25$ μm, are depicted as squares in Fig. 3b.

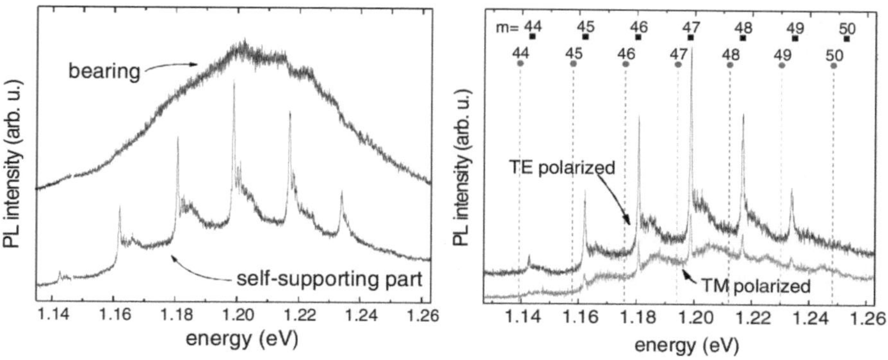

Fig. 3. (**a**) Micro-PL spectra of the self-supporting part (*lower curve*) and of a bearing (*upper curve*, *vertically shifted* for clarity) of the microtube bridge. (**b**) TE (*upper graph*, *vertically shifted*) and TM (*lower graph*) polarized spectra of the microtube bridge. The symbols indicate calculated mode positions (without any fitting) labeled with their azimuthal mode number m. The squares (circles) represent the waveguide (exact) approach

The exact positions of the calculated modes are afflicted with some uncertainty because they are very sensitive to the assumed radius. Therefore the mode spacing is the important quantity to compare to the experiment. This comparison exhibits striking accordance. In the second, so-called exact approach, we solve Maxwell's equations for a dielectric disk with a hole in its center [25]. The dots in Fig. 3b represent the results obtained from this solution. The deviation to our first approach is very small. Especially the mode spacings fit perfectly. This shows that the first approach, which is easier to calculate, delivers sufficient accurate results.

Besides the sharp resonances identified as the first TE modes we observe broader signals on the high energy side of every TE mode, sometimes exhibiting a fine structure. These signals are regular with the TE modes and therefore cannot be attributed to radially higher TE modes. Furthermore we do not observe any distinct modes in TM polarization. The absence of both TM modes and higher TE modes can be explained with their weaker confinement inside the tube wall and, consequentially, with their higher susceptibility to imperfections of the waveguide surface, especially to the stricture [10].

Figure 4a shows TE spectra obtained for different positions of the exciting laser on the microtube axis. Here, $z = 0$ is somewhere in the middle of the self-supporting tube. The distance between two adjacent positions on the sample is less than 1 µm. The PL intensity is encoded in a gray scale, where dark means high intensity. In the displayed energy range each spectrum exhibits two groups of peaks representing two optical modes with neighboring azimuthal mode numbers m. Over a distance of about 20 µm along the microtube, the optical mode positions shift less than 2 meV. A variation of the microtube radius of just 0.3% would lead to a larger shift. This impressively

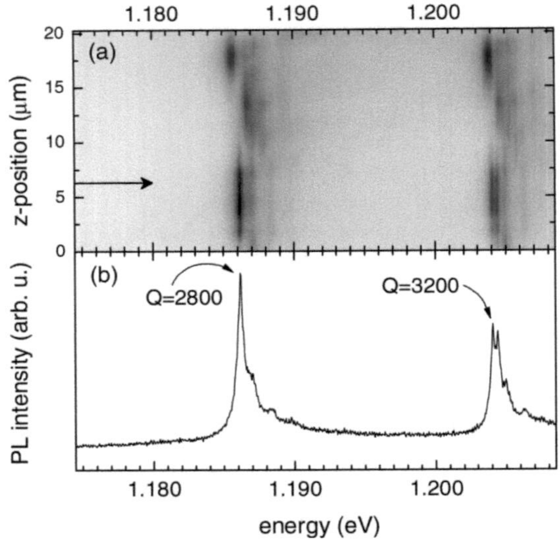

Fig. 4. (**a**) Gray scale plot of PL spectra measured at different positions on the microtube along its axis. Dark regions represent strong intensities. The spectrum marked with the *arrow* is depicted in (**b**)

demonstrates the homogeneity of the microtube and of its underlying self-rolling mechanism. Figure 4b shows the spectrum indicated with an arrow in Fig. 4a at about $z = 6$ µm. Here, the fine structure on the high-energy side of the modes is clearly visible. If we fit the peaks by multiple Lorentzians, we receive quality factors defined by $Q = E/\Delta E$ of 2800 and 3200 for the modes at 1.186 eV and 1.204 eV, respectively. We now want to address the signals at the high-energy side of the modes. In a perfect homogeneous and infinite long microtube, only light traveling perpendicular to the tube axis, i.e., having no wave vector component k_z, propagates in discrete modes. A non-zero k_z leads to spiral-shaped orbits with continuous mode energy. A finite length of the microtube would allow only discrete values of k_z leading to fully discretized modes. Therefore one might interpret the strong peaks in Fig. 4 as modes with $k_z = 0$, whereas signals on their high-energy side represent modes with finite k_z. Following this model, from the fine structure of the broad signal we can approximately determine the confining length L_z in z direction. For the spectrum depicted in Fig. 4b this leads to $L_z \approx 10$µm, which is much shorter than both, the length of the whole tube (120 µm) and the length of its self-supporting part (50 µm). Interestingly, L_z is comparable to the length over which the peak positions are nearly constant see Fig. 4a. This finding strongly suggests that light is confined also along the tube axis on a scale of about 10 µm. We will show later in this paper that this interpretation is consistent with our measurements on a QW microtube. Nevertheless, until now it is an open question, what mechanism exactly causes the confinement along z. SEM pictures do not resolve any inhomogeneities like decreasing radius or deformation of the tube.

4 Results on QW Microtubes

The SEM image in Fig. 5 depicts the specific microtube on which all measurement presented in this section have been performed. It is slightly more than twofold rolled in its self-supporting part, correspondingly, the wall thickness is mostly 130 nm except for the small region along the tube axis where three strained sheets sum up to a 195 nm thick wall (see *non-scaled* cross-sectional sketch of the microtube in Fig. 5). The self-supporting part has a diameter of about 6.4 μm and a distance to the substrate of about 1 μm.

Figure 5 shows (a) TE and (b) TM polarized spectra of the microtube. For these measurements we focussed the exciting laser spot centrally on the microtube but collected the light from a position slightly shifted in radial direction as indicated by the two vertical arrows in the cross-sectional sketch in Fig. 5. This improves the collection of resonant light with respect to light emitted into leaky modes at the position of the laser spot [13]. Similarly to the QD microtube case, we observe a regular sequence of TE polarized optical modes superimposed on a background, which is in this case the emission of the QW. No modes are observed in TM polarization (note that we used a setup with a much better polarization selectively than in the measurement on the QD microtube in Fig. 3).

The strong peaks at about 939 nm, 951 nm, and 963.5 nm clearly show shoulders on their high energy side. Their origin is not unambiguously clear. They might arise due to the lifting of the degeneracy of a perfectly symmetric cylindrical resonator by the inner and outer edges of the real structure (see sketch in Fig. 5). Another explanation for the double peak structure might be the existence of axially confined modes of slightly different energy which spatially overlap or at least can not be separated by our experimental setup. Such axially localized modes will be thoroughly addressed later in this paper. At about 977 nm a further mode with low intensity can be seen (marked by an arrow and magnified by a factor of 20 in Fig. 5a). By trying to approximate the spectrum by a superposition of Lorentzians for the resonant modes and their high-energy shoulders together with a homogeneous Gaussian curve at about 933 nm for the residual intensity from leaky modes, one can clearly identify a further, rather broad mode on the high energy side of the QW emission at about 929 nm. This peak can clearly be observed only in the described excitation and detection configuration. Collecting the PL light at the excitation position would strongly overlay this peak by leaky modes. It is broadened by a factor of about ten compared to the sharpest peaks on the low energy side of the QW emission because of reabsorption inside the QW.

Figure 6a shows a magnified SEM picture of a part of the investigated self-supporting microtube bridge. As previously mentioned, the tube has rolled-up slightly more than two times. For this particular microtube the small region where the wall consists of three rolled-up strained layers was orientated on top of the tube, just like sketched in Fig. 5. Using a rather high acceleration voltage of 20 kV at the SEM, not only the outside edge of the microtube wall, but also the inside edge can be resolved. Since the microtube wall in the region between the inside (left) and outside (right) edge consists of three rolled-up layers it appears slightly brighter in the SEM picture than the material besides this region. We find that applying our preparation technique to the layer system described above, both edges of the microtube tend to randomly fray over some microns instead of forming straight lines. The fraying occurs predominantly along the ⟨110⟩ direction of the crystal, whereas the rolling direction of the microtube is along ⟨100⟩. The spectrum in Fig. 5 proves that optical modes still can develop despite the frayed edges since it is actually obtained from exactly the microtube shown in Fig. 6a.

In order to investigate the influence of the frayed edges on the mode spectrum, we performed micro PL measurements scanning along the tube axis (z direction) which are shown in Fig. 6b. The graph's horizontal axis gives the wavelength of the detected light, whereas its vertical axis gives the axial position on the microtube which can directly be related to the SEM picture in Fig. 6a. The PL intensity is encoded in a gray scale where dark means high signal. Note that the gray scale is chosen to be logarithmic in order to better resolve in one and the same graph both, intense peaks on a

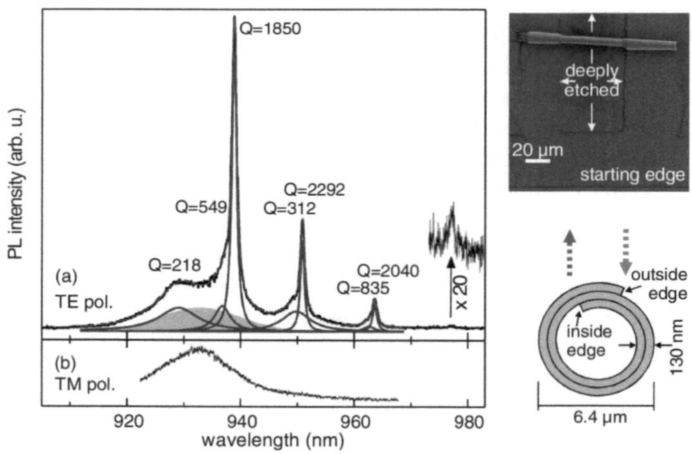

Fig. 5. Measured PL spectrum (black line) of the microtube (SEM picture on the top right). The spectrum is approximated by the sum (red line) of Lorentzian curves for the resonant modes *(blue) (online color)* and a Gaussian curve for the leaky modes (grey shaded). Q factors have been obtained by fitting. Additionally, a non-scaled cross-sectional sketch of the microtube together with the excitation and collection configuration is shown.

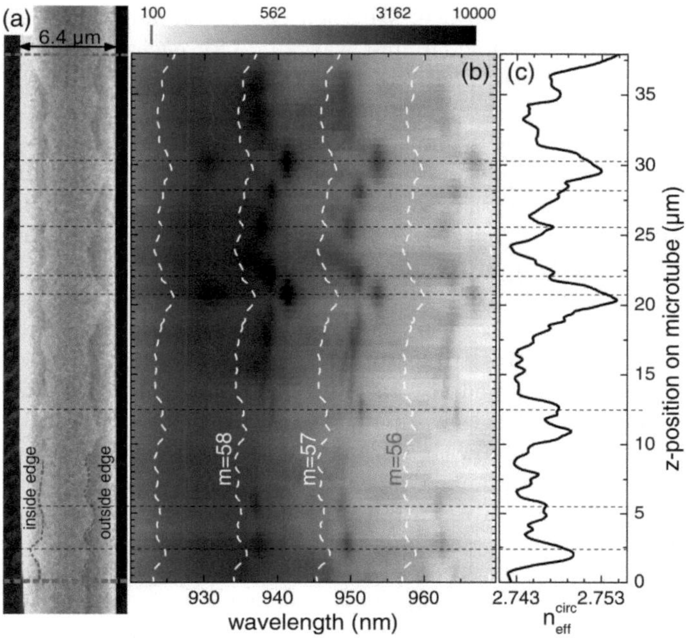

Fig. 6. (**a**) SEM picture of a part of the microtube. Inside and outside edges are marked with *broken lines* for clarification. The tube wall is thicker between these edges see sketch in Fig. 5. (**b**) Scanning micro PL spectra show three dimensionally confined optical modes. The *white broken* lines represent calculated mode energies for $m = 59$ to 56. (**c**) Calculated effective refractive index vs. position. *Horizontal broken lines* mark positions of pronounced localized modes.

comparatively intense background and small peaks on a weak background. The microtube was scanned in 80 steps over a length of 35 µm (between the two broken horizontal lines in Fig. 6a.

For nearly every position on the microtube we observe three or four optical modes. Their energies are shifting along the z direction, but interestingly this shifting is not continuous: The mode energies sometimes seem to be spatially pinned. First, we want to explain the shifting by an expanded model of a closed waveguide. For each position in z direction we regard a cross section perpendicular the tube axis as a circular waveguide, having a diameter of $d = 6.4$ µm. The waveguide thickness abruptly changes at the edges from 130 nm to 195 nm. For each region we translate the thickness in a related effective refractive index n_{eff}^{130nm} and n_{eff}^{195nm} respectively. For this calculation we assumed the average refractive index of the layer system to be $n(E) = 0.3225 \times E[\text{eV}] + 2.98232$ by linearly approximating values given in [24] and [23]. From the micrograph in Fig. 6a we determine the distance $L^{195nm}(z)$ from the inside to the outside edge on the tube surface by taking into account that the picture is a projection of a curved surface. We then can define an overall effective refractive index for the whole circular waveguide for each z position: $n_{\text{eff}}^{\text{circ}}(z) = n_{\text{eff}}^{195nm} L^{195nm}(z)/(\pi d) + n_{\text{eff}}^{130nm}[1 - L^{195\text{nm}}(z)/(\pi d)]$. The periodic boundary condition for a resonant mode then reads $n_{\text{eff}}^{\text{circ}}(z)\pi d = m\lambda$, with the vacuum wavelength λ and the azimuthal mode number $m \in N$. The resonances calculated with this model have azimuthal mode numbers around $m = 57$ and are depicted in Fig. 6b as bright broken lines. As a striking result, the calculation reveals that the overall z dependency of the measured resonant wavelengths is nicely approximated by our model.

In the following we want to concentrate on the spatially pinned resonances. These resonances can be seen in Fig. 6b as isolated dark spots with no significant wavelength shift over 1–2 μm. This implies that these resonances are localized modes, confined also in z direction along the tube axis. Comparing Figs. 6a and b it seems that some of these localized modes can be attributed to positions on the tube with a broad distance between inside and outside edge. To work out this point in more detail, Fig. 6c shows $n_{\text{eff}}^{\text{circ}}(z)$ calculated for $\lambda = 940$ nm, which essentially reflects the distance between the edges in dependency of the z position. Comparing now (b) and (c), it becomes obvious that optical modes are predominantly localized in regions of local maxima of $n_{\text{eff}}^{\text{circ}}(z)$ representing local maxima of $L^{195nm}(z)$. To better visualize this behavior, all pronounced localized modes are indicated by horizontal lines in Fig. 6b. Confinement of light along the z direction is achieved by a variation of $n_{\text{eff}}^{\text{circ}}(z)$. This important result can be qualitatively explained within the waveguide model. Until now we implied in our model light having no wave vector component along the tube axis, because, for a perfect infinite microtube, this light would just run away along the tube axis. If we now regard light with a finite but small wave vector component along the axis, this light can experience total internal reflection also in z direction, which leads to a three dimensional confinement. Our model using an averaged effective refractive index for a circular cross-sectional area of a microtube explains our measured data astonishingly well, despite this model neglects the abrupt changes of the inside and outside radius. The experimental data shows that it is possible to confine light along the axis of a microtube ring resonator by a slight change of the wall geometry along the axis.

5 Conclusions

In summary, we fabricated optical microtube ring resonators by exploiting the self-rolling mechanism of strained layer systems. As an optically active material, we embedded either QDs or QWs inside the tube wall. Polarized optical modes are observed, with modes spacings in very good agreement to our theoretical modelings. In our QW microtube, confinement of these modes on a length of 1–2 μm along the tube axis is provoked by spatial variations of the inside and outside edges along this axis. We also observe such a confinement in our QD microtube, but with a larger confining length of about 10 μm.

The Q factors of our structures are about 2000–3000. The Q factors of state-of-the-art semiconductor microcavities used in cavity quantum electrodynamic (CQED) experiments range from 9000 to 20000 [18–20]. Thus, a further improvement of optical quality will make the novel microtube ring resonator a good candidate for CQED experiments, too. Because of the high-selective lift-off process during the preparation, the surfaces of our microtubes should be epitaxially smooth, suppressing scattering on surface roughness. However, quantification of losses induced by the discontinuous inside and outside radius of microtubes will be subject of future investigations. As an intrinsic advantage of microtube resonators, the optically active material is located very close to the optical field intensity. This is important for both CQED experiments and for a future utilizing of microtubes in optoelectronic devices like lasers.

Acknowledgements

The author would like to thank Ch. Strelow, H. Rehberg, C. M. Schultz, H. Welsch, S. Mendach, Ch. Heyn, and D. Heitmann. Financial support of the Deutsche Forschungsgemeinschaft via the SFB 508 "Quantum Materials" and the Graduiertenkolleg 1286 "Functional Metal-Semiconductor Hybrid Systems" is gratefully acknowledged.

References

[1] V. Y. Prinz, V. A. Seleznev, A. K. Gutakovsky, A. V. Chehovskiy, V. V. Preobrazhenskii, M. A. Putyato, and T. A. Gavrilova: Free-standing and overgrown InGaAs/GaAs nanotubes, nanohelices and their arrays, Physica E **6**, 828 (2000)

[2] O. G. Schmidt and K. Eberl: Thin solid films roll up into nanotubes, Nature **410**, 168 (2001)

[3] C. Deneke, C. Müller, N. Jin-Phillipp, and O. G. Schmidt: Diameter scalability of rolled-up In(Ga)As/GaAs nanotubes, Semicond. Sci. Technol. **17**, 1278 (2002)

[4] M. Grundmann: Nanoscroll formation from strained layer heterostructures, Appl. Phys. Lett. **83**, 2444 (2003)

[5] A. B. Vorob'ev and V. Y. Prinz: Directional rolling of strained heterofilms, Semicond. Sci. Technol. **17**, 614 (2002)

[6] S. Mendach, O. Schumacher, C. Heyn, S. Schnüll, H. Welsch, and W. Hansen: Preparation of curved two-dimensional electron systems in InGaAs/GaAs-microtubes, Physica E **23**, 274 (2004)

[7] S. Mendach, T. Kipp, H. Welsch, C. Heyn, and W. Hansen: Interlocking mechanism for the fabrication of closed single-walled semiconductor microtubes, Semicond. Sci. Technol. **20**, 402 (2005)

[8] M. Hosoda, Y. Kishimoto, M. Sato, S. Nashima, K. Kubota, S. Saravanan, P. O. Vaccaro, T. Aida, and N. Ohtani: Quantum-well microtube constructed from a freestanding thin quantum-well layer, Appl. Phys. Lett. **83**, 1017 (2003)
[9] T. Fleischmann, K. Kubota, P. O. Vaccaro, T.-S. Wang, S. Saravanan, and N. Saito: Self-assembling GaAs mirror with electrostatic actuation using micro-origami, Physica E **24**, 78 (2004)
[10] T. Kipp, H. Welsch, C. Strelow, C. Heyn, and D. Heitmann: Optical modes in semiconductor microtube ring resonators, Phys. Rev. Lett. **96**, 77403 (2006)
[11] S. Mendach, R. Songmuang, S. Kiravittaya, A. Rastelli, M. Benyoucef, and O. G. Schmidt: Light emission and wave guiding of quantum dots in a tube, Appl. Phys. Lett. **88**, 111120 (2006)
[12] R. Songmuang, A. Rastelli, S. Mendach, and O. G. Schmidt: SiO_x/Si radial superlattices and microtube optical ring resonators, Appl. Phys. Lett. **90**, 091905p (2007)
[13] C. Strelow, C. M. Schultz, H. Rehberg, H. Welsch, C. Heyn, D. Heitmann, and T. Kipp: Three dimensionally confined optical modes in quantum well microtube ring resonators, arXiv:0704.3971
[14] S. L. McCall, A. F. J. Levi, R. E. Slusher, S. J. Pearton and R. A. Logan: Whispering-gallery mode microdisk lasers, Appl. Phys. Lett. **60**, 289 (1992)
[15] J. M. Gérard, D. Barrier, J. Y. Marzin, R. Kuszelewicz, L. Manin, E. Costard, V. Thierry-Mieg, and T. Rivera: Quantum boxes as active probes for photonic microstructures: The pillar microcavity case, Appl. Phys. Lett. **69**, 449 (1996)
[16] O. Painter, R. K. Lee, A. Scherer, A. Yariv, J. D. OBrien, P. D. Dapkus, and I. Kim: Two-dimensional photonic band-gap defect mode laser, Science **284**, 1819 (1999)
[17] K. J. Vahala: Optical microcavities, Nature **424**, 839 (2003)
[18] E. Peter, P. Senellart, D. Martrou, A. Lemaître, J. Hours, J. M. Gérard, and J. Bloch: Exciton-photon strong-coupling regime for a single quantum dot embedded in a microcavity, Phys. Rev. Lett. **95**, 67401 (2005)
[19] J. P. Reithmaier, G. Sek, A. Löffler, C. Hofmann, S. Kuhn, S. Reitzenstein, L. V. Keldysh, V. D. Kulakovskii, T. L. Reinecked, and A. Forchel: Strong coupling in a single quantum dot-semiconductor microcavity system, Nature **432**, 197 (2004)
[20] T. Yoshie, A. Scherer, J. Hendrickson, G. Khitrova, H. M. Gibbs, G. Rupper, C. Ell, O. B. Shchekin, and D. G. Deppe: Vacuum Rabi splitting with a single quantum dot in a photonic crystal nanocavity, Nature **432**, 200 (2004)
[21] G. Khitrova, H. M. Gibbs, M. Kiraz, S. W. Koch, and A. Scherer: Vacuum Rabi splitting in semiconductors, Nature Physics **2**, 81 (2006)
[22] H. Kogelnik: *Integrated Optics.* Topics in Applied Physics, vol. 7 (Springer Verlag, Berlin 1975) Chap. 2, pp. 13–81
[23] A. N. Pikhtin and A. D. Yas'kov: Dispersion of the refractive index of semiconducting solid solutions with sphalerite structure, Sov. Phys. Semicond. **14**, 389 (1980)
[24] T. Takagi: Refractive index of $Ga_{1-x}In_xAs$ prepared by vapor-phase epitaxy, Jpn. J. Appl. Phys **17**, 1813 (1978)
[25] M. Hentschel: *Mesoscopic Wave Phenomena in Electronic and Optical Ring Structures*, Ph.D. thesis, Technische Universität Dresden (2001)

Wide-Bandgap Quantum Dot Based Microcavity VCSEL Structures

K. Sebald[1], H. Lohmeyer[1], J. Gutowski[1], C. Kruse[2], T. Yamaguchi[2], A. Gust[2], D. Hommel[2] J. Wiersig[3], N. Baer[3], F. Jahnke[3]

[1] Semiconductor Optics, Institute of Solid State Physics,
[2] Semiconductor Epitaxy, Institute of Solid State Physics,
[3] Institute of Theoretical Physics,
University of Bremen, P.O.Box 33 04 40, 28334 Bremen, Germany
ksebald@ifp.uni-bremen.de

Abstract. In this contribution we report on the optical properties of planar and pillar structured GaN- and ZnSe-based monolithic microcavities. These structures reveal three-dimensional confined optical modes with high quality factors and potentially small mode volumes especially for the ZnSe-based samples. The measurements are completed with theoretical calculations. Furthermore, the optical emission properties of CdSe quantum dots embedded into microcavities have been studied. The Purcell effect demonstrated by means of the pronounced enhancement of the spontaneous emission rate of quantum dots coupled to the discrete optical modes of the cavities. This enhancement depends systematically on the pillar diameter and thus on the Purcell factor of the individual pillars.

In the past decade, it became possible to confine light on the wavelength scale by means of semiconductor microcavities (MCs) [1]. The resulting reduced mode volume can lead to laser devices with a strongly reduced threshold and allows for a control of the light-matter interaction. The implementation of semiconductor quantum dots (QDs) in solid state MCs is promising for a variety of classical and quantum-optical devices leading to improved properties and new applications. Cavity quantum electrodynamics phenomena, such as a strong enhancement of the spontaneous emission rate of QDs coupled to a discrete mode of the MC (Purcell effect), have been observed with semiconductor quantum dots (QDs) embedded [2, 3]. A single-mode solid-state single-photon source based on isolated QDs in pillar MCs has been demonstrated [4, 5]. Recently, strong coupling in this system has been reported [6]. These results have been achieved almost exclusively with the technologically well developed (Al,Ga,In)As system. However, due to the comparatively shallow electron confinement in conventional InAs QDs, the applicability of these structures is limited to low temperatures.

By utilization of the selenide and nitride wide-bandgap system devices for the UV to green spectral region can be realized. For the II–VI based material system there are only few reports in literature for monolithic MCs. CdTe-based pillar MCs emitting at λ=767 nm have been realized by *Obert et al.* [7], and the achievement of ZnSe-based MCs is presented in [8]. Other re-

R. Haug (Ed.): Advances in Solid State Physics,
Adv. in Solid State Phys. **47**, 29–41 (2008)

ports treat MCs with dielectric Bragg mirrors [9] and ZnSe-based micro-disc resonators with quantum wells [10] and CdSe QDs embedded [11], for the latter stimulated emission was observed as well. Furthermore, there are only few publication concerning the realization of high-quality and small-mode-volume cavities for the important nitride system emitting in the blue-green spectral region [12–14]. This is mainly due to the difficulties arising from the lattice mismatch in the Al(Ga)N/GaN system which for a long time hampered the epitaxial growth of high-quality distributed-Bragg reflectors (DBRs). AlN/GaN DBR mirrors with $R \geq 99\%$ have been reported [15, 16]. Results like optically pumped lasing [17, 18] and normal-mode coupling with InGaN quantum wells (QWs) at room temperature [19] have been achieved using, however, at least one dielectric mirror to form a complete MC. Recently, monolithic fully-epitaxial VCSEL structures with quality factors approaching their hybrid counterparts have been reported [20, 21].

Furthermore, QDs being formed on base of these material systems are characterized by the high temperature stability of their emission making them good candidates for operation at elevated temperatures. In the II–VI system emitting in the green spectral region single-photon emission of CdSe QDs up to 200 K [22] and a high quantum efficiency for CdSe QDs with MgS barriers up to room temperature [23] as well as a green QD laser [24] have been demonstrated. The emission of single InGaN QDs can be traced up to 150 K and these samples are characterized by a low QD density [25] as it is preferable for single-photon emitters. Moreover, the larger oscillator strengths of the II–VI and nitride based QDs make them promising to achieve the control of the light-matter interaction at elevated temperatures.

1 Experimental Details

The lack of suitable II–VI based low-index materials for the green spectral region has been overcome by the use of a short-period MgS/ZnCdSe superlattice (SL) as low-index material in the DBRs. In the case of the ZnSe-based cavities, $ZnS_{0.06}Se_{0.94}$ (lattice matched to the GaAs substrate) is used as the high-index material (layer thickness 48 nm) whereas a SL of typically 24 periods of alternating layers of 1.9 nm MgS and 0.9 nm ZnCdSe serves as the low-index material. The samples are grown by molecular beam epitaxy at a growth temperature of 280 °C (for growth details, see [26]). This SL approach leads to fully strained structures with reflectivities beyond 0.99 and additionally provides the possibility of a DBR doping [27]. The λ cavity for the ZnSe material system is formed by ZnSSe with three ZnCdSSe QWs or a single layer of CdSe QDs embedded, respectively. The CdSe QDs (density $\approx 5 \cdot 10^{10} cm^{-2}$) were grown by migration-enhanced epitaxy [28]. This cavity is positioned between an 18-period bottom and a 15-period top DBR stack each of the type described above.

For the growth of the GaN-based DBRs a comparable SL approach is used to obtain a suitable low-index material. In this case each SL is composed of 19.5 periods of AlN (thickness 1.6 nm) and $InGa_{0.25}N_{0.75}$ layers (thickness 0.75 nm) grown by plasma-assisted molecular-beam epitaxy at a growth temperature of 620 °C on GaN/sapphire template layers provided by metal-organic vapour-phase epitaxy. The high-index material layer of the DBR consists of 42 nm GaN. The nitride based VCSEL consists of a 20.5-periods bottom DBR, a λ cavity containing three InGaN quantum wells (QWs) at the antinode position of the optical field, and an 18-periods top DBR.

Cylindrically shaped pillar MCs etched down to the substrate and with diameters between 500 nm and 2.6 μm were prepared from the planar cavities by focused-ion-beam etching using a FEI Nova NanoLab system [8]. Scanning electron microscope (SEM) images of resulting structures with 500 nm and 1 μm diameter are shown in Fig. 1a for a ZnSe and in Fig. 3a for a nitride based pillar structure, respectively. The smooth sidewalls of the pillars reflect the good structural quality realized by FIB etching.

A micro-photoluminescence (μ-PL) setup has been used employing a microscope objective to excite individual pillar structures with a continuous wave or pulsed laser system and to collect their emission at variable temperatures (the measurements were performed at 4 K unless otherwise stated). The PL emission was detected either by a liquid-nitrogen cooled charged-coupled device camera or a multichannel plate after dispersion in a monochromator. Time-correlated single-photon counting was used for lifetime measurements with spectral and temporal resolutions of 1.8 meV and of about 30 ps, respectively (for experimental details, see [8, 29, 30]).

2 II–VI Based Quantum Well Microcavities

The VCSEL structures have been designed such that the longitudinal resonance of the cavity and the emission maximum of the embedded QWs coincide at room temperature. In order to study the optical modes of the cavities and to acquire quality (Q) values representative for the empty cavity, we measured μ-PL at low temperatures. Due to the blueshift of the QW emission with decreasing temperature only emission from the low-energy tail of the QWs acts as an internal light source to reveal the mode structure of the cavity in a regime of weak QW absorption (details are given in [8]). μ-PL spectra of pillars with various diameter are shown in Figs. 1c–g. A series of sharp emission peaks is observed which are, for decreasing diameter, increasingly blueshifted with respect to the resonance of the planar cavity at normal incidence (2.43 eV). The individual peaks are resonances of the discrete optical modes due to the optical confinement provided by the DBRs in longitudinal direction and by total internal reflection at the sidewalls of the pillars in lateral direction. The relative height of the peaks decreases when going to higher energies due to increasing absorption. Well-known experimental trends [31, 32] are clearly identified for decreasing diameters: The energy position of the fundamental mode as well as the splitting between the individual modes increase monotonously due to the stronger confinement of the light in the smaller pillars.

In the μ-PL setup, the sample surface is imaged on the entrance slit of the spectrometer yielding a spectrally resolved quasi one "dimensional" cut of the farfield emission pattern of the pillar at the position of the CCD array detector [33]. Figure 1b shows the spatial images recorded by the CCD detector corresponding to the μ-PL spectrum in Fig. 1c. The images provide information about the spectral width of the individual modes (abscissa) together with a cut through their farfield emission pattern (ordinate). The outermost left peak belongs to the fundamental mode of the pillar having a simple Gaussian-type intensity profile. For modes with higher energies, an increasing number of intensity minima and maxima is observed corresponding to nodes and antinodes of the transverse electric field pattern in the pillar.

To identify the observed cavity modes, we compute the modes of the three-dimensional pillars using a vectorial transfer-matrix method [34]. In this approach, the electric and magnetic fields in each layer of the pillar are expanded with respect to the modes of a cylindrical optical waveguide. Expansion coefficients of adjacent layers are related by the continuity of the transverse components of the electric and magnetic fields at the interface between these layers. Then, light propagation through the pillar is described by a $2N \times 2N$ transfer matrix, where N is the number of waveguide modes included in the expansion. Waveguide modes which are unbound in the transverse direction are ignored. This approximation is well justified for pillars with not too small diameters because radiation in the transverse direction is then negligible. The dotted lines in Figs. 1d–g show the calculated cavity spectra. Most of the modes are two-fold degenerate. It is the overall number of modes contributing to each peak which determines the peak height of the calculated cavity spectrum. For some experimental peaks several modes with very similar resonance wavelengths contribute. The second experimental peak (from the left), e.g., occurring as one band for pillars with diameters $\geq$2.5 μm (Figs. 1c–e) consists of three modes becoming experimentally resolved just in the case of the 1.5 μm pillar (Fig. 1f) due to increasing mode splitting. The calculated resonance wavelengths show a convincing agreement with the experiment which confirms the nature of the peaks as different transverse modes in the pillar. Slight deviations are probably due to imperfections of the microcavity and to a limited knowledge of the dispersion entering the calculations. For further illustration, calculated transverse electric field patterns for the indicated modes of the 3.5 μm pillar are shown on top of Fig. 1b. The numbers of field maxima in radial direction corresponds to those experimentally observed in the farfield intensity patterns (vertical double lines indicate the cut in the experiment).

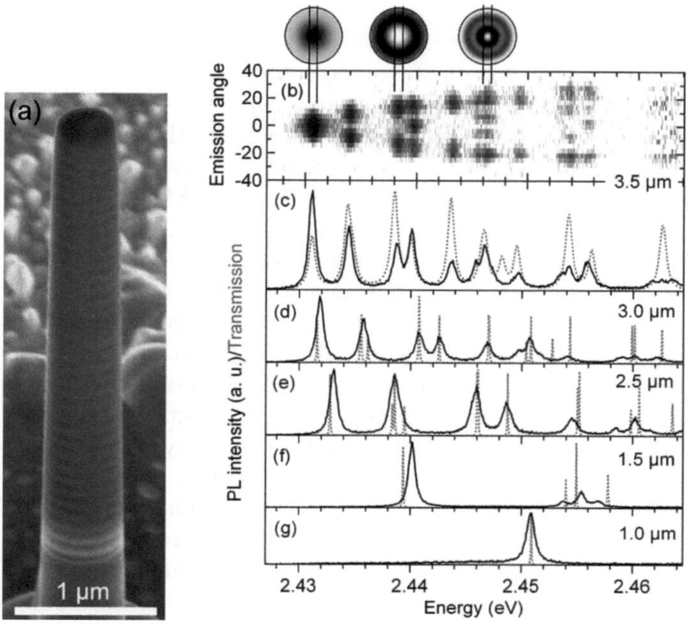

Fig. 1. (**a**) SEM picture of a ZnSe-based VCSEL pillar structure with a diameter of 500 nm. (**b**) Spectrally resolved vertical cuts of farfield emission pattern of a d=3.5 μm diameter pillar (intensity encoded in a logarithmic gray scale). Examples of calculated transverse electric-field patterns of the indicated modes are shown on top. (**c**) Corresponding μ-PL spectrum (*solid line*) and calculated pillar modes (*dotted line*) (calculation with background absorption). (**d**) to (**g**) μ-PL spectra (*solid line*) and calculated modes (*dotted line*, without background absorption) of pillars with decreasing diameter. After [8]

The ability of the cavity to confine the light is described by the quality factor which is defined as $Q=\lambda_c / \Delta\lambda_c$ with λ_c being the peak wavelength and $\Delta\lambda_c$ being the FWHM of the resonance. For pillars with diameters exceeding 1.5 μm the Q factor of the fundamental mode is nearly constant approaching the value Q=3500 of the planar structure, which is the highest value reported so far for ZnSe-based monolithic MCs. For smaller pillars, the measured Q starts to decrease. For pillar diameters of 1 μm and 700 nm, Q≈1800 and 1100 can be determined, respectively. The decrease of Q has to be attributed mainly to scattering losses due to the residual roughness of the pillar sidewalls becoming more important for smaller pillars [35]. In general, there is a difference in the spectral width of the calculated and measured resonant modes. Considering additional background absorption in the calculation the spectral width can be reproduced as well as shown exemplarily in Fig. 1c for a 3.5 μm pillar.

2.1 Stimulated Emission

In order to investigate the influence of the mode volume on the threshold of stimulated emission, excitation dependent measurements were performed on the pillars with different diameters. To limit the heating of the micropillars during optical pumping experiments, a short-pulse laser system providing 120 fs wide pulses has been chosen for the investigation of the laser threshold of the VCSEL structures at T=280 K. In Fig. 2a the μ-PL spectra of a 1.2 μm structure in the spectral range of the fundamental resonator mode is exemplarily shown for different excitation densities. The numbers given in Fig. 2a denote the incident energy per pulse in relation to the threshold pump energy of $E_{th} = 0.23$ pJ. The linewidth of the fundamental mode for the 1.2 μm pillar below threshold is broadened to an FWHM of about 7 meV due to the overlap with the QW emission peak, and its spectral position of 2.424 eV remains almost unchanged when the excitation energy is raised from 0.15 E_{th} to 0.8 E_{th}. However, an additional peak at 2.430 eV arises close to threshold. When the pump power is further increased, stimulated emission sets in at that particular spectral position, i.e., shifted by roughly 6 meV to higher energy compared to the position of the emission maximum below threshold at low pump power. Above threshold, the laser line does not show a significant shift when the pump energy is raised further. In the inset of Fig. 2a the development of the PL intensity in dependence on the pump energy is depicted at the position of the dominating mode (2.430 eV), confirming the existence of a sharp rise in intensity which is characteristic for the onset of stimulated emission. For pillars with a larger diameter the free spectral range between the modes is reduced, therefore, stimulated emission occurs also for higher transversal modes (not shown here). Below a pillar diameter of 2 μm single mode lasing is obtained up to 1.5 E_{th}. The lowest energy lasing line above threshold shows the expected blueshift with pillar diameter, confirming its origin to lie in the fundamental mode of the pillar. This effect is observed for the first time, to the best of our knowledge, in ZnSe-based monolithic MCs.

The pronounced blueshift of the stimulated emission peak amounting to about 6 meV when compared to the spectral position of the luminescence maximum at weak excitation is attributed to a breakup of the normal-mode coupling (details are given in [36]). The threshold excitation power density systematically decreases with decreasing diameter of the cavities. The results obtained for different pillar diameters with respect to the threshold pulse energy is summarized in Fig. 2b, the dashed line is a guide to the eye. Obviously the increased spacing between the modes and the smaller mode volumes for pillars with decreasing diameter lead to an improved coupling efficiency of the emission into the fundamental mode of the VCSEL and consequently to a reduction of the threshold density. This effect seems to overcompensate the slightly reduced Q values (Q = 2600–2800) and the increased mismatch between gain maximum and resonator mode of the MCs with smaller diameter.

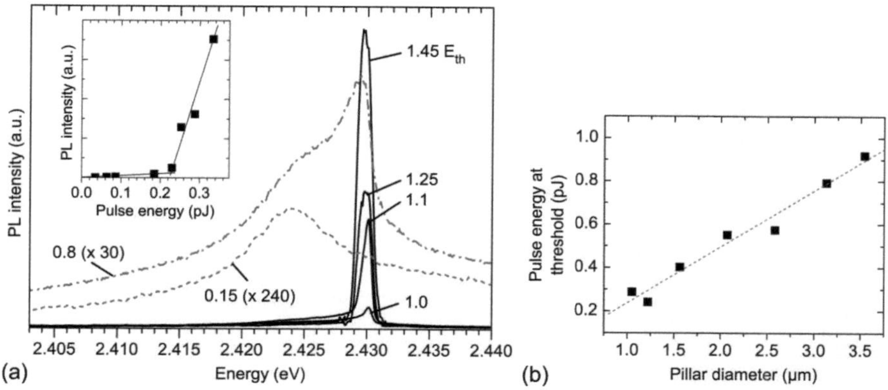

Fig. 2. (**a**) Spectra of a VCSEL structure with 1.2 µm diameter for increasing excitation power density at T=280 K. The numbers denote the incident energy per pulse in the relation to the lasing threshold (E_{th}=0.22 pJ). Inset: PL intensity in dependence on the excitation pulse energy. (**b**) Lasing threshold in dependence on the pillar diameter

3 Nitride Based Quantum Well Microcavities

Up to now GaN-based structures considerably suffered from a limited applicability or even failure of methods like (selective) wet or reactive ion etching to prepare small structures with large aspect ratios. In the last years, a photoelectrochemical etching process has been developed which can be applied to slab-like structures like microdisk resonators [13] or photonic crystal defect cavities. We show that nitride based pillar MCs can be successfully realized by FIB etching starting from a monolithically grown VCSEL structure. FIB structuring allows for production of pillars with smooth sidewalls and high aspect ratios as depicted in Fig. 3a.

Micropillars with different diameters were structured by FIB (for details, see [30]). PL spectra of pillars with diameters between 800 nm and 2.1 µm are shown in Fig. 3b (solid lines). Although not reaching the distinct mode separation obtained for the II–VI VCSELs, the pillar spectra reflect clear modulations becoming more pronounced for smaller pillar diameters which can be attributed to the modes resulting from the three-dimensional optical confinement in the pillar structures. To clarify the observations, we again calculated the transmission spectrum of the resonant modes for the three-dimensional pillars by using the vectorial transfer-matrix method. The SL was approximated by an $Al_{0.4}Ga_{0.6}N$ layer of the same total thickness and dispersion, and the pillar diameter was measured independently by SEM. The dashed lines in Fig. 3b show the calculated mode spectra. The spectral positions of the calculated individual modes are indicated by arrows. The calculation yields a Q factor of about 500 for the empty cavity which corresponds to an FWHM of the resonances accounting to 6 meV. The spectral

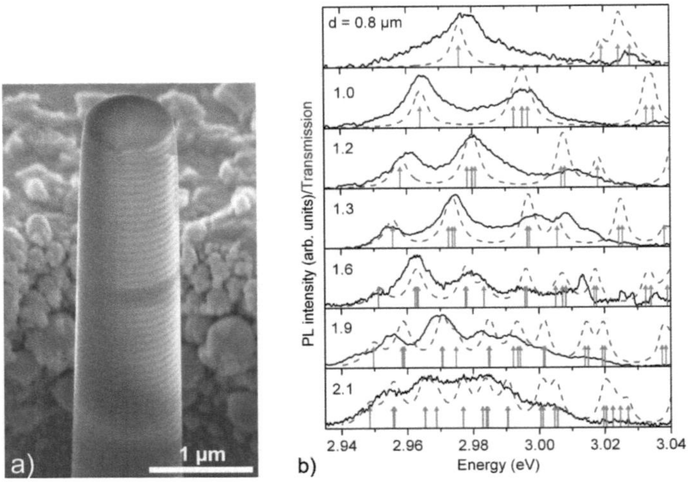

Fig. 3. (**a**) SEM image of a 1 µm nitride-based pillar structure. (**b**) *(online color)* PL spectra of pillars (*solid lines*) with different diameters and calculated transmission spectra (*dashed lines*) of the nitride based MCs. *Arrows* mark the calculated spectral positions of the individual modes. After [30]

position of the calculated resonances and the characteristic trend of the mode shifts for decreasing pillar diameter corresponds very well to the measured spectra which confirms the nature of the observed resonances as the different lowest-lying transverse modes of the pillars. For the fundamental mode of the 1 µm pillar lying at 2.965 eV, an FWHM of 12.5 meV and, thus, a Q factor of 240 can be determined which is comparable to the value measured for the planar structure.

In comparison to the II–VI based QW MCs the Q factor of the nitride based MCs is lower. Thus, the spectral width of resonant modes is increased leading to partly overlapping resonances of the modes. The lower Q is mainly due to the smaller index contrast within the nitride based DBRs and, additionally, due to background absorption in the mirrors. Nevertheless, in this material system, no other method than FIB is capable of producing structures of the demonstrated quality and aspect ratios making the observation of a resonant mode spectrum possible. The realization of InGaN QD samples with a low QD density and rather temperature stable QD emission [25] as well as of nitride based MC samples are promising results with respect to the intended implementation of QD layers into MCs. This may lead in future to efficient single photon emitters in the blue spectral region operating at elevated temperatures and could be explored for QD VCSELs as well.

4 CdSe Quantum Dots in ZnSe-based Microcavities

Cavity quantum electrodynamics phenomena, such as a strong enhancement of the spontaneous emission (SE) rate, can only be observed in systems with high quality factor and small mode volume V, and for emitters with a small spectral width like QDs. Therefore CdSe QDs were embedded into monolithic ZnSe-based MCs for the first time in this kind of material system. The longitudinal cavity resonance was set on the low energy tail of the QD emission band centered at 2.50 eV at 4 K (for more details, see [29]). In spectra taken at 5 K, partly resolution limited emission lines are observed superimposed to and spectrally in between the modes and being attributed to emission of individual QDs (see Fig. 4a for the spectral region around the FM of a 1.2 μm pillar). We attribute the dominant part of the PL detected on the position of the fundamental mode (FM) to emission from a weak continuous background due to coupling of QDs to phonons which would hardly be detected without the cavity. However, if QDs couple resonantly to the FM they may show up as enhanced single lines, a fact which reflects the much larger collection efficiency of the PL emitted into the pillar modes. For the 1.2 μm pillar such a situation can be reached by temperature tuning, as shown in Fig. 4a. With increasing temperature the pronounced QD line at 510.9 nm (10 K) shifts to longer wavelengths due to the band gap shrinkage and becomes exactly resonant with the FM at about 54 K. Due to the enhanced collection efficiency of PL emitted into the FM the emitter can be clearly traced up to 60 K. At higher temperatures the QD emission lines become broadened due to coupling to acoustic phonons [22] and can no longer be resolved individually for the present structures. From the spectral width of the modes the theoretical expected Purcell factor $F=3Q\lambda_c^3/4\pi^2 n^3 V$ can be calculated (n refractive index at the position of the emitter, λ_c resonance wavelength in vacuum, V effective mode volume). Mode volumes of the FMs have been calculated by an extended transfer-matrix method [29]. For an 0.7 μm pillar a mode volume of $V=6.1(\lambda/n)^3$ was found resulting in a Purcell factor of F=19.5, for example.

To probe the action of the Purcell effect, time-resolved PL measurements have been performed on a series of pillars with varying diameters. For individual pillars PL transients as obtained for detection at the spectral position of the fundamental mode (FM) are compared to those of leaky modes (LMs) of the pillars.

As shown in Fig. 5a, the PL decay characteristics clearly differ when detected at the FM and LMs. For the LMs (spectral positions labeled U1) the transients can be fitted within the experimental accuracy using a single exponential decay (including an offset for dark counts). The obtained decay times (380–500 ps for the larger pillars and 210 ps for the 0.5 μm pillar, see Fig. 5b, triangles) are comparable to data of the QD ensemble in a ZnSe matrix without cavity at the respective spectral position. Such a behavior is expected for QDs coupling to a continuum of low-Q LMs. PL spectrally selected at the

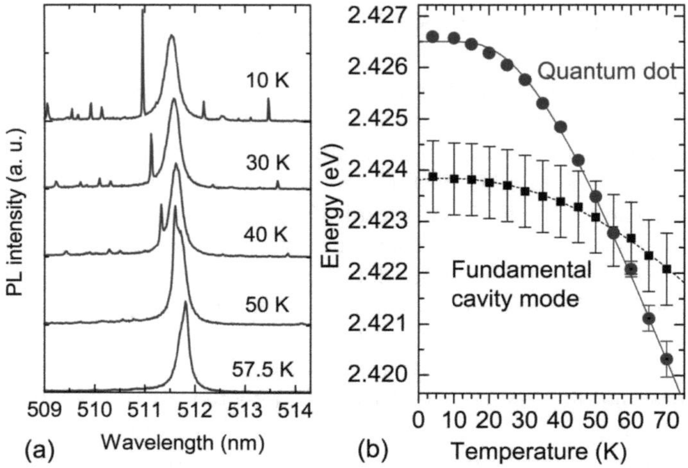

Fig. 4. (**a**) PL spectra of a ZnSe-based QD pillar with a diameter of 1.2 μm for different temperatures. (**b**) Spectral positions of the FM and the QD emission line of the 1.2 μm pillar shown in (a) for different temperatures.

FM of the pillars exhibits a clearly faster but nonexponential decay dynamics. A quite satisfying agreement with the experimental transients is obtained by using a biexponential decay. For example, a fast decay time of (120±10) ps and a long decay time of (700±50) ps are found for the 0.7 μm pillar (Fig. 5a). As summarized in Fig. 5b, the determined fast decay component (boxes) becomes systematically shorter when going to smaller pillar diameters. The slow decay component being in the range of 600–950 ps (±50 ps) shows only a weak correlation to the pillar diameter (more details [29]), its origin is unclear up to now. With respect to the findings in [38] we note that the PL transients do not change significantly with excitation density.

We approximately quantify an achieved effective SE enhancement by the shortening of the extracted fast decay time τ_f, similar to what has been done in [2]. For pillars with 1.0, 0.7, and 0.5 μm diameters, a τ_f of 205, 120, and 75 ps is obtained, respectively. In relation to the lifetime of QDs in a ZnSe matrix (460 ps for 1.0 and 0.7 μm and 250 ps for 0.5 μm diameter) an SE enhancement value F_{eff}=2.3, 3.8, and 3.3 can thus be stated for the respective pillars. Although for a given structure the PL detected at the spectral position of the FM may be highly dominated by contributions of few QDs, one can explain these numbers in terms of averaging about many QDs with different spectral detuning from the mode and different spatial location with regard to the field maximum coupling to the mode. For this situation it has been estimated [2, 3] that the spatial and spectral averaging in the QD ensemble with respect to the FM leads to a PL transient exhibiting a roughly twelvefold reduced SE enhancement. Thus, for the twofold degenerate FM $F_{eff} = \tau_0/\tau \approx 2F/12 + \tau_0/\tau_{LM}$ should hold, where the second term on

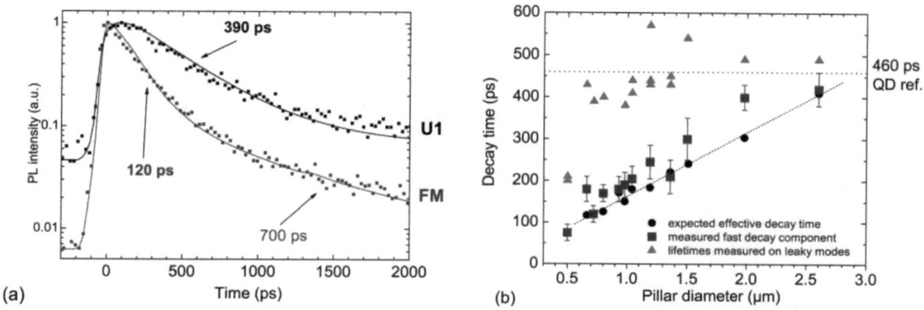

Fig. 5. (**a**) PL transients for the 0.7 μm diameter structure. The lifetimes obtained from the biexponential fit for the FM data and from the monoexponential fit for the U1 data are given in the figure. (**b**) Data in dependence on the pillar diameter: fast decay time measured spectrally in resonance with the FM (*boxes*), decay times measured off resonant to the FMs (*triangles*), and expected decay time as effect of average enhancement (*circles*, *solid line* as guide to the eye). The decay time of about 460 ps measured on the reference sample is indicated as a *dotted line*. After [29]

the right accounts for the emission into leaky modes, and $\tau_0 \approx \tau_{LM}$. For the pillars with 0.7 μm and 1.0 μm diameter (see above) expected effective enhancements F_{eff} result to amount to 4.25 and 2.75, respectively, which are reasonably close to the values calculated from the experimental transients. The circles in Fig. 5b corresponds to the calculated effective lifetime $\tau_e = 460\ \text{ps}/F_{eff}$ for the different pillar diameters. The observed fast decay times follow remarkably well the expected trend for decreasing pillar diameter.

5 Conclusion

Pillars yielding three-dimensional confined optical modes and possessing high quality factors and potentially small mode volumes have been successfully FIB etched from GaN- and ZnSe-based monolithic planar microcavities. The resulting experimentally observed fundamental and higher pillar modes could rather perfectly modelled by theoretical calculations based on a vectorial transfer-matrix method. CdSe quantum dots could be successfully embedded into the microcavities resulting in the Purcell effect to occur, proved through the pronounced enhancement of the spontaneous emission rate of quantum dots coupling to the discrete optical modes of the cavities with an effective Purcell factor of up to four. These are promising results for classical and quantum optical device applications of wide-bandgap based vertical emitting microcavities in future.

Acknowledgements

The work on GaN based MCs was supported by the Deutsche Forschungsgemeinschaft (Research Group 506, "Physics of Nitride-Based Nanostructured Light-Emitting Devices"). K. Sebald acknowledges support by the BFK of the University of Bremen. The theory group acknowledges a grant for CPU time at the NIC, FZ Jülich.

References

[1] K. J. Vahala, Nature (London) **424**, 839 (2003).
[2] J. M. Gérard, B. Sermage, B. Gayral, B. Legrand, E. Costard, and V. Thierry-Mieg, Phys. Rev. Lett. **81**, 1110 (1998).
[3] For a review see J. M. Gérard, Top. Appl. Phys. 90, 269 (2003).
[4] E. Moreau, I. Robert, J. M. Gérard, I. Abram, L. Manin, and V. Thierry-Mieg, Appl. Phys. Lett. **79**, 2865 (2001).
[5] M. Pelton, C. Santori, J. Vučković, B. Zhang, G. S. Solomon, J. Plant, and Y. Yamamoto, Phys. Rev. Lett. **89**, 233602 (2002).
[6] J. P. Reithmaier, G. Sek, A. Löffler, C. Hofmann, S. Kuhn, S. Reitzenstein, L. V. Keldysh, V. D. Kulakovskii, T. L. Reinecke, and A. Forchel, Nature (London) **432**, 197 (2004).
[7] M. Obert, B. Wild, G. Bacher, A. Forchel, R. André, and L. S. Dang., Appl. Phys. Lett. **80**, 1322 (2002).
[8] H. Lohmeyer, K. Sebald, C. Kruse, R. Kröger, J. Gutowski, D. Hommel, J. Wiersig, N. Baer, and F. Jahnke, Appl. Phys. Lett. **88**, 051101 (2006).
[9] I. C. Robin, R. André, A. Balocchi, S. Carayon, S. Moehl, and J. M. Gérard, Appl. Phys. Lett. **87**, 233114 (2005).
[10] M. Hovinen, J. Ding, A. V. Nurmikko, D. C. Grillo, J. Han, L. He, and R. L. Gunshor, Appl. Phys. Lett. **63**, 3128 (1993).
[11] J. Renner, L. Worschech, A. Forchel, S. Mahapatra, and K. Brunner, Appl. Phys. Lett. **89**, 091105 (2006).
J. Renner, L. Worschech, A. Forchel, S. Mahapatrad, and K. Brunner, Appl. Phys. Lett. **89**, 231104 (2006).
[12] S.X. Jin, J. Li, J.Z. Li, J. Y. Lin, and H.X. Jiang, Appl. Phys. Lett. **76**, 631 (2000).
[13] E. D. Haberer, R. Sharma, C. Meier, A. R. Stonas, S. Nakamura, S. P. DenBaars, and E. L. Hu, Appl. Phys. Lett. **85**, 5179 (2004).
[14] M. Kneissl, M. Teepe, N. Miyashita, N. M. Johnson, G. D. Chern, and R. K. Chang, Appl. Phys. Lett. **84**, 2485 (2004).
[15] H. M. Ng, T. D. Moustakas, and S. N. G. Chu, Appl. Phys. Lett. **76**, 2818 (2000).
[16] T. Ive, O. Brandt, H. Kostial, T. Hesjedal, M. Ramsteiner, and K.-H. Ploog, Appl. Phys. Lett. **85**, 1970 (2004).
[17] T. Someya, R. Werner, A. Forchel, M. Catalano, R. Cingolani, and Y. Arakawa, Science **285**, 1905 (1999).
[18] A. V. Nurmikko, and J. Han, Progress in Blue and Near- Ultraviolet Vertical-Cavity Emitters: A Status Report, edited by H. Li, K. Iga, VCSEL Devices (Springer, Berlin, 2002).

[19] T. Tawara, H. Gotoh, T. Akasaka, N. Kobayashid, and T. Saitoh, Phys. Rev. Lett. **92**, 2564021 (2004).
[20] E. Feltin, R. Butté, J.-F. Carlin, J. Dorsaz, N. Grandjean, and M. Ilegems, Electronics Lett. **41**, 94 (2005).
[21] G. Christmann, D. Simeonov, R. Butté, E. Feltin, J.-F. Carlin, and N. Grandjean, Appl. Phys. Lett. **89**, 261101 (2006).
[22] K. Sebald, P. Michler, T. Passow, D. Hommel, G. Bacher, and A. Forchel, Appl. Phys. Lett. **81**, 2920 (2002).
[23] R. Arians, T. Kümmell, G. Bacher, A. Gust, C. Kruse, and D. Hommel, Appl. Phys. Lett. **90**, 101114 (2007).
[24] M. Klude, T. Passow, H. Heinke, and D. Hommel, Phys. Stat. Sol. (b) **229**, 1029 (2002).
[25] K. Sebald, H. Lohmeyer, J. Gutowski, T. Yamaguchi, and D. Hommel, Phys. Stat. Sol. (b) **243** 1661 (2006).
[26] C. Kruse, S.M. Ulrich, G. Alexe, E. Roventa, R. Kröger, B. Brendemühl, P. Michler, J. Gutowski, and D. Hommel, Phys. Stat. Sol. (b) **241**, 731 (2004).
[27] K. Otte, C. Kruse, J. Dennemarck, and D. Hommel, phys. stat. sol. (c) **3**, 1217 (2006).
[28] T. Passow, K. Leonardi, H. Heinke, D. Hommel, D. Litvinov, A. Rosenauer, D. Gerthsen, J. Seufert, G. Bacher and A. Forchel, J. Appl. Phys. **92**, 6546 (2002).
[29] H. Lohmeyer, C. Kruse, K. Sebald, J. Gutowski, and D. Hommel, Appl. Phys. Lett. **89**, 091107 (2006).
[30] H. Lohmeyer, K. Sebald, J. Gutowski, R. Kröger, C. Kruse, D. Hommel, J. Wiersig, and F. Jahnke, European Physical Journal B **48**, 291 (2005).
[31] J.M. Gérard, D. Barrier, J.Y. Marzin, R. Kuszelewicz, L. Manin, E. Costard, V. Thierry-Mieg, and T. Rivera, Appl. Phys. Lett. **69**, 449 (1996).
[32] J.P. Reithmaier, M. Röhner, H. Zull, F. Schäfer, A. Forchel, P.A. Knipp, and T.L. Reinecke, Phys. Rev. Lett. **78**, 378 (1997).
[33] A. Baas, O. El Daïf, M. Richrad, J.P. Brantut, G. Nardin, R. Idrissi Kaitouni, T. Guillet, V. Savona, J.L. Staehli, F. Morier-Genoud, and B. Deveaud, phys. stat. sol. (b) **243**, 2311 (2006).
[34] D. Burak and R. Binder, IEEE J. Quantum Electron. **33**, 1205 (1997).
[35] T. Rivera, J.-P. Debray, J.M. Gérard, B. Legrand, L. Manin-Ferlazzo, and J. L. Oudar, Appl. Phys. Lett. **74**, 911 (1999).
[36] C. Kruse, H. Lohmeyer, K. Sebald, J. Gutowski, D. Hommel, J. Wiersig, and F. Jahnke submitted to xxx.
[37] M. Röhner, J.P. Reithmaier, A. Forchel, F. Schäfer, and H. Zull, Appl. Phys. Lett. **71**, 488 (1997).
[38] M. Schwab, H. Kurtze, T. Auer, T. Berstermann, M. Bayer, J. Wiersig, N. Baer, C. Gies, F. Jahnke, J.P. Reithmaier, A. Forchel, M. Benyoucef, and P. Michler, Phys. Rev. B **74**, 045323 (2006).

Momentum Space Wave Functions in InAs Quantum Dots Mapped by Capacitance Voltage Spectroscopy

Dirk Reuter

Lehrstuhl für Angewandte Festkörperphysik, Ruhr-Universität Bochum,
Universitätstrasse 150, D-44780 Bochum, Germany
dirk.reuter@ruhr-uni-bochum.de

Abstract. We have investigated the charging of conduction as well as the valence band states of InAs quantum dots by magneto capacitance voltage spectroscopy. Whereas measurements in magnetic fields perpendicular to the base plane of the quantum dots reveal the dispersion of the single particle levels and help to identify the corresponding charging peaks, measurements in parallel magnetic fields allow the mapping of the momentum space wave functions corresponding to the individual charging peaks. The wave function maps directly visualize anisotropies in the confinement potential. For electrons, a sequential filling of a shell-like energy level structure is observed and the results can be explained well by using a harmonic approximation for the lateral confinement potential and a single effective mass. For holes, the charging behaviour is more complicated. The dispersion measurements in conjunctions with wave function maps show that the orbital angular momentum cannot be treated as good quantum number and more complex theories have to be used. Electron as well as hole wave function reveal an anisotropy in the confinement potential with the low energy axis along the [0-11] direction.

1 Introduction

InAs quantum dots (QDs) have been studied intensively in the past as model systems for strong three-dimensional carrier confinement [1]. The interest is rooted in basic research questions as well as many envisioned novel devices. Arguably the most fundamental property making InAs QDs so interesting, are their atomically sharp energy levels. In addition to the single particle spectrum determined by the three-dimensional confinement potential, the carrier-carrier interactions become important, if the QD is multiply charged.

It is known for quite a while that capacitance voltage (C-V) spectroscopy is a versatile tool to determine the addition spectra for InAs QDs [2–4]. These measurements give information on the single particle levels and especially on carrier-carrier interaction energies. Whereas the conduction band states have been investigated intensively for over a decade [2–5], high quality charging spectra for holes became available only in recent years [6–8]. To assign the individual charging peaks to the corresponding single particle energy level, measurements in perpendicular magnetic field (field along the growth direction) are of invaluable help [4, 5, 8]. For electrons, the identification of the

R. Haug (Ed.): Advances in Solid State Physics,
Adv. in Solid State Phys. **47**, 43–54 (2008)

charging peaks was possible by using a harmonic lateral confinement potential and assume sequential shell filling [5]. For holes, the situation is more complicated and to explain the dispersion measurements in perpendicular magnetic field using the approach of *Warburton* and co-workers [5] one has to evoke an non-sequential shell filling sequence [8]. Based on an idea developed by *Patané* and co-workers [9] for magneto-tunneling spectroscopy, the application of C-V spectroscopy was extended to the mapping of momentum space wave functions [10, 11]. With this method it was possible to visualize anisotropies in the confinement potential directly.

In this paper, I will discuss how the wave functions maps of conduction and valence band states have extended our understanding of InAs QD charging spectra. This contribution is organized as follows: I will introduce the measurement principle for the wave function mapping in Section 2. In Section 3, the wave function mapping experiments for conduction band states will be discussed with respect to the dispersion of the individual charging peaks in perpendicular field. For holes, I will do the same in Section 4, before I close with a brief summary in Section 5.

2 Wave Function Mapping by Capacitance Voltage Spectroscopy

Since C-V spectroscopy has been used for quite a while to determine the charging spectra of InAs QDs, I will only briefly describe the basic measurement principle and focus on the new aspect of wave function mapping. For more details, the reader is referred to [2, 5].

For charging spectroscopy, the InAs QDs have to be embedded into the undoped part of a Schottky diode. For the investigation of conduction band states an n-type sample is required whereas for valence band charging p-type doping has to be employed. For our samples, the relevant part of the layer sequence is as follows: A doped (2-4$\times 10^{18}$ $\mathtt{cm}^{-3}$) GaAs back contact is followed by a thin GaAs layer that serves as a tunnelling barrier. For n-type samples the typical barrier thickness t is $\sim$ 25 nm but for wave function mapping experiments, t $\sim$ 40 nm has to be chosen because otherwise the measurement frequencies would be impracticable high. For valence band state spectroscopy, the values for t are between 17 nm and 21 nm. After the tunnelling barrier, a single layer of InAs QDs is inserted. It is important to use growth conditions that result in a narrow size and composition distribution. The growth conditions we use result in QDs that show at 300 K ground state emission between 1250 nm and 1275 nm. The whole layer sequence is terminated by a 10 nm thick GaAs cap layer. The QDs are overgrown by 30 nm GaAs which is followed by a 3 nm AlAs/1 nm GaAs superlattice of variable thickness. The thickness of the superlattice is chosen so that the overall distance between back contact and the surface is approximately eight (n-type) to eleven times (p-type) larger than t. From these samples, Schottky diodes are fabricated employing Au or Cr/Au gates. For typical gate areas of 300×300 μm^2, a single device contains about 10^7 QDs.

The capacitance of the devices described above, is measured at 4.2 K, as function of an applied DC voltage V_G by superimposing a small AC component onto the DC bias and recording the charging current. If one of the QD energy levels is in resonance with the Fermi level in the back contact, the QD is periodically charged and de-charged, which results in a peak in the capacitance (see Fig. 1 for an example). A detailed discussion of the resonance conditions [5] shows that any degeneracy present in the single particle energy spectrum is lifted due to Coulomb interactions.

The height of the capacitive signal depends on the number of QDs charged during one AC cycle. If the AC frequency is much lower than the tunnelling rate, each QD is charged and de-charged during every AC cycle, the charging current scales with the frequency and the signal height depends only on the number of QDs under the gate. If the frequency becomes larger than the tunnelling rate, the charging current scales not longer with the frequency and the height of the capacitive signal decreases. This has been investigated in detail by *Luyken* and co-workers [12], *Medeiros-Ribeiro* and co-workers [13] as well as *Horiguchi* and co-workers [14] for conduction band state charging and by us for hole charging [7].

To get access to the QD momentum space wave functions corresponding to the individual charging peaks, we have adopted an idea developed by *Patané* and co-workers for magneto tunnelling spectroscopy [9]. They showed that to a good approximation the tunnelling rate $1/\tau$ as function of an in-plane magnetic field is related to the momentum space wave functions by

$$1/\tau(B_x, B_y) \propto |\phi_{QD}(k_x, k_y)|^2 \,. \tag{1}$$

In (1), B_x and B_y are the components of the in-plane magnetic field (parallel to the QD base plane) and $\phi_{QD}(k_x,k_y)$ is the in-plane momentum space wave function of the single particle level through which tunnelling occurs. The magnetic field and the in-plane momentum are related by the following expression:

$$k_{||} = \frac{et}{\hbar} B_{||} \,. \tag{2}$$

Here t is the thickness of the tunnelling barrier and e the elementary charge. The tunnelling current and $B_{||}$ form with $k_{||}$ a right angle trihedral, i.e., $B_{||}$ and $k_{||}$ are orthogonal to each other.

To transfer this idea to C-V spectroscopy, one takes advantage of the fact that for an AC frequency in the range of the tunnelling rate $1/\tau$, the height of the capacitance signal is a measure for the tunnelling rate. In first approximation, the following relation is valid:

$$C_{QD}(B_x, B_y) \propto |\phi_{QD}(k_x, k_y)|^2 \,. \tag{3}$$

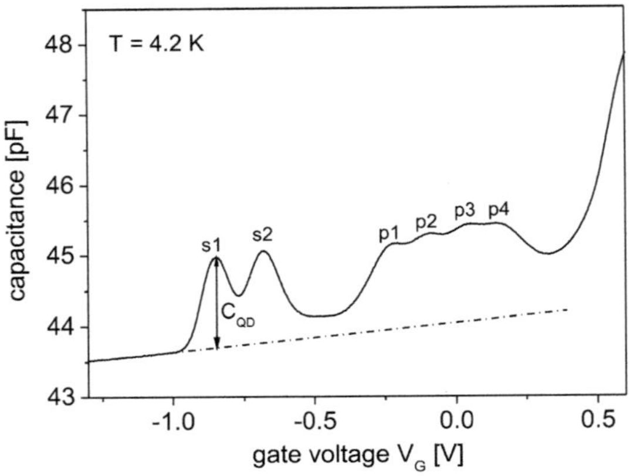

Fig. 1. The capacitance as function of an applied gate voltage is shown for the electron charging of InAs quantum dots. The signal caused by the charging of the quantum dots C_{QD} is indicated by an *arrow*. The dashed line visualizes the background caused by the capacitance of the Schottky diode. Please note that the gate voltage scale corresponds to a grounded back contact

Here, C_{QD} is the height of the charging peak with the background subtracted (see Fig. 1). If C_{QD} varies strongly with the magnetic field, a more complicated relation [12] between $1/\tau$ and C_{QD} has to be used for a quantitative analysis to determine the square of the wave function. To reveal the basic character of the wave functions, their symmetries and perhaps anisotropies, it is sufficient to plot C_{QD} for the individual charging peaks as function of the in-plane magnetic $B_{||}$ field.

Compared to magneto-tunnelling spectroscopy, it is not directly evident that with C-V spectroscopy really the single particle wave functions are mapped: The resonance conditions [5] show that – strictly speaking – to each charging peak corresponds the difference between the N-carrier and the (N-1)-carrier state. *Rontani* and *Molinari* discussed this problem in detail [15]. They showed that for strongly confined QDs, as our InAs QDs, the measured wave functions reflect quite well the single particle wave function to which the charging peaks belongs.

3 C-V Spectroscopy for Conduction Band States

In Fig. 1, a high quality charging spectrum for the conduction band states of InAs QDs is shown. The large background under the charging peaks is due to the voltage dependent capacitance of the Schottky diode. This background has to be subtracted to obtain the height of the capacitive signal C_{QD} required for the wave function mapping as discussed above. As already discussed by other authors [2–5], the first two charging peaks belong to a twofold

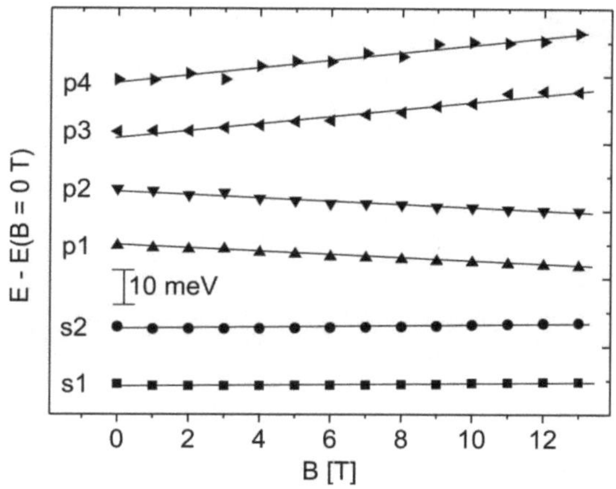

Fig. 2. Energy shift with respect to the 0 T value of the individual electron charging peaks seen in Fig. 1 as function of a perpendicular magnetic field B (field along growth direction). The *straight lines* are guides to the eyes

degenerated s-like ground state whereas the peaks p1–p4 reflect the charging of a fourfold degenerated p-shell. This assignment of the individual charging peaks to single particle levels is based on the dispersion of the individual charging peaks in a perpendicular magnetic field as shown in Fig. 2 [5]. The first two peaks do not shift at all, which is consistent with the charging of an s-like ground state with no orbital angular momentum. The peaks p1 and p2 shift downward in energy whereas p3 and p4 shift upwards in energy with approximately the same slope, which is consistent with charging p-like levels ($L = \pm 1$). It seems that the filling within the p-shell is for our QDs not according to Hunds first rule which would favour a filling sequence (-1, 1, -1, 1). This points to a small asymmetry in the confinement potential lifting the fourfold degeneracy of the p-shell expected for rotational symmetry in the confinement potential but allowing still the classification of the single particle levels by their orbital angular momentum as good approximation. For a conclusive decision on the filling sequence at 0 T, the dispersion measurements offer probably not enough resolution. At higher magnetic fields, the field itself enforces a filling sequence that deviates from Hunds rule. Calculations by *Warburton* and co-workers [5] show that this might already happen at fields as low as 2 T and the corresponding level crossing might not be resolved in the experiments. In Fig. 3, wave function maps for the s1 and s2 charging peaks are shown. One clearly sees that the signal is maximum for $k_{||} = 0$ and decreases monotonically towards larger k-values. This is exactly what one expects for an s-like ground state with no orbital angular momentum. One also sees the elliptical shape of the wave function which proves an anisotropy in the confinement potential already suspected on basis of the measurements in perpendicular magnetic field. The wave functions

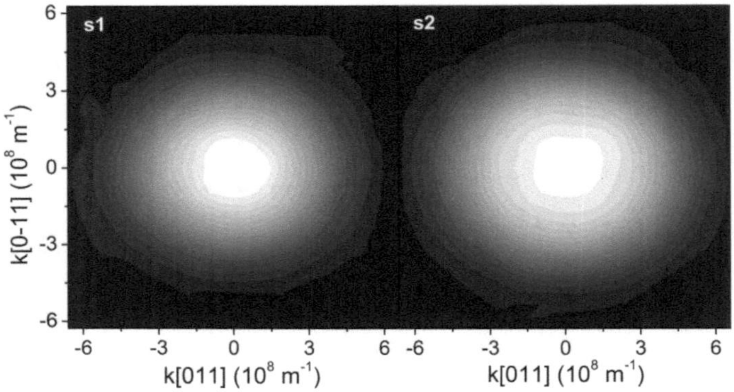

Fig. 3. C_{QD} of the electron charging peaks s1 and s2 as function of an in-plane momentum generated by an in-plane magnetic field B (field perpendicular to the growth direction). *Bright* represents high C_{QD} values whereas *dark* corresponds to low capacitive signals. As discussed in the text, C_{QD} is a measure for the momentum space wave functions of the corresponding charging peak. One clearly sees that the wave functions are elongated along the [0-11] direction in real space representation

are in k-space representation elongated along the [011] direction, i.e., in real space representation in [0-11] direction. Although the maps of the s-state wave functions reveal an anisotropic confinement potential, it is not yet clear if the resulting p-shell splitting is strong enough to enforce a filling violating from Hunds rule. The wave functions for the excited states p1/p2 and p3/p4 show a bone like structure (see Fig. 4). This again points to an asymmetry in the lateral confinement potential because otherwise the p-like wave functions would show circular symmetry. The fact that p1 and p2 show the same symmetry proves that they belong to the same orbital angular momentum number (L $= -1$), i.e. in our QDs the filling sequence of the p-shell is indeed not according to Hunds rule. For p3/p4 the bone pattern is orthogonal to the one for p1/p2 which is consistent with the charging of an L $= +1$ level.

In summary, the wave functions maps for the electron system directly visualize the anisotropy in the confinement potential and reveal that the filling of the p-shell is not according to Hunds rule. All experimental results presented above, can at least qualitatively be explained by using a harmonic approximation for the lateral confinement potential, employing a single effective mass for all single particle levels and taking the orbital angular momentum as good quantum number [5]. This is different for the valence band states as discussed in the next section.

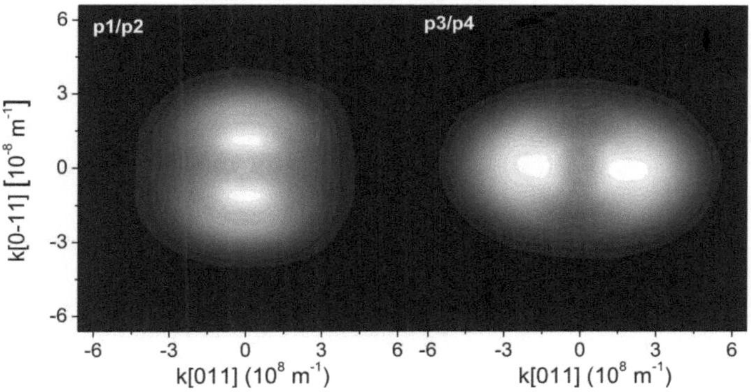

Fig. 4. C_{QD} of the electron charging peaks p1/p2 and p3/p4 as function of an in-plane momentum generated by an in-plane magnetic field B. *Bright* represents high C_{QD} values and *dark* corresponds to low capacitive signals. As discussed in the text, C_{QD} is a measure for the momentum space wave functions of the corresponding charging peak. One clearly sees a bone like shape in both maps which is consistent with a p-like excited state

4 Charging Spectroscopy for Holes in InAs Quantum Dots

The charging spectra for valence band states look at first glance somewhat different from the electron C-V spectra (see Fig. 5). The charging peaks seem to come in pairs and no fourfold degenerated p-shell can be identified directly. Also, up to eight charging peaks can be identified. In Fig. 6, the dispersion in a perpendicular magnetic field for the individual charging peaks is shown. It can be seen that peaks 1 and 2 do not shift which makes the assignment to the charging of an s-like ground state reasonable. Peaks 3 to 6 shift downward and upwards in energy in an alternating fashion. Arguing in analogy to the electron system, one is tempted as first try to assign these four peaks to the filling of a fourfold degenerated p-shell according to Hunds rule. A quantitative analysis shows that this is no satisfying explanation: Peaks 5 and 6 shift twice as strong as 3 and 4, which contradicts, at least in the harmonic approximation with a single effective mass, the assignment to a level with the same orbital angular momentum (L $= \pm 1$). But one knows [5], that in this simple model, d-like states shift approximately twice as strong as p-states. Based on this, we proposed an incomplete shell filling model [8], where peak 3 and 4 belong to p-shell charging and d-shell charging is responsible for peaks 5 and 6. This means, the p-shell is not completely filled before the d-shell filling starts. Under this assumption, one can explain the dispersion behaviour in perpendicular magnetic field quite well but severe contradictions exist to the measured wave functions, as discussed in the following. Figure 7 shows the wave function maps for peak 1 and 2. The signal is maximum for $k_{||} = 0$ and decreases monotonically towards larger $k_{||}$-values. This behaviour is in

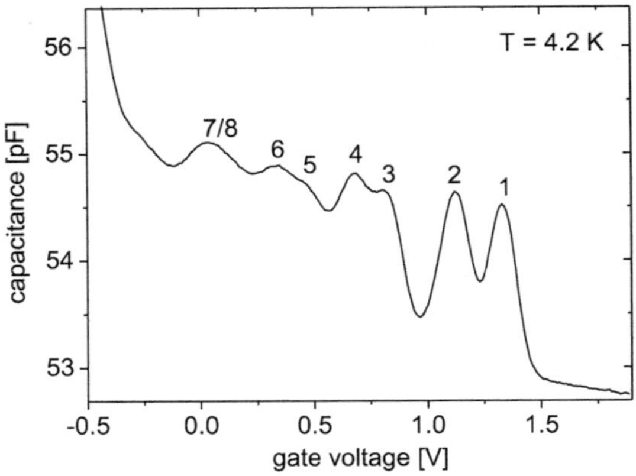

Fig. 5. Addition spectrum for holes in InAs quantum dots. Six charging peaks are clearly resolved and two more can be seen as one broad feature. Please note that the gate voltage scale corresponds to a grounded back contact

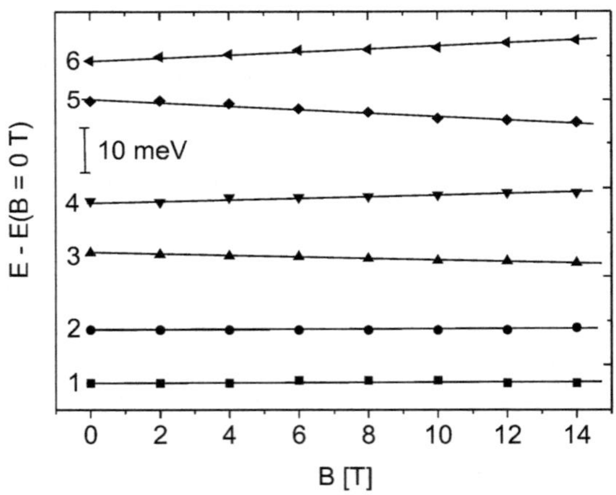

Fig. 6. Energy shift with respect to the 0 T value of the individual valence band charging peaks seen in Fig. 5 as function of a perpendicular magnetic field B (field along growth direction). The *straight lines* are guides to the eyes

agreement with our assignment as an s-like ground state. As for the electron ground state, the wave function has an elliptical shape with the long axis along the [0-11] direction in real space orientation. This means, electron and hole wave functions show the same anisotropy and are not oriented orthogonal to each other as it might be the case if a strong piezoelectric contribution to the confinement potential is present [16]. The wave functions for peaks 3 to 6 show a node like structure around $k_{||} = 0$ (see Fig. 8). This is consistent with the charging of exited states with finite orbital angular momentum. Peaks 3

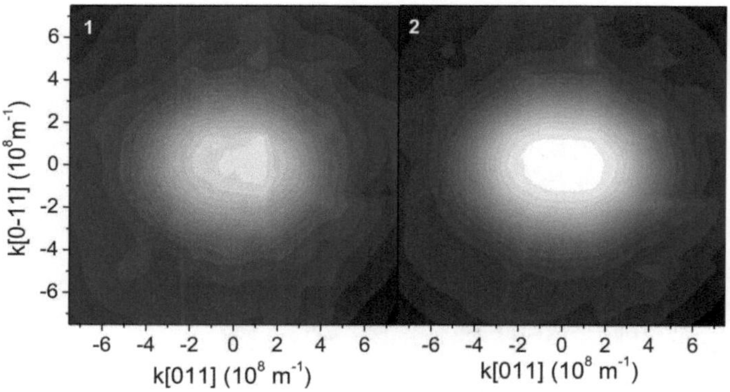

Fig. 7. C_{QD} of the hole charging peaks 1 and 2 as function of an in-plane momentum generated by an in-plane magnetic field B. *Dark* corresponds to low C_{QD} values whereas *bright* corresponds to high capacitive signals. As discussed in the text, C_{QD} is a measure for the momentum space wave functions of the corresponding charging peak. One clearly sees that the wave functions are elongated along the [0-11] direction in real space representation

and 4 look very similar and show two maxima at finite $k_{||}$-values along the [0-11] direction. Although the shape itself is compatible with a p-like wave function for a slightly asymmetric confinement potential as discussed for the electron system, it is surprising that both peaks are oriented in the same direction: Based on the dispersion behaviour, we have assigned peak 3 to an L $= -1$ state and 4 to an L $= 1$ state. Under the assumptions made so far, especially that the orbital angular momentum number is a good quantum number, p-states with L $= -1$ and L $= 1$ should show orthogonal symmetry as indeed observed for the electron system. More complex theoretical treatments show that the orbital angular momentum is no longer a good quantum number [17, 18] and that also light hole and heavy hole mixing has to be taken into account. Here only a quantitative explanation should be given that is mainly based on the work of *Climente* and co-workers [18]: They argue that the overall angular momentum number F_z composed from an orbital component and the Bloch character of the hole J_z can be treated as good quantum number. In their model, peaks 3 and 4 belong to the single particle level with $| F_z | = 1/2$, where peak 3 (4) has the orbital quantum number L $= -1$ (L $= +1$). Because both peaks belong to charging into the same twofold degenerated single particle level, the wave functions are the same. In perpendicular magnetic field, the dispersion is mainly determined by the orbital angular momentum part which results in a downward shift in energy for peak 3 (L $= -1$) and in an upward shift for peak 4 (L $= 1$). In principle, the same arguments hold for peaks 5 and 6. Of course, the arguments, which led us to propose above an incomplete shell filling are with this new model not longer valid. Due to heavy-hole-light-hole mixing also single particle states

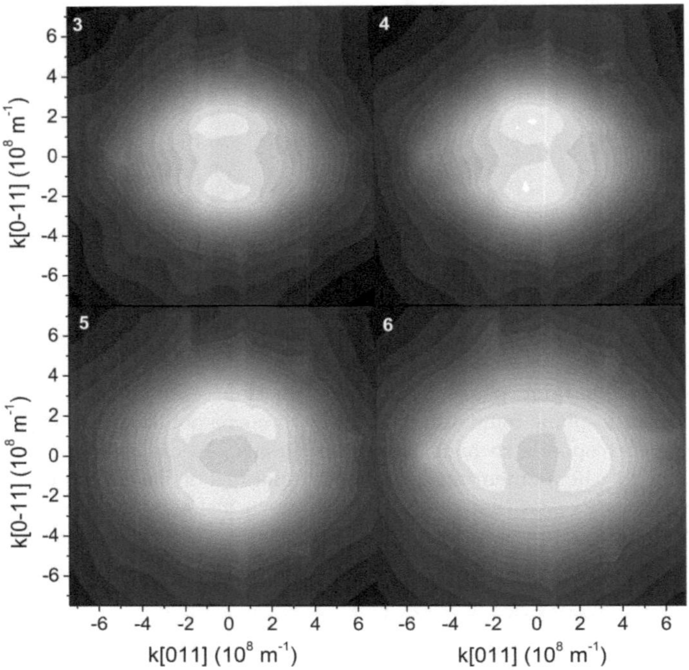

Fig. 8. C_{QD} of the valence band charging peaks 3 to 6 as function of an in-plane momentum generated by an in-plane magnetic field B. The colour coding is the same as in Fig. 7. One clearly sees a node like structure for all 4 peaks with is consistent with exited state wave functions with a finite orbital angular momentum component

with predominately p-character can show different dispersions [18]. On the other hand, a non-sequential filling sequence cannot be excluded on the basis of our data. Whereas *Climente* and co-workers [18] predict a sequential shell filling, *He* and co-workers predict a non-sequential shell filling [17]. It seems that the wave function shape for peaks 5 and 6 differs significantly from that of peak 3 and 4. This seems to point to a different character of the wave function supporting more the assignment of a d-like state for peak 5 and 6, i.e., suggesting a non-sequential shell filling as proposed in [17]. However, this question cannot be answered conclusive based on our experimental results and further work is necessary.

5 Summary

I have shown that C-V spectroscopy can today determine the charging spectra for conduction and valence band states of InAs QDs with high resolution. Measurements in perpendicular magnetic field help to identify the peaks with respect to the corresponding single particle level. Additionally, C-V spectroscopy at frequencies comparable to the tunnelling rate can be used to map the momentum space wave functions corresponding to the individual charging peaks by employing a parallel magnetic field [10, 11]. For the electrons a shell like energy structure is observed and the dispersion measurements as well as the wave function can be understood assuming a slightly asymmetric harmonic confinement potential, a single effective mass and treating the orbital angular momentum as good quantum number. For the valence band states this is not longer valid and more sophisticated theories have to be used. Also it seems that the filling of the single particle valence band levels occurs in a non-sequential manner.

Acknowledgement

I gratefully acknowledge the contribution of P. Schaafmeister, P. Kailuweit, M. Richter, R. Roescu, O. S. Wibbelhoff, A. Lorke, U. Zeitler, J. C. Maan, and A. D. Wieck for their contributions to the work presented above. Financial support by the DFG and the BMBF is gratefully acknowledged.

References

[1] D. Bimberg, M. Grundmann, and N.N. Ledentsov: *Quantum Dot Heterostructures*, (Wiley, New York, 1999), *Semiconductor Quantum Dots*, Eds. Y. Masamoto and T. Takagahara, (Springer, Berlin, 2002).
[2] H. Drexler, D. Leonard, W. Hansen, J.P. Kotthaus, and P.M. Petroff: Phys. Rev. Lett. **73**, 2252 (1994).
[3] M. Fricke, A. Lorke, J.P. Kotthaus, G. Medeiros-Ribeiro, and P.M. Petroff: Europhys. Lett. **36**, 197 (1996).
[4] B.T. Miller, W. Hansen, S. Manus, R. J. Luyken, A. Lorke, and J.P. Kotthaus: Phys. Rev. B **56**, 6764 (1997).
[5] R. J. Warburton, B.T. Miller, C.S. Dürr, C. Bdefeld, K. Karrai, J.P. Kotthaus, G. Medeiros-Ribeiro, and P.M. Petroff: Phys. Rev. B **58**, 16221 (1998).
[6] C. Bock, K.H. Schmidt, U. Kunze, S. Malzer, and G. Döhler: Appl. Phys. Lett. **82**, 2071 (2003).
[7] D. Reuter, P.Schafmeister, P. Kailuweit, and A.D. Wieck: Physica E **21**, 445 (2005).
[8] D. Reuter, P. Kailuweit, A.D. Wieck, U. Zeitler, O. Wibbelhoff, C. Meier, A. Lorke, and J.C. Maan: Phys. Rev. Lett. **94**, 26808 (2005).
[9] A. Patané, R.J.A. Hill, L. Eaves, P.C. Main, M. Henini, M.L. Zambrano, A. Levin, N. Mori, C. Hamaguchi, Yu. V. Dubrovskii, E.E. Vdovin, D.G. Austing, S. Tarucha, and G. Hill: Phys. Rev. B **65**, 165308 (2002),
E.E. Vdovin, A. Levin, A. Patané, L. Eaves, P.C. Main, Yu. N. Khanin, Yu. V. Dubrovskii, M. Henini, and G. Hill, Science **290**, 122 (2000).
[10] O.S. Wibbelhoff, A. Lorke, D. Reuter, and A.D. Wieck: Appl. Phys. Lett. **86**, 92104 (2005), Appl. Phys. Lett. **88**, 129901 (2006).

[11] P. Kailuweit, D. Reuter, A. D. Wieck, O. Wibbelhoff, A. Lorke, U. Zeitler, and J. C. Maan: Physica E **32**, 159 (2006),
D. Reuter, P. Kailuweit, R. Roescu, A. D. Wieck, O. S. Wibbelhoff, A. Lorke, U. Zeitler, and J. C. Maan: Phys. Stat. Sol.(b) **243**, 3942 (2006).
The directions for the hole wave function anisotropy are wrong in these publications. After careful experimental checks, we are sure that the directions as given in this contribution are the correct ones.
[12] R. J. Luyken, A. Lorke, A. O. Govorov, J. P. Kotthaus, G. Medeiros-Ribeiro, and P. M. Petroff: Appl. Phys. Lett. **74**, 2486 (1999).
[13] G. Medeiros-Ribeiro, J. M. Garcia, and P. M. Petroff: Phys. Rev. B **56**, 3609 (1997).
[14] N. Horiguchi, T. Futatsugi, Y. Nakata, and N. Yokoyama: Jpn. J. Appl. Phys. **36**, L1247 (1997).
[15] M. Rontani and E. Molinari: Jpn. J. Appl. Phys. **45**, 1966 (2006).
[16] O. Stier, M. Grundmann, and D. Bimberg: Phys. Rev. B **59**, 5688 (1999).
[17] L. He, G. Bester, and A. Zunger: Phys. Rev. Lett. **95**, 246804 (2005).
[18] J. I. Climente, J. Planelles, M. Pi, and F. Malet: Phys. Rev. Lett. **72**, 233305 (2005).

Coupling Phenomena in Dual Electron Waveguide Structures

Saskia F. Fischer[1]

Werkstoffe und Nanoelektronik, Ruhr-Universität Bochum,
D-44780 Bochum, Germany
saskia.fischer@rub.de

Abstract. Quantum transport in dual one-dimensional electron systems (1DES) is reviewed for different types of dual electron waveguides prepared from GaAs/AlGaAs-heterostructures by nanolithography. Tunneling and mode coupling phenomena can be observed measuring the conductance in applied electric and magnetic fields. This work classifies (a) laterally tunnel-coupled 1DES, (b) vertically-stacked tunnel-coupled 1DES and (c) spatially-coincident 1DES and reports on the control of quantum wire superposition states, which is envisioned for electronic waveguide devices.

1 Introduction

Nanoscale semiconductors reveal quantum properties in the charge carrier transport such as the conductance quantization for quantum wires, transmission through tunneling barriers or interference effects as the Aharonov–Bohm effect in quantum rings [1]. Along with the advances in the field of nanotechnology we gain increasing control in the preparation and manipulation of quantum objects, and transport properties of single quantum objects have been widely investigated. However, whether our deepened understanding of quantum transport will impact future nanoelectronics depends strongly on the implementation of complex quantum structures and networks. Therefore, fundamental properties of quantum transport in *interacting* quantum objects are of high interest. This review deals with aspects of coupling phenomena which occur in the electron transport through adjacent quantum wires, also referred to as dual waveguide structures.

Various concepts exist for nanoelectronic devices involving electron waveguides, some of which lie within the framework of solid-state quantum information processing, e.g., quantum logic gates [2–5] or resonant quantum waveguide networks [5–9]. The realization of quantum wire *systems* and *networks* in which coherent wave propagation persists is a non-trivial task, however, much progress has been achieved for one-dimensional electron systems (1DES) in dual waveguide structures as sketched in Fig. 1.

Typical for electron waveguides is the discrete energy spectrum of 1D subbands due to the discrete modes of standing waves that form in the quantum wire confinement [1]. All electronic waveguide confinements that are

R. Haug (Ed.): Advances in Solid State Physics,
Adv. in Solid State Phys. **47**, 55–66 (2008)

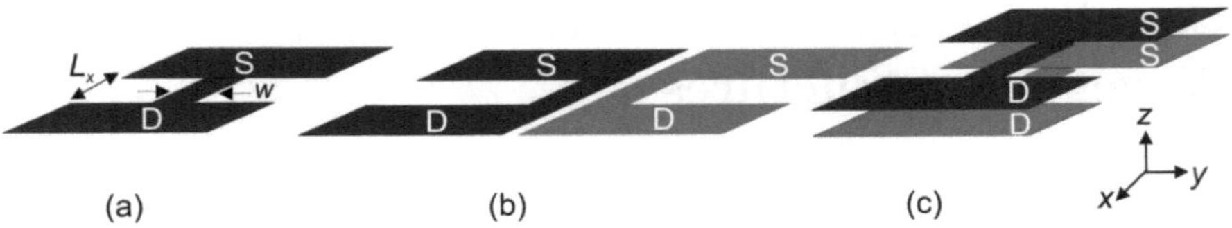

Fig. 1. Schematics of electron waveguides with source (S) and drain (D) contacts of (**a**) a single wire with characteristic lateral confinement width w and longitudinal length L_x, and of dual electron waveguides which are (**b**) laterally aligned and (**c**) vertically stacked

discussed below can be described in a first approximation by a (nearly) harmonic lateral confinement leading to the properties of 1D quantum harmonic oscillators [10]. Thus, mode coupling can be understood by the coupling of quantum harmonic oscillator modes.

To date, the prerequisites for a full control of mode coupling have been fulfilled for dual electron waveguides: First, a single-mode operation, second, a spectroscopy of the 1D subband structures of each 1DES in the uncoupled and coupled case, and third, a mode identification and the energetic variation of the involved 1D subband ladders.

2 Quantum Transport

Electrons at the Fermi energy E_F move ballistically with the Fermi velocity v_F if their mean free path $l_\mathrm{m} = v_\mathrm{F}\tau_\mathrm{m}$ exceeds the wire length L, here τ_m is the elastic scattering time. If in a 1D constriction only elastic scattering events take place phase coherence may persist [11]. Furthermore, if electrons experience a free movement in the longitudinal direction (x) but are confined in the lateral (y) and vertical (z) directions such that (see Fig. 2a) standing electron waves evolve as transverse and vertical modes, then a discrete 1D subband structure will result [1].

The confining potential $V(x, y, z)$ determines the energy eigenvalues E of the 1DES from the time-independent Schroedinger equation. If $V(x, y, z)$ is separable the electron wavefunction can be given by a product as $\Psi_{n,s}(x, y, z) = e^{-ik_x x}\phi_{n,s}(y)\chi_{n,s}(z)$, with the quantum numbers s and n of the vertical and lateral confinement, respectively. The discrete 1D energy spectra can be written as

$$E_{s,n}(k_x) = \frac{\hbar^2 k_x^2}{2m^*} + E_s + E_n \,.$$

The subband edges are given by $E_s + E_n$. Here, m^* is the effective mass and $\hbar$ is the reduced Planck constant. For every lateral mode n the 1D density of states $D_n(E) = 1/\pi \frac{dk}{dE}$ and the electron velocity $v_n = 1/\hbar \frac{dE}{dk}$ cancel. Therefore the current density I depends only on the difference of the chemical

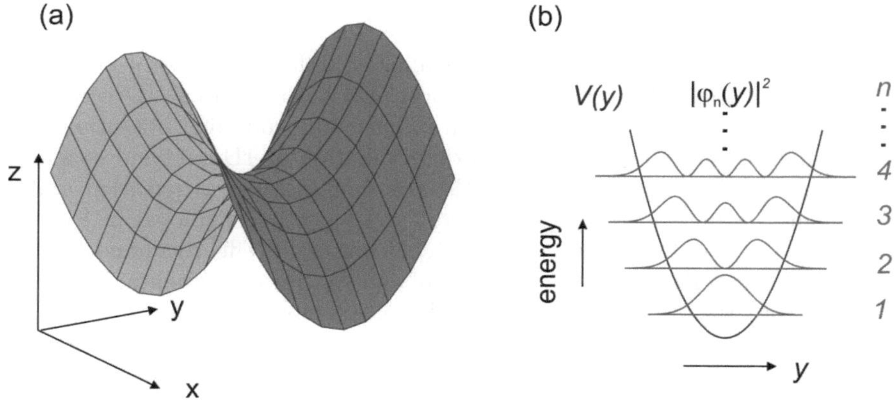

Fig. 2. (**a**) Saddle-point potential for short quantum wires or quantum point contacts allowing for a free electron movement along x. (**b**) A lateral harmonic confinement $V(y)$ at the center of a saddle-point potential but also valid for many elongated quantum wires. 1D subband edges are given by the discrete 1D harmonic oscillator eigenenergies of modes n. Plotted is the square of the corresponding transversal part of the electron wavefunction $|\phi_n(y)|^2$

potentials $\mu_b - \mu_a = eV_{ab}$ and we will measure a quantized conductance [12, 13], as is depicted, e.g., in Fig. 3c and given by $G = I/V_{ab}$:

$$I = \sum_n \int_{\mu_a}^{\mu_b} eD_n(E)v_n(E)dE = \frac{2e}{h}NeV_{ab} \quad \text{and} \quad G = \frac{2e^2}{h}N.$$

Short ballistic 1D constrictions are often well-described in a first approximation by a saddle-point potential, see (see Fig. 2a) [10] :

$$V(x,y) = V_0 + \frac{1}{2}m^*y^2\omega_y^2 - \frac{1}{2}m^*x^2\omega_x^2.$$

Here, V_0 is the electrostatic energy at the saddle point and ω_x, ω_y represent the harmonic oscillator strengths. Considering adiabatic constrictions [14] and only one vertical mode contribution s the subbands are given by

$$E_{s,n}(k_x) = \frac{\hbar^2 k_x^2}{2m^*} + V_s + \left(n - \frac{1}{2}\right)\hbar\omega_y$$

with equal subband spacings $\Delta E_{n+1,n} = \hbar\omega_y$, as the 1D quantum harmonic oscillator, see Fig. 2b.

The advantage of the lateral harmonic confining potential is immediately evident: The transverse eigenenergies E_n define the 1D subband edges and we exactly know the eigenfunctions of transverse modes represented by 1D quantum harmonic oscillator eigenstates

$$\phi_n(y) = (2^n n!\sqrt{\pi}y_0)^{-\frac{1}{2}} \left[\exp -\frac{1}{2}\left(\frac{y}{y_0}\right)^2\right] H_n\left(\frac{y}{y_0}\right)$$

with $y_0 = \sqrt{\frac{\hbar}{m^* \omega_y}}$, and the Hermite polynomials are given by $H(y) = (-1)^n \exp\left(y^2\right) \frac{d^n}{dy^n} \exp\left(-y^2\right)$. Orthogonality of different modes n simplifies the interpretation of mode coupling for 1DES with identical 1D confining potentials [22, 25, 29]. The assumption of harmonic confining potentials can be compared with the experimental determination of 1D subband spacings by transport spectroscopy. Hereby, the transconductance is recorded under finite source-drain bias voltage and grey-scale plots, as depicted in Fig. 4a reveal the subband spacings [14, 15]. A decrease of the effective nanochannel width leads to a linear increase of 1D-subband spacings by lithographical [16] or electrostatical [17] variation of the 1D constriction.

The temperature dependence of the quantized conductance is primarily determined by that of the Fermi–Dirac-distribution function $f_{\mathrm{FD}}(E - E_{\mathrm{F}}) = [1 + exp((E - E_{\mathrm{F}})/k_{\mathrm{B}}T)]^{-1}$. With increasing temperature T the conductance steps smear out according to the the thermal broadening of f_{FD}:

$$G(E_{\mathrm{F}}, T) = \int G(E, 0) \left(-\frac{f_{\mathrm{FD}}}{dE} \right) dE = 2e^2/h \sum_n f_{\mathrm{FD}}(E_n - E_{\mathrm{F}}).$$

3 Fabrication of Dual Electron Waveguides

Dual electron waveguide structures can be prepared by nanolithography from high-mobility two-dimensional electron gases (2DEGs) in GaAs-AlGaAs-heterostructures. Typically modulation-doped field-effect transistor structures are fabricated by conventional photolithography and wet-etching. Additionally, nanolithography must be employed to define a 1D channel, for example, by means of electron-beam lithography or scanning-probe lithography. In general, a 1DES can be formed by electrostatic depletion of the 2DEG, which has been performed for dual electron waveguides by split-gates [18–25] or by etch patterning [26–29] .

Interaction between one-dimensional electron systems may take place via two generic mechanisms. First, waveguides can be *spatially separated* by a barrier. Non-transparent barriers can lead to the observation of Coulomb drag between the adjacent 1D currents [30]. For barriers with a finite transmission, tunneling between the 1D wires will prevail which plays a central role in the observations of mode coupling. Second, within the same waveguide structure two (or more) energetically displaced 1D subband ladders can be formed. Such 1DES are classified as *spatially coincident* and they exist, if electrons are injected into the same geometrically defined 1D constriction, from two or more occupied subbands of the two-dimensional electron gas reservoirs.

Tunnel-coupled dual electron waveguide structures are particularly interesting for device concepts. They can be formed *laterally aligned* from one single 2DEG or *vertically stacked* from double-quantum well heterostructures. Electron waveguides arranged in lateral configuration, as for example shown

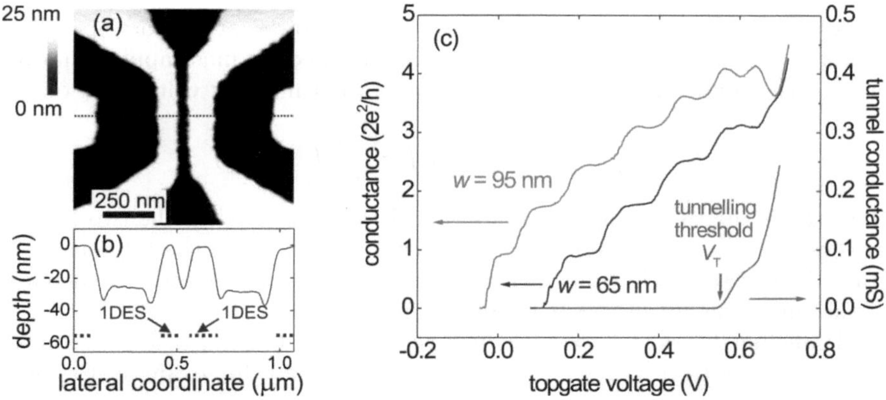

Fig. 3. (**a**) Atomic force micrograph of the inner part of a lateral 1D-1D tunneling device (from [26]). Dark areas represent etched regions where the electron gas underneath is depleted. The waveguides are 400 nm long and of geometric widths $w = 65$ nm and $w = 95$ nm. The image was taken before deposition of a global top gate. (**b**) Cross section of the structure along the *grey dashed line* in (**a**) The *dotted line* depicts schematically the depletion of the 2DEG in 55 nm depth. (**c**) Differential conductance of the two waveguides and across the barrier. A series resistance of about 800 Ω was corrected. For details see [26]

in Fig. 3a, can easily be contacted by individual source-drain reservoirs, and additionally a the tunnel transmission may be controlled by a gate voltage. However, the preparation of a homogeneous lateral tunneling barrier presents a considerable technological challenge. Instead, vertically-stacked electron waveguides in general exhibit atomically flat tunneling barriers as they are prepared by molecular-beam epitaxy and, thus, have a high-potential for fundamental investigations. However, such barriers have a distinct barrier height and width which defines a fixed tunnel transmission and the preparation of individual source-drain contacts to each quantum wire requires non-standard processing which is incompatible with lateral integration technology.

4 Laterally Coupled Electron Waveguides

Tunnel-coupled electron waveguides enable to investigate directional electron waveguide coupler [2–4] and resonant quantum waveguide networks or potentially even charge qubits [5–9] if each waveguide can be addressed via source and drain contacts. A lateral configuration would additionally allow to include the waveguides in integrated circuits by top-down fabrication technology.

In tunneling transport the wavevector component parallel to the barrier $k_{\|,l} = k_{\|,r} = k_{\|,\mathrm{barrier}}$, and the total energy of the tunneling electron $E_l = E_r \equiv E$ remains unchanged. The incident current density on both sides of the barrier is $j_{l,r} = -eD(\mathbf{k}_{l,r})f(\mathbf{k}_{l,r})v_\perp(\mathbf{k}_{l,r})d\mathbf{k}_{l,r}$ in the infinitesimal range of $d\mathbf{k}_{l,r}$. $D(\mathbf{k}_{l,r})$ and $D(\mathbf{k}_{l,r})$ depict the density of states for the corresponding contacts on the left and right side [11]. In the effective mass-approximation we find the tunneling current density [31] at the tunneling voltage V_{T} of

$$j_{\mathrm{T}} = \frac{2e}{h} \int dE[f_r(E) - f_l(E + eV_{\mathrm{T}})] \int \frac{d^2k_{\|}}{(2\pi)^2} T^*(E, \mathbf{k}_{\|}) \,.$$

Here, T^* is the tunneling probability. In 1D $k_{\|}$ is reduced to the two Fermi-wavevectors of the quantum wire and the discrete nature of the 1D-density of states is represented in tunneling spectroscopy in a non-monotonous increase of the tunneling current I_{T} with increasing V_{T}.

Tunneling barriers have been implemented by top-down fabrication as finger-gate electrodes [18, 19, 32], or via lithography for example by local barriers by local anodic oxidation [33]. Lithographically defined thinner barriers are achieved by wet-etching [26]. By usage of side-gate and finger-gate electrodes waveguides and barrier were defined and investiated in 2D-2D-[32, 34, 35] and 2D-1D-configuration [18, 19]. 1D-1D lateral waveguide structures were also prepared by splitgate structures in combination with finger gates [20, 36, 37]. In order to achieve ultra-homogenous tunneling barriers heterostructures were used, and quantum wires contacted indiviually in a *flip-chip*-technique [38, 39] or using *cleaved-edge-overgrowth*-processing [40, 41]. However, both fabrication techniques can hardly be incorporated in standard top-down processing and do not allow for a variable tunnel transmission.

Etched waveguide structures that exhibit a strong confinement with 1D-subband spacings exceeding 10 meV are shown in Fig. 3a. Here, a GaAs/AlGaAs-heterostructure hosts a 2DEG about 55 nm beneath the surface with an electron density of $n = 4.4 \times 10^{11}$ cm^{-2} and electron mobility of $\mu = 1.1 \times 10^6$ cm$^2V^{-1}s^{-1}$ (measured in the dark at T= 4.2 K). For experimental details in processing and measurements see [26]. Both waveguides exhibit clear conductance quantization in units of $2e^2/h$ as depicted in Fig. 3b. The tunneling conductance measured *across* the barrier reflects the transmission through the potential barrier.

In Fig. 4b, the tunnel transmission increases while raising the applied voltage of the common top gate, and therefore simultaneously, higher 1D subbands become occupied in the waveguides. The tunneling threshold coincides with significant losses in the waveguide conductance of the wide wire which indicate a tunneling leakage. Transport spectroscopy allows to determine subband spacings of $\Delta E_{1,2} = 8.0 \pm 0.5$ meV (see Fig. 4a) and $\Delta E_{1,2} =$ 10.0 $\pm$0.5 meV for the two waveguides [26].

Figure 4b shows a series of tunnel spectra for different gate voltages. Conductance oscillations appear for top-gate voltages larger than the transmission threshold mapping the density of states of the coupled 1D electron systems. The oscillation period measured for large top-gate bias corresponds to an energy interval which is of the same order of magnitude as the subband energy separations found in each waveguide. The amplitude of the oscillations increases with increasing top gate voltage as the effective tunnel barrier

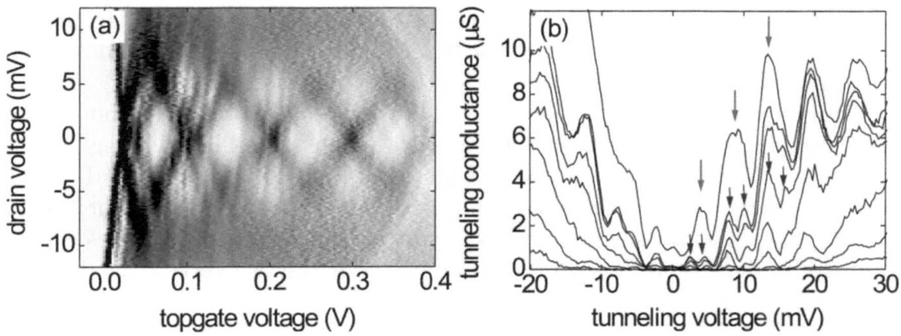

Fig. 4. Lateral 1D-1D tunneling device at 4.2 K: Grey scale plots of the calculated transconductance from the measured conductance of the 400 nm long and 95 nm wide quantum wire in Fig. 3. Tunneling conductance vs. bias voltage for different top gate voltages from 0.5 V to 0.646 V in intervals of 0.08 V. Details are given in [26]

height is decreased. Additionally, the periodicity of the tunneling oscillations depends on the top-gate voltage as depicted by arrows in Fig. 4b. While such phenomena might reflect 1D mode coupling effects between the two electron waveguides variations of waveguide widths and barrier inhomogeneities must be excluded first. Similarly, tunneling spectroscopy of 1D subband edges in waveguide split gate structures has been reported [20] as well as in *cleaved-edge-overgrowth*-processing [41].

On one hand, it follows that advances in top-down processing enable us to fabricate elongated electron waveguides with 1D subband spacings of more than 10 meV which allow high-resolution transport and tunneling spectroscopy at temperatures of 4.2 K and higher. On the other hand, recent experiments [36, 37] and numerical simulations [42] state that potential asymmetries and local minima must be avoided in waveguides thought to demonstrate a directional electron waveguide coupling [2, 3, 5]. This puts considerable constraints, in particular a nanometer-scale precision, on the fabrication process. The visionary concept of charge qubits furthermore demands the experimental realization of the entanglement of pairs of tunnel-coupled waveguides [9, 43].

5 Vertically-Stacked Electron Waveguides

Mode coupling phenomena are observed in vertically-stacked waveguides: The subband structure of each waveguide can be significantly modified when 1D-subband edges of the two subladders become degenerate, $E_{n,s} = E_{n^*,s^*}$. If the 1D modes (n, s) and (n^*, s^*) are coupled [44] by the matrix element $M = \langle \Psi_{s,n} | V(z) | \Psi_{s^*,n^*} \rangle$, where $V(z)$ is the potential between the coupled wires, superposition states $\Psi_{a,s} = \Psi_{s,n} \pm \Psi_{s^*,n^*}$ are formed and the degeneracy is lifted by $\Delta E_{(n,s),(n*,s*)}$. The symmetry of the confining potential determines whether mode coupling occurs and the coupling strength is given by the overlap of the electron wavefunctions of the two degenerate modes [22,44,45].

Tunnel-coupled vertically-stacked 1DES have been prepared from GaAs double-quantum well heterostructures with a thin tunnel barrier (e.g., 1 nm AlGaAs) [22,28,29]. An example is depicted in Figs. 5a and b. Instead, GaAs quantum wells with two occupied subbands lead to spatially coincident 1DES. If the vertical confinement is slightly asymmetric a difference in the electrons probability density in vertical direction exists for the two vertical modes and these systems can also be classified as vertically stacked, see [25,27].

In dual 1DES with common source-drain reservoirs mode identification is of particular importance. An energetic shift of the 1D subladders relative to each other may be established by a variation of the confinement due to applied magnetic or electric fields. Figure 5c shows the effect of different backgate voltages on the conductance vs. topgate voltage characteristic. Here, double steps in the conductance of $4e^2/h$ signal near-coincidences or degeneracies of subband edges. The backgate voltage changes the vertical asymmetry in the confining potential. If the conductance is quantized at $2e^2/h$ but exibits unusual small plateau widths with increasing gate voltage, this indicates split subbands due to mode coupling. A mode spectrum is best visualized by plotting the transconductance maxima (see Fig. 4d) for applied gate voltages (for example top- and backgate [28] or splitgates [22]).

Magnetic fields can be applied to dual 1DES in order to identify the modes of each 1D subsystem on behalf of distinct magnetodispersions of the two 1DES [48]. Mode spectra have been recorded in transversal and longitudinal magnetic fields for spatially coincident [25, 27] and tunnel-coupled systems [21–24, 29]. Subband degeneracies are displayed in mode spectra as a level crossings and mode coupling induced by wavefunction hybridization as level anticrossings. 1D subband anticrossings are reported from various investigations [21, 23, 25].

Splitting energies due to mode coupling can be estimated correcting the applied gate voltages with by the gate efficiency [22, 24, 27]. However, also direct transport spectroscopy can be applied successfully as shown in Fig. 6a in the case of mode-coupling of vertically-stacked tunnel-coupled 1DES with nearly identical 1D confining potentials [28]. In this case, the energy separations between the first and second 1D subband of the top and bottom 1DES amount to 12.3 and 10.8 meV, respectively. Here, all subband edges of identical mode index n become mode-coupled at zero back gate voltage and splitting energies of 5.4 meV, 3.9 meV and 2.9 meV occur for $n = 1, 2$, and 3, respectively.

Mode coupling persists for temperatures above 10 K, as shown at zero source-drain voltage in Fig. 6b: The transconductance peaks of the split levels are unambiguously distinguishable. The thermal dephasing time decreases from 6 ps at 4.2 K to 2.4 ps at 10 K and is in the investigated heterostructures above 4 K the dominating time scale [28]. A scattering time of 9 ps (from mea-

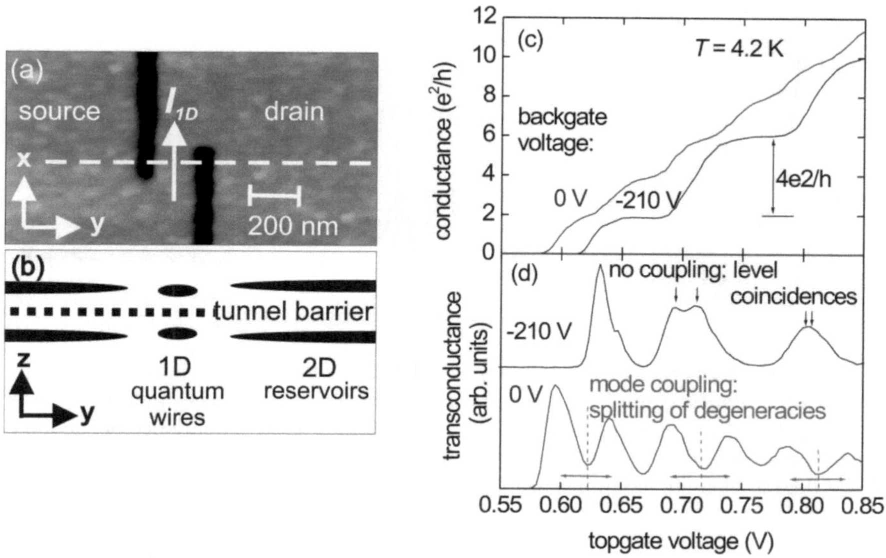

Fig. 5. Vertically-stacked 1DES prepared from a 30 nm wide GaAs double-quantum well with a 1 nm centered AlGaAs barrier: (**a**) atomic force topgraphy micrograph of etched nanogrooves (*dark*) which define the lateral constriction width. (**b**) Schematic of cross-sectional view of 2D and 1D charge distribution (*black*) separated by a tunnelbarrier (*dotted line*). Typical (**c**) conductance and (**d**) transconductance measurements versus top gate voltage at 4.2 K which show effects of coupled (backgate: 0 V) and decoupled (backgate: -210 V) modes. Details see [28,29,49]

surements of the electron mobility) and phase-coherence times of about 60 ps in single 1DES ensure the so-called coherent quantum wire states. Therefore, strong indications exists that the coherence in the transverse modes meets the requirements for superposition states formed by coupling.

Our ability to control coupled modes has been demonstrated in particular by the magneto-oscillations of energy splittings in longitudinal magnetic fields as shown in Fig. 6c and detailed in [29]. The experimental findings are well reproduced by analytical calculation in the framework of the simple harmonic oscillator model [50] and by numerical simulations [42].

6 Summary and Outlook

To date, quantum transport in interacting one-dimensional electron systems is investigated in a variety of dual electron waveguide structures, such as laterally aligned or vertically stacked, spatially separated or coincident. In each case the electrostatic confining potentials forming the waveguides play a decisive role: The confinement strength and geometry determine the 1D subband structure and the symmetry influences the degree of mode coupling.

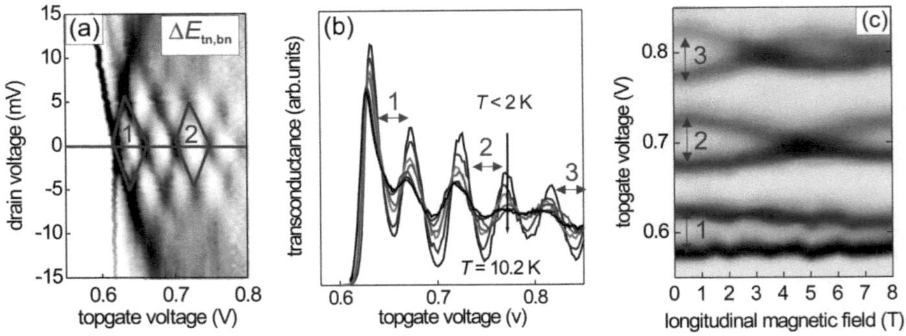

Fig. 6. Vertically-stacked tunnel-coupled 1DES as in Fig. 5 in the case of mode-coupling for identical modes n: (**a**) Transport spectroscopy of the first two energy splittings labelled 1 and 2, details see [28]. (**b**) Temperature dependence of the measured transconductane (*line* along zero drain voltage in (a), see [28]. (**c**) Variation of the degree of mode coupling in longitudinal magnetic fields [29]

Commonly, the vertical component of the confining potential is defined by the molecular-beam epitaxy of band-edge engineered GaAs/AlGaAs-heterostructures hosting high-mobility electron gases in quantum wells. However, future material choices that guarantee quantum coherent transport and additionally exhibit different effective masses, valley degeneracies or large g-factor might be of interest.

In general, the lateral confinement can be achieved in a large geometrical variety by means of advanced nanolithography allowing for a single-mode operation at liquid-helium temperature and above. The opportunity of stable high temperature operation might be crucial for potential applications. High-resolution of mode coupling in transport spectroscopy demonstrates a high degree of coherence in the quantum superposition states generated in the investigated 1DES.

The results indicate that we have access to electron waveguide structures which should allow the preparation and manipulation of coherent electronic states in the quantum transport. Combining dual electron waveguides into larger quantum-wire systems or network, however, represents a considerable challenge, both, in technological implementation as well as in fundamental aspects relating to the field of solid-state quantum information processing. Novel phenomena are to be expected from investigations of coherent electron transport in coupled quantum wire systems in combination with single-electron control and spin-related phenomena.

Acknowledgements

I gratefully acknowledge the scientific contributions of my collaborators Gabriela Apetrii, Sven S. Buchholz, Jean-Laurent Deborde, and Ulrich Kunze from Werkstoffe und Nanoelektronik, Ruhr-Universität Bochum, Germany. The GaAs/AlGaAs heterostructures were partly grown by Dirk Reuter and Andreas D. Wieck, Angewandte Festkörpphysik, Ruhr-Universität Bochum and Dieter Schuh, Universität Regensburg and Gerhard Abstreiter, Walter-Schottky Institut, Technische Universität München, Germany. Part of this work was supported by the Bundesministerium für Bildung und Forschung under grants no. 01BM920 and NanoQUIT 01BM454.

References

[1] C. W. J. Beenakker, H. van Houten: Solid State Physics **44**, 1–228 (1991)
[2] N. Tsukuda, A. D. Wieck, K. Ploog: Appl. Phys. Lett. **56**, 2527 (1990)
[3] A. Bertoni, P. Bordone, R. Brunetti, C. Jacoboni, S. Reggiani: Phys. Rev. Lett. B **84**, 5912 (2001)
[4] J. Harris, R. Akis, D. K. Ferry: Appl. Phys. Lett. **79**, 2214 (2001)
[5] M. J. Gilbert, R. Akis, D. K. Ferry: Appl. Phys. Lett. **81**, 4284 (2002)
[6] M. Sabathil, D. Mamaluy, P. Vogl: Semicond. Sci. Technol. **19**, S137 (2004)
[7] G. B. Akguc, L. E. Reichl, A. Shaji, M. G. Snyder: Phys. Rev. A **69**, 042303 (2004)
[8] M. G. Snyder, L. E. Reichl: Phys. Rev. A **70**, 052330 (2004)
[9] R. Ioniciou, G. Amaratunga, F. Udrea: Int. J. Mod. Physics B **15**, 125 (2001)
[10] M. Büttiker: Phys. Rev. B **41**, 7906 (1990)
[11] D. K. Ferry, S. M. Goodnick: *Transport in Nanostructures*, Cambridge Studies in Semiconductor Physics and Microelectronic Engineering, (Cambridge University Press, 1997)
[12] B. J. van Wees, et al.: Phys. Rev. Lett. **60**, 848 (1988)
[13] D. A. Wharam, et al.: J. Phys. C: Solid State Phys. **21**, L209 (1988)
[14] L. I. Glazman and A. V. Khaetskii: Europhys. Lett. **9**, 263 (1989)
[15] N. K. Patel, et al.: PRB **44**, 13549 (1991)
[16] G. Apetrii, S. F. Fischer, U. Kunze, D. Reuter, A. D. Wieck: Semicond. Sci. Technol. **17**, 735 (2002)
[17] S. F. Fischer, S. Skaberna, G. Apetrii, U. Kunze, D. Reuter, A. D. Wieck: Appl. Phys. Lett. **81** (15), 2779–2781 (2002)
[18] C. G. Eugster and J. A. del Alamo: Phys. Rev. Lett. **67**, 3586 (1991)
[19] C. G. Eugster, J. A. del Alamo, M. R. Melloch, M. J. Rooks: Phys. Rev. B **48**, 15057 (1993)
[20] C. G. Eugster, J. A. del Alamo, M. J. Rooks, M. R. Melloch: Appl. Phys. Lett. **64**, 3157 (1994).
[21] I. M. Castleton,A. G. Davies, A. R. Hamilton, J. E. F. Frost, M. Y. Simmons, D. A. Ritchie, M. Pepper: Physica B **249–245**, 157 (1998)
[22] K. J. Thomas, J. T. Nicholls, M. Y. Simmons, W. R. Tribe, A. G. Davies, M. Pepper: Phys. Rev. B **59**, 12252 (1999)
[23] M. A. Blount, J. S. Moon, J. A. Simmons, S. K. Lyo, J. R. Wendt, J. L. Reno: Physica E **6**, 689 (2000)
[24] K. J. Friedland, T. Saki, Y. Hirayama, K. H. Ploog: Physica E **11**, 144 (2001)
[25] G. Salis, T. Heinzel, K. Ensslin, O. J. Herman, W. Bchtold, K. Maranowski, A. C. Gossard: Phys. Rev. B **60**, 7756 (1999)

[26] J. L. Deborde, S. F. Fischer, U. Kunze, D. Reuter, A. D. Wieck: ICPS-28, AIP Conf. Proc. **893**, 725 (2007)
[27] S. F. Fischer, G. Apetrii, S. Skaberna, U. Kunze, D. Schuh, G. Abstreiter: Phys. Rev. B **71**, 195330 (2005)
[28] S. F. Fischer, G. Apetrii, S. Skaberna, U. Kunze, D. Schuh, G. Abstreiter: Nature Physics **2**, 91 (2006)
[29] S. F. Fischer, G. Apetrii, S. Skaberna, U. Kunze, D. Schuh, G. Abstreiter: Phys. Rev. B **74**, 115324-1-7 (2006)
[30] M. Yamamoto, M. Stopa, Y. Tokura, Y. Hirayama, S. Tarucha: Science **313**, 204.207 (2006).
[31] E. Burstein (ed.): *Tunneling Phenomena in Solids*, Stig Lundqvist, (Plenum Press, New York, 1969)
[32] S. J. Manion, L. D. Bell, W. J. Kaiser, P. D. Maker, R. E. Muller: Appl. Phys. Lett. **59**, 213 (1991)
[33] U. Keyser, H. W. Schuhmacher, U. Zeitler, R. J. Haug, K. Eberl: Appl. Phys. Lett. **76**, 457 (2000)
[34] S. K. Lyo, J. A. Simmons: J. Phys. Condens. Matter **5**, L299 (1993)
[35] G. Rainer, J. Smoliner, E. Gornik, G. B"ohm, G. Weimann: Phys. Rev. B **51**, 17642 (1995)
[36] P. Pingue, et al.: Appl. Phys. Lett. **86**, 05102 (2005)
[37] A. Ramamoorthy, et al.: Appl. Phys. Lett. **89**, 013118 (2006)
[38] E. Bielejec, J. A. Seammons, J. L. Reno, M. P. Lilly: Appl. Phys. Lett. **86**, 083101-1-3 (2005)
[39] E. Bielejec, J. A. Seammons, J. L. Reno, S. K. Lyo, M. P. Lilly: Physica E **34**, 433–436 (2006)
[40] O. M. Ausländer, et al.: Science **295**, 825 (2002)
[41] F. Ertl, S. Roth, D. Schuh, M. Bichler, G. Abstreiter: Physica E **22**, 292 (2004)
[42] T. Ziebold, P. Vogl: private communication, 2007; Numerical calculation with NextNano, S. Birner, P. Vogl: http://www.wsi.tum.de/nextnano3/tutorial/2Dtutorial_CoupledQWRsMagnetic.htm
[43] L. E. Reichl, M. G. Snyder: Phys. Rev. A **72**, 032330 (2005)
[44] E. Merzbacher, Treatments of mode coupling, in: *Quantum Mechanics*, 2nd ed (Wiley, New York, 1970), pp. 428–429 or C. Cohen-Tannoudji, B. Diu, F. Lalo e: *Quantum Mechanics Vol. 1* (John Wiley & Sons, 2005) ISBN 0-471-16433-X(v.1), pp. 406.
[45] N. Mori, P. H. Beton, J. Wang, L. Eaves: Phys. Rev. B **51**, 1735 (1995)
[46] S. F. Fischer, G. Apetrii, U. Kunze, D. Schuh, G. Abstreiter: Phase Transitions **79** (9–10), 815–825 (2006)
[47] S. F. Fischer, G. Apetrii, U. Kunze, D. Schuh, G. Abstreiter: ICPS-28, AIP Conf. Proc., in print
[48] S. F. Fischer: Int. Journal of Modern Physics B, Issue 8 March 30 (2007)
[49] S. F. Fischer G. Apetrii, U. Kunze, D. Schuh and G. Abstreiter: Physica E **34**, 568–571 (2006)
[50] L. G. Mourokh, A. Y. Smirnov, S. F. Fischer: Appl. Phys. Lett. **90**, 132108-1-3 (2007)

Part II

Correlation Effects

Electronic Correlations in Electron Transfer Systems

Ralf Bulla[1], Sabine Tornow[1], and Frithjof Anders[2]

[1] Theoretische Physik III, Elektronische Korrelationen und Magnetismus, Institut für Physik, Universität Augsburg, 86135 Augsburg, Germany
Ralf.Bulla@Physik.Uni-Augsburg.De
[2] Fachbereich Physik, Universität Bremen, 28334 Bremen, Germany

Abstract. Electron transfer processes play a central role in many chemical and biological systems. Already the transfer of a single electron from the donor to the acceptor can be viewed as a complicated many-body problem, due to the coupling of the electron to the infinitely many environmental degrees of freedom, realized by density fluctuations of the solvent or molecular vibrations of the protein matrix. We focus on the quantum mechanical modelling of two-electron transfer processes whose dynamics is governed by the Coulomb interaction between the electrons as well as the environmental degrees of freedoms represented by a bosonic bath. We identify the regime of parameters in which concerted transfer of the two electrons occurs and discuss the influence of the Coulomb repulsion and the coupling strength to the environment on the electron transfer rate. Calculations are performed using the non-perturbative numerical renormalization group approach for both equilibrium and non-equilibrium properties.

1 Introduction

1.1 Electron Transfer

The transfer of an electron from a donor to an acceptor via a bridge is an essential ingredient of numerous reactions occuring in biological and chemical systems [1, 2]. The transfer process itself is typically a non-equilibrium process: the donor is prepared in an excited state, for example through irradiation with light or by accepting an electron from another source. An interesting physical quantity in this respect is the occupation probability of the donor site by the excess (or excited) electron – we call this quantity $P(t)$ – which typically decays exponentially with time due to the transfer of the electron to the acceptor. The occupation probability $P(t)$ – or occupation $P(t)$, for short – can be measured experimentally (via femtosecond spectroscopy, see, for example, [3]) and is the focus of the theoretical calculations we present in this paper.

Electron transfer is a quantum mechanical process since the electron has to tunnel through a barrier (the bridge). The quantum mechanical nature is best seen experimentally when the temperature dependence of the electron transfer rate is followed from physiological temperatures down to a few

R. Haug (Ed.): Advances in Solid State Physics,
Adv. in Solid State Phys. **47**, 69–78 (2008)

Kelvin, where electron transfer reactions are still possible, with a reduced or even an enhanced rate [4].

Using Fermionic creation and annihilation operators $c^\dagger_{\mathrm{D/A}}$, $c_{\mathrm{D/A}}$ for the electrons at donor/acceptor sites, we can write the tunneling part of the Hamiltonian as

$$H_\Delta = -\frac{\Delta}{2}\left(c^\dagger_{\mathrm{D}} c_{\mathrm{A}} + c^\dagger_{\mathrm{A}} c_{\mathrm{D}}\right) , \tag{1}$$

with Δ the tunneling amplitude. This is, however, only part of the story. As first realized by *Marcus* [5], the electron transfer process is accompanied by a reorganization of the environment, since the volume occupied by the donor and acceptor ions depends on their charge. Furthermore, the change of the dipole moment due to the transfer leads to a rearrangement of polar molecules in the environment. Within the celebrated Marcus theory, such a reorganization is modeled by a single reaction coordinate, and a semiclassical analysis of the resulting model leads to a highly successful quantitative description of electron transfer processes.

Nevertheless, the reduction to a single reaction coordinate is an approximation, and if we wish to keep the coupling to the full spectrum of bath modes in our model, we arrive at the following description:

$$H_{\mathrm{b}} = \sum_n \omega_n b^\dagger_n b_n \ , \quad H_{\mathrm{c}} = \frac{1}{2}(n_{\mathrm{D}} - n_{\mathrm{A}}) \sum_n \lambda_n \left(b^\dagger_n + b_n\right) \ , \tag{2}$$

where H_{b} describes a collection of harmonic oscillators with frequency ω_n. For the term H_{c}, we assume a linear coupling of the charge difference, $n_{\mathrm{D}} - n_{\mathrm{A}}$, to the displacements of the bath modes, $b^\dagger_n + b_n$, with a coupling strength λ_n.

1.2 Models

The Hamiltonian $H = H_\Delta + H_{\mathrm{b}} + H_{\mathrm{c}}$ corresponds to the well-known spin-boson model [6], provided we have a single electron in the system. Much is known about the real-time dynamics of the spin-boson model (at least for the Ohmic case) and we show in Fig. 1 the result from a recent calculation using the time-dependent numerical renormalization group method [7, 8].

In these calculations, the system is prepared such that the donor is occupied by one electron at time $t = 0$, therefore we have for the occupancy at the donor site $P_{\mathrm{D}}(t = 0) = 1$. Results are shown here for different values of α, which parameterizes the coupling strength to the bath [6], and for zero bias (same on-site energies at donor and acceptor).

For increasing values of α, we can identify three regions: (i) coherent oscillations for small α, (ii) exponential relaxation for intermediate values of α, and (iii) localization of the electron at the donor site above a critical α. For the details of this calculation we refer the reader to [8,9], but we want to stress here that already for a single electron, the system donor/acceptor/bath is a

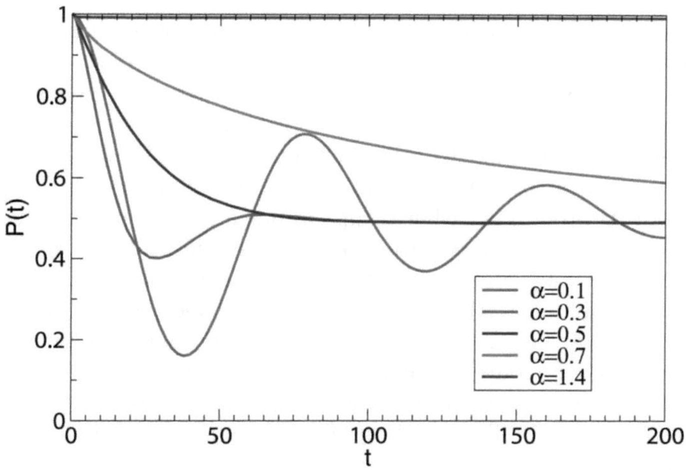

Fig. 1. Time-dependence of the occupation at the donor site, $P(t)$, with the system prepared such that at time $t = 0$ the electron sits at the donor. Results are shown for different coupling strengths α between the donor/acceptor system and the environment

complicated many-body problem for which exact results are available only for a few limiting cases (for example, the real-time dynamics in the case of a sub-Ohmic bath has only recently been investigated non-perturbatively, see [10]).

Let us now turn to the description of *two*-electron transfer processes, for which a description in terms of the spin-boson model is clearly not applicable. A suitably generalized model, as proposed in [11,12] takes the following form

$$\begin{aligned} H = & \sum_{\sigma,i=\mathrm{A,D}} \varepsilon_i c^\dagger_{i\sigma} c_{i\sigma} - \frac{\Delta}{2} \sum_{\sigma} \left(c^\dagger_{\mathrm{D}\sigma} c_{\mathrm{A}\sigma} + c^\dagger_{\mathrm{A}\sigma} c_{\mathrm{D}\sigma} \right) \\ & + U \sum_{i=\mathrm{A,D}} c^\dagger_{i\uparrow} c_{i\uparrow} c^\dagger_{i\downarrow} c_{i\downarrow} + \sum_n \omega_n b^\dagger_n b_n \\ & + (g_\mathrm{A} n_\mathrm{A} + g_\mathrm{D} n_\mathrm{D}) \sum_n \frac{\lambda_n}{2} \left(b^\dagger_n + b_n \right) \ . \end{aligned} \quad (3)$$

Here we add the spin degree of freedom to the fermionic operators $c^{(\dagger)}_{D/A\sigma}$ and also include a Coulomb repulsion U between electrons at donor and acceptor sites.

When we now prepare the system such that, at time $t = 0$, there are two electrons at the donor site, these two electrons will interact both with the bosonic bath and with each other. As we will show below, the Coulomb repulsion between the electrons as well as the interaction with the bath generates correlations which significantly influence the dynamics and the type of the electron transfer process; in other words, when going from single- to two-electron transfer, the difference is more than just a factor of two.

A few remarks are in order on actual realizations of two- (in general: multi-) electron transfer reactions. There are, in fact, numerous examples for such reactions (see, for example, [13]), but for concreteness, let us briefly discuss electron transfer in the protein 'respiratory arsenite oxidase' [14]. The biological function of this protein is rather complex, but a certain part of the reaction involves a two-electron transfer from a molybdenum center to iron-sulfur centers. There is strong evidence that both electrons are transfered simultaneously, or that, at least, the rate of the second electron is much faster than that of the first electron. This is essential for many biological electron transfer systems, as a purely single-electron transfer would lead to intermediate reaction products, e.g., superoxide, which are very reactive and therefore harmful to the cell [15]. The question now is how nature manages to avoid this harmful intermediate stage–and based on the results we present for the model (3) we claim that electronic correlations are the key to understand this remarkable feature of multi-electron transfer reactions in proteins.

Although our calculations are performed in the nuclear tunneling regime our results are still applicable as long as the temperature is smaller than the Coulomb matrix elements. Electron pair transfer is a tunneling process which we also expect at room temperature similar as the superexchange transfer or transfer in the inverted region [1].

2 Numerical Renormalization Group

The dynamics of the Hamiltonian (3) is investigated with the numerical renormalization group method (NRG). This technique has been originally developed by *Wilson* for the Kondo problem in which a magnetic impurity couples to the electrons of a conduction band [16]. The basic technical steps of the NRG are: (i) a logarithmic discretization of the bath continuum, (ii) a mapping of the discretized model onto a semi-infinite chain, and (iii) an iterative diagonalization of the chain-Hamiltonian (for a review, see [17]).

Applications of the NRG have focused on quantum impurity models, where a small system (such as a magnetic impurity) couples to the infinitely many degrees of freedom of a non-interacting bath (such as the electrons in a conduction band or vibrational modes in a molecule). We can also view the model (3) as a quantum impurity model because the donor-acceptor system contains only a small number of degrees of freedom whereas the bosonic environment is characterized by a continuous bath covering a broad range of energies.

Most of the NRG applications have been for impurities in fermionic baths, and the method has only recently been generalized to bosonic baths [9, 18]. Even more recent is the generalization of the NRG for the calculation of time-dependent quantities, such as the occupation $P(t)$ as plotted in Fig. 1. For the details we refer the reader to [8].

At this point one might ask why it is at all reasonable to use a method as sophisticated as the NRG for electron transfer problems as introduced above. First of all, the dissipation strength in many biological molecules has to be large enough to avoid oscillations between donor and acceptor (this would reduce the efficiency of the electron transfer process). This requires an approach which is non-perturbative in the coupling strength α. Second, the spectral function of the bath, $J(\omega)$, covers a broad spectrum of modes, from an upper cut-off of the order of 10–100 meV, down to very low energies. This can be seen in molecular dynamics simulations [19], in which the lower cut-off is only limited by the size of the system that can be handled numerically. The modes from all energies are relevant for the dissipation and, apparently, renormalization group approaches are designed to cover such a broad range of energies.

3 Electron Transfer: Results

3.1 Equilibrium

Let us start with the calculated phase diagram of the two-site model (3). For these calculations, we used a zero bias, $\epsilon = \epsilon_{\rm A} - \epsilon_{\rm D} = 0$, as this enables a clear determination of a phase boundary–for any finite ϵ, the phase transition turns into a smooth crossover. In Fig. 2 we plot the phase boundaries between localized and delocalized phases in the α-U-plane, both for single- and two-electron subspaces. When we have only a single electron in the system, the Coulomb repulsion does not play a role, therefore the phase boundary does not depend on U. The value of the critical coupling strength, $\alpha_{\rm c}$, is identical to the corresponding spin-boson model. For $\Delta \to 0$, a value of $\alpha_c = 1$ has been found [6]; the deviation from $\alpha_c = 1$ in Fig. 2 is caused by the finite value of the tunneling rate Δ used here (see also, for example, Fig. 8 in [9]).

On the other hand, the phase boundary for the two-electron subspace *does* depend on U, which has drastic consequences for the electron transfer process. Imagine that the parameters are such that the system is placed between the two phase boundaries, for example, for $U > 2.2$ above the single-electron and below the two-electron phase boundary. This means that the system is localized in the single-electron subspace, but the addition of a second electron immediately places the system in the delocalized phase, and one or even both electrons can be transfered.

Note also the different values of the $\alpha_{\rm c}$'s for $U = 0$. This is because the interaction of the donor/acceptor system with a common bosonic bath induces an effective (attractive) interaction between the two electrons, so that they are correlated even without an explicit Coulomb interaction. See [11] for more details about the equilibrium properties of the model (3).

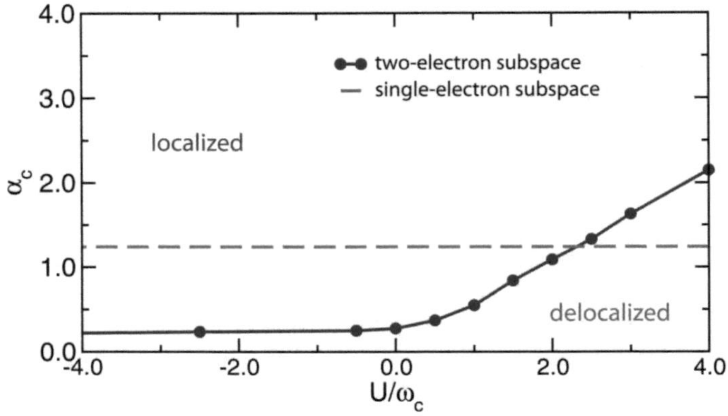

Fig. 2. Phase diagram of the two-site model (3) for both single- and two-electron subspaces. The phase boundaries separate delocalized phases, in which electron transfer of one or two electrons from donor to acceptor is allowed, from localized phases in which no electron transfer occurs

3.2 Non-Equilibrium

Phase diagrams as shown in Fig. 2 already provide us with some information about the physics of the two-site model, but for a deeper understanding of the dynamics of the electron transfer process, we have to discuss time-dependent quantities such as the occupation $P(t)$ shown in Fig. 1. In Fig. 3, we show the real-time dynamics of the double occupancies $P^{\rm d}_{\rm D/A}(t)$ for the two-site model with parameters, $U = \omega_c$, $\Delta = 0.1$, and different values of α. The system is prepared such that, at time $t = 0$, both electrons (with spin $\uparrow$ and $\downarrow$) are located at the donor site: $P^{\rm d}_{\rm D}(t = 0) = 1$, $P^{\rm d}_{\rm A}(t = 0) = 0$.

For $\alpha = 0$, the donor/acceptor system is decoupled from the bosonic bath and the two electrons perform coherent oscillations with frequencies $\omega_1 = U$ and $\omega_2 = 4\Delta^2/U$, as given by the analytically exact solution. Note that the results shown here are from the numerical calculations and we find perfect agreement between numerical and analytical results.

The dynamics for $\alpha = 0.2$ shows a relaxation of $P^{\rm d}_{\rm D}(t)$ to a finite value and a simultaneous increase of $P^{\rm d}_{\rm A}(t)$ (on the same time scale but with a smaller amplitude). The limiting values of $P^{\rm d}_{\rm D}$ and $P^{\rm d}_{\rm A}$ do not add up to 1, because there is also an increase of the probability of having one electron on both donor and acceptor. The value of $\alpha = 0.2$ for $U/\omega_{\rm c} = 1$ is located below both phase boundaries as depicted in Fig. 2.

Finally, we also show the result for a value $\alpha > \alpha_{\rm c}$, for which the system is localized. For the double occupancies plotted in Fig. 3, we see only a very small deviation of $P^{\rm d}_{\rm D}(t)$ from its initial value, and almost no increase in $P^{\rm d}_{\rm A}(t)$.

For intermediate values of α, we observe an exponential relaxation of the occupation at the donor site: $P^{\rm d}_{\rm D}(t) \propto \exp(-kt)$, which defines the electron

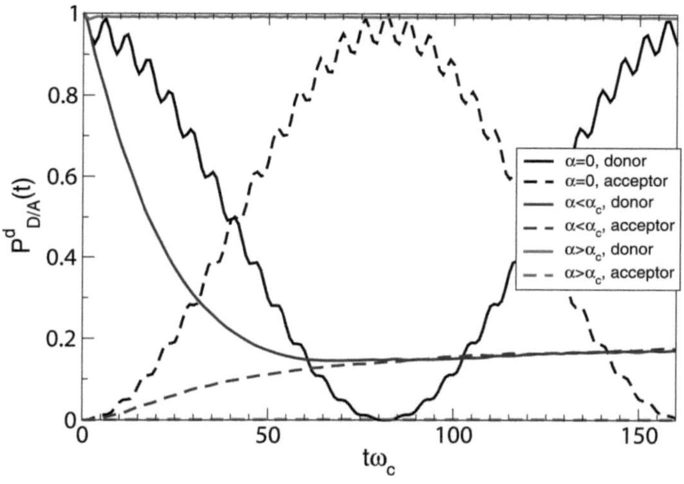

Fig. 3. Time evolution of double occupancies at donor and acceptor sites, $P^{\mathrm{d}}_{\mathrm{D/A}}(t)$, for different values of the coupling strength α. Undamped, coherent oscillations are observed for $\alpha = 0$, while for intermediate α, the double occupancies show exponential relaxation. Above a critical α_{c}, the system is localized and no electron transfer occurs

transfer rate k. In the upper panel of Fig. 4, we show a sequence of results for single and double occupancies for $\alpha = 0.2$ and different values of U. For these parameters, the probability of the singly occupied state rises at short times while the double occupancy at the acceptor stays very small, $P^{\mathrm{d}}_{\mathrm{A}}(t) \ll 1$, so we are in a regime of sequential transfer. For larger values of U (only these are plotted in Fig. 4, upper panel) we find that the single electron transfer rate (starting from the doubly occupied donor) is decreasing with increasing U, as plotted in the lower panel of Fig. 4. However, for the whole range of U-values, we find a non-monotonic behaviour of k, with an initial increase for small values of U. This is very similar to the inverted region as predicted by Marcus for single-electron transfer reactions and is, in fact, due to very similar reasons. As the single-electron transfer rate becomes very small, pair tunneling is the most probable process for large U. This is seen in the inset of Fig. 4 (lower panel) where we plot the occupancies for $U = 5\omega_{\mathrm{c}}$; note that for these parameters the *single* occupancies are suppressed while the double occupancy at the acceptor site increases and even performs (together with $P^{\mathrm{d}}_{\mathrm{D}}(t)$) coherent oscillations. A detailed discussion including inter-site Coulomb interactions will be given in [20].

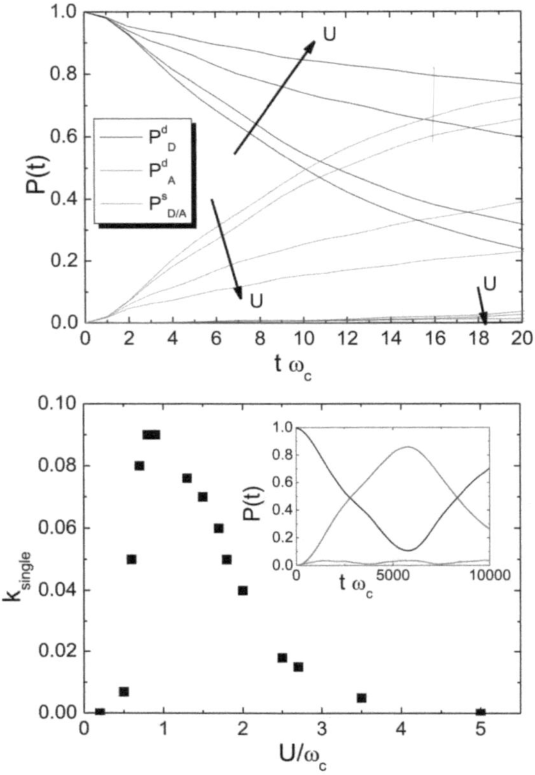

Fig. 4. *Upper panel*: time dependence of the single occupancy and the double occupancy at donor and acceptor sites. For increasing values of U, the electron transfer slows down, i.e., the rate decreases. *Lower panel*: dependence of the electron transfer rate k on U. An inverted region is visible beyond the maximum of $k(U)$. *Inset*: time dependence of the single occupancy and the double occupancy at donor and acceptor sites for $U = 5\omega_c$, where pair transfer dominates

4 Summary and Discussion

We have presented numerical renormalization group calculations for the real-time dynamics of electron transfer systems, focusing on situations where the donor is initially occupied by two electrons. We find significant differences in the electron transfer dynamics between the one- and two-electron sector, which is caused by the Coulomb interaction U and an effective interaction mediated by the coupling to a common bosonic bath, absent in the one-electron sector.

This correlation is already visible in the equilibrium phase diagram, Fig. 2, calculated with the bosonic NRG [9]. We have also presented results from the time-dependent NRG [8], for the time-evolution of single- and double occupancies at the donor and acceptor sites. This information helps us to identify conditions under which the systems performs

a) concerted two-electron transfer, or
b) uncorrelated, step-by-step single-electron transfer.

The single-occupancy of donor and acceptor, $P^{\mathrm{s}}_{\mathrm{D/A}}(t)$, allows us to distinguish between these two scenarios. Due to the specific preparation – both electrons at the donor site at time $t = 0$ – the single-occupancy starts from zero: $P^{\mathrm{s}}_{\mathrm{D/A}}(t = 0) = 0$. If the single-occupancy remains close to zero during the electron transfer process, the states $|\uparrow, \downarrow\rangle$ and $|\downarrow, \uparrow\rangle$ are only virtually occupied, and we are in the regime of concerted two-electron transfer. If, on the other hand, $P^{\mathrm{s}}_{\mathrm{D/A}}$ increases significantly, we take this as an indication of step-by-step single-electron transfer.

A virtual occupation of the singly occupied states can be understood by simple energetic arguments for the decoupled donor/acceptor system ($\alpha = 0$). The difference in the energies of the singly and doubly occupied states is given by $|U|$; the larger this difference, the smaller the probability for finding the intermediate, singly occupied state. Of course, there is a change of the energetics by the coupling to the bosonic bath, but we see already from this simplified picture that it is the Coulomb interaction between the two electrons which decides about the character of the transition. In other words, electronic correlations are essential to understand the dynamics of multi-electron transfer events.

The calculations described in this paper can be easily extended by adding the degrees of freedom for an intervening bridge between donor and acceptor. In future, we are planning to study donor/bridge/acceptor systems related to the charge transfer in DNA [21], which shows a very interesting distance dependence of the electron transfer rate, that is a crossover from a superexchange mechanism for short distances, to sequential hopping for larger distances. For charge transfer reactions in DNA, a single-electron picture is generally not applicable. Multi-electron transfer processes have recently been suggested [22–24] and are – in our opinion – essential to understand radiative damages and their repair in DNA as well as the connection between electron transfer rates and molecular conductivity.

Acknowledgements

We acknowledge helpful discussions with A. Nitzan and A. Schiller. S. Tornow is grateful to the School of Chemistry of the Tel Aviv University for the kind hospitality during her stay. This research was supported by the DFG through SFB 484 (RB,ST). FBA acknowledges funding of the NIC, Forschungszentrum Jülich under Project No. HHB000, and DFG funding under project No. AN 275/5-1.

References

[1] J. Jortner, M. Bixon: *Electron Transfer – from isolated Molecules to Biomolecules*, Adv. in Chem. Phys. **106/107** (1999)
[2] V. May, O. Kühn: *Charge and Energy Transfer Dynamics in Molecular Systems*, (Wiley-VCH, Weinheim 2004)
[3] C. Wan, T. Fiebig, O. Schiemann, J. K. Barton, A. H. Zewail: Proc. Natl. Acad. Sci. U.S.A. **97**, 14052 (2000)
[4] D. DeVault: *Quantum-mechanical tunneling in biological systems*, (Cambridge University Press, Cambridge 1981)
[5] R. A. Marcus: J. Chem. Phys. **24**, 966 (1956); Rev. Mod. Phys. **65**, 599 (1999)
[6] A. J. Leggett, S. Chakravarty, A. T. Dorsey, M. P. A. Fisher, A. Garg, W. Zwerger: Rev. Mod. Phys. **59**, 1 (1987)
[7] F. B. Anders, A. Schiller: Phys. Rev. Lett. **95**, 196801 (2005)
[8] F. B. Anders, A. Schiller: Phys. Rev. B **74**, 245113 (2006)
[9] R. Bulla, H.-J. Lee, N.-H. Tong, M. Vojta: Phys. Rev. B **71**, 045122 (2005)
[10] F. B. Anders, R. Bulla, M. Vojta: preprint cond-mat/0607443 (2006)
[11] S. Tornow, N.-H. Tong, R. Bulla: Europhys. Lett. **73**, 913 (2006)
[12] L. Mühlbacher, J. Ankerhold, A. Komnik: Phys. Rev. Lett. **95**, 220404 (2005)
[13] L. D. Zusman, D. V. Beratan: J. Chem. Phys. **105**, 165 (1996)
[14] K. R. Hoke, N. Cobb, F. A. Armstrong, R. Hille: Biochemistry **43**, 1667 (2004)
[15] A. Naqui, B. Chance, E. Cadenas: Ann. Rev. Biochem. **55**, 137 (1986)
[16] K. G. Wilson: Rev. Mod. Phys. **47**, 773 (1975)
[17] R. Bulla, T. Costi, Th. Pruschke: preprint cond-mat/0701105 (2007)
[18] R. Bulla, N.-H. Tong, M. Vojta: Phys. Rev. Lett. **91**, 170601 (2003)
[19] D. Xu, K. Schulten: Chem. Phys. **182**, 91 (1994).
[20] S. Tornow, R. Bulla, F. Anders, A. Nitzan: in preparation
[21] H.-A. Wagenknecht: *Charge Transfer in DNA – From Mechanism to Application*, (Wiley-VCH, Weinheim 2005)
[22] G. Pratviel, B. Meunier: Chem. Eur. J. **12**, 6018 (2006)
[23] E. B. Starikov: Philos. Mag. Lett. **83**, 699 (2003)
[24] S. Tornow, R. Bulla, F. Anders, E. Starikov: in preparation

Criticality and Correlations in Cold Atomic Gases

Michael Köhl[1,2], Tobias Donner[1], Stephan Ritter[1], Thomas Bourdel[1,3], Anton Öttl[1,4], Ferdinand Brennecke[1], and Tilman Esslinger[1]

[1] Institute for Quantum Electronics, ETH Zürich, Schafmattstrasse 16, 8093 Zürich, Switzerland
[2] Department of Physics, University of Cambridge, JJ Thomson Avenue, Cambridge, CB3 0HE, United Kingdom
mk540@cam.ac.uk
[3] Laboratoire Charles Fabry de l'institut d'Optique, CNRS, Univ Paris-Sud, Campus Polytechnique, RD128, 91127 Palaiseau cedex, France
[4] Department of Physics, University of California, Berkeley, CA 94720, United States

Abstract. We study the phase transition of Bose–Einstein condensation in a dilute atomic gas very close to the critical temperature. The critical regime we enter is governed by fluctuations extending far beyond the length scale of thermal de-Broglie waves. Using matter-wave interference we measure the correlation length of these critical fluctuations as a function of temperature. From this we determine the critical exponent of the correlation length for a trapped, weakly interacting Bose gas to be $\nu = 0.67 \pm 0.13$.

1 Ultracold Quantum Gases as Model Systems for Condensed Matter Physics

In 1925 *S. N. Bose* [1] and *A. Einstein* [2] predicted that below a critical temperature a macroscopic fraction of bosons occupies the lowest-lying single-particle quantum state. The perception of *F. London* that this concept is applicable to condensed matter at ultralow temperatures [3] has initiated the understanding of quantum matter, such as superfluid Helium. In 1995 a new form of quantum matter has been realized by cooling a dilute atomic vapor through the phase transition temperature of Bose–Einstein condensation [4]. It could be witnessed how a gas of neutral atoms condensed into a single zero momentum quantum state displaying the behaviour of one macroscopic quantum mechanical wave function. This achievement of Bose–Einstein condensation has initiated a wave of research on quantum degenerate gases. The route towards zero temperature in a gas of atoms is a narrow path which circumvents solidification by keeping the atomic cloud at very low densities. Spin-polarized ultracold dilute vapors are cooled by employing the mechanical effect of laser light to slow down atoms, followed by evaporative cooling of the atoms in a magnetic trap. Upon reaching temperatures of 100 Nanokelvin

R. Haug (Ed.): Advances in Solid State Physics,
Adv. in Solid State Phys. **47**, 79–88 (2008)

and below, the experimental efforts are rewarded with a quantum many-body system of ultimate controllability and access to microscopic understanding.

The conceptual links between the physics of Bose–Einstein condensates in dilute gases and superfluid Helium are manyfold. Experimentally this has been highlighted by the observation of frictionless flow [5] and the generation of quantized vortices [6,7] in dilute atomic Bose–Einstein condensates. In contrast to superfluid Helium, in which the Bose–Einstein condensed quantum state is depleted by the strong interatomic interactions, in dilute atomic gases nearly pure Bose–Einstein condensates can be realized. The weak interactions between neutral atoms are through s-wave collisions via the van-der-Waals interaction. The strength of the interaction is parameterized by the s-wave scattering length a, and the diluteness of the gas implies $n^{1/3}a \ll 1$ with n being the particle density. In this regime, a mean-field description – for example provided through the Gross–Pitaevskii-equation – has been successfully used to explain the physics of many experiments with atomic Bose–Einstein condensates. Only in very few exceptional cases beyond mean field effects could be observed with cold atoms. While strong interactions alone are not sufficient to observe beyond mean-field effects [8, 9], interacting systems in reduced dimensionality have offered a first glimpse of this exciting regime [10–15]. In this work we follow a different experimental route to progress into a strongly correlated regime, namely the progress into the critical regime of the Bose–Einstein phase transition [16]. The critical regime of a weakly interacting Bose gas offers an intriguing possibility to study physics dictated by fluctuations and their correlations [17].

2 Criticality and Critical Exponents

The transition between two phases of matter is driven by fluctuations. For example, at a liquid-gas phase transition this leads to critical opalescence because large scale density fluctuations induce strong light scattering off the sample and the liquid-gas mixture becomes opaque. In general, for second order phase transitions, to which also Bose–Einstein condensation of gases or the λ-transition in ^{4}He belong, strong fluctuations of the order parameter are encountered close to the phase transition. When approaching the critical temperature T_c the fluctuations of the order parameter rather than its mean value dictate the behaviour of the system.

The nature of fluctuations is characterized by the density matrix of the system which measures the two-point correlation function of the order parameter Ψ: $\langle \Psi^\dagger(r)\Psi(0)\rangle$. The separation of the two probed locations is r. As pointed out by *Penrose* and *Onsager*, a Bose–Einstein condensate manifests as off-diagonal long-range order of the density matrix, whereas a thermal gas far above the transition temperature exhibits only short-range correlations [18]. Their characteristic length scale is determined by the thermal

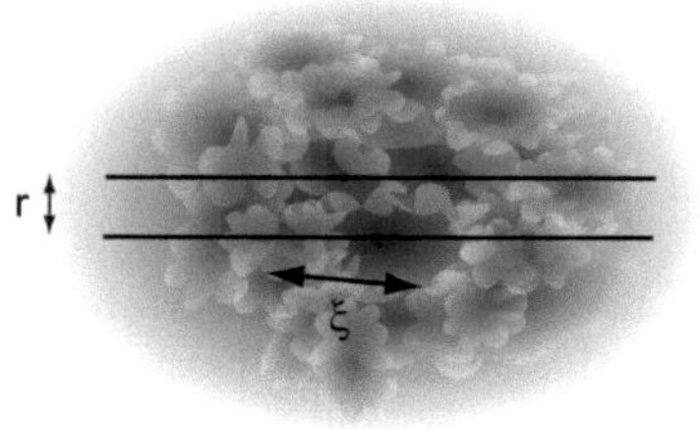

Fig. 1. Fluctuations of a trapped atom cloud close to the critical temperature. The characteristic length scale of the fluctuations is the correlation length ξ and we probe the correlations at two planes intersecting the atom cloud. The planes are chosen symmetric about the center of the cloud and are spaced by a variable distance r

de Broglie wavelength $\lambda_{dB} = \sqrt{2\pi\hbar^2/(mk_BT)}$ with $\hbar$ being Planck's constant divided by 2π, m the mass of the particles, k_B Boltzmann's constant, and T the temperature. In this regime the correlation function decays as $\langle\Psi^\dagger(r)\Psi(0)\rangle \propto \exp(-\pi r^2/\lambda_{dB}^2)$. Critical fluctuations of the order parameter Ψ close to the critical temperature become visible when their length scale becomes larger than thermal fluctuations. For long length scales $r > \lambda_{dB}$ the correlation function of a Bose gas can be expressed as [19]

$$\langle\Psi^\dagger(r)\Psi(0)\rangle \propto \frac{1}{r^{d-2+\eta}}\exp(-r/\xi)\,. \tag{1}$$

Here ξ denotes the correlation length of the order parameter which is the typical length scale over which its fluctuations are correlated (see Fig. 1), d is the dimensionality and η is a critical exponent whose numerical value is on the order of 10^{-2}. The correlation length ξ is a function of temperature and diverges as the system approaches the phase transition. This results in the algebraic decay of the correlation function with distance $\langle\Psi^\dagger(r)\Psi(0)\rangle \propto 1/r^{d-2+\eta}$ at the phase transition. The theory of critical phenomena predicts a power law behavior of the divergence of the correlation length ξ of the order parameter vs. temperature according to

$$\xi \propto |(T-T_c)/T_c|^{-\nu}\,, \tag{2}$$

where ν is the critical exponent of the correlation length. The value of the critical exponent depends only on the universality class of the system. This universal behaviour implies that microscopic details are irrelevant but critical fluctuations dominate the system at a phase transition.

Determining critical exponents has a long history in studying phase transitions. For non-interacting systems the critical exponents can be easily calculated. For the noninteracting homogeneous Bose gas one finds $\nu = 1$. The effect of the harmonic trapping potential – which is inevitable in experiments with cold atoms – modifies the critical exponent for the non-interacting gas to $\nu = 1/2$. The difference arises from the relation $\mu \propto 1/\xi^2$ in combination with the different dependence of the chemical potential μ on $T - T_c$ for the homogeneous and the trapped case. The non-interacting gas, however, is a peculiar case from the theoretical point of view and its relation to the interacting system is limited.

The presence of interactions adds richness to the system. Determining the value of the critical exponent using Landau's theory of phase transitions results in a value of $\nu = 1/2$ for the homogeneous system. This value is the result of both a classical theory and a mean field approximation to quantum systems. However, calculations initially by *Onsager* [20] and later the techniques of the renormalization group method [21] showed that mean field theory fails to describe the physics at the phase transition and the resulting critical exponent is $\nu \simeq 0.67$ [38–40]. Very close to the critical temperature – in the so-called critical regime – the fluctuations of Ψ become more dominant than its mean value. The onset of this regime can be determined by the Ginzburg criterion $\xi > \lambda_{dB}^2/(\sqrt{128}\pi^2 a)$ [22]. The numerical coefficient may be different for a trapped gas and has been omitted in reference [23]. The enhanced fluctuations are also responsible for a nontrivial shift of the critical temperature of Bose–Einstein condensation [24–28]. However, despite previous attempts [29] the critical regime of the phase transition has been accessed experimentally only recently [16].

3 Measuring the Spatial Correlation Function

In the experiments reported here, we perform a direct measurement of the density matrix of an ultracold atom cloud and determine the temperature dependence of the correlation length very close to the critical temperature. To prepare the ultracold atom gas we load 10^9 laser cooled ^{87}Rb atoms into a magnetic trap [30]. The sample is evaporatively cooled to a temperature slightly below the critical temperature being $T_c = 146\,\mathrm{nK}$ and we end up with a cloud of 4×10^6 atoms in the $|F = 1, m_F = -1\rangle$ hyperfine ground state. The characteristic frequencies of the harmonic magnetic trap are (ω_x, ω_y, ω_z) $= 2\pi\times$ (39, 7, 29) Hz, where z denotes the vertical axis. The apparatus and the methods have been described in detail previously [31].

We obtain the spatial correlation function through a measurement of the visibility of an interference signal of two atomic beams (see Fig. 2). The atomic beams originate from two distinct regions inside the trapped atomic cloud with the distance r of these regions being adjustable [32,33]. For output coupling the atomic beams we employ microwave frequency fields to spin-flip the atoms into a magnetically non-trapped state $|F = 2, m_F = 0\rangle$. The resonance condition for this transition is given by the local magnetic field and the released atoms propagate downwards due to gravity. The regions of output coupling can be approximated by horizontal planes oriented orthogonally

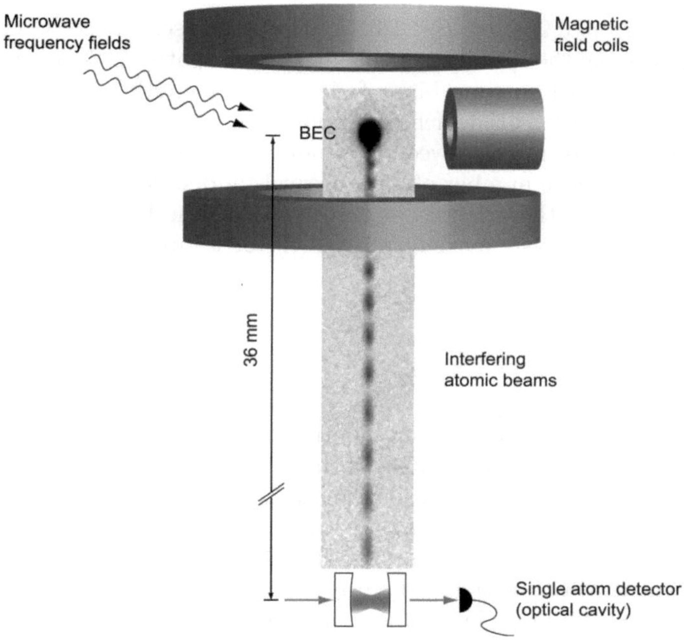

Fig. 2. Schematic of the experiment. The Bose–Einstein condensate (BEC) is trapped in the inhomogeneous magnetic field produced by the magnetic field coils. Using two microwave fields we release atomic beams from two different locations inside the trapped cloud. The beams propagate downwards and interfere. We detect the interference pattern with a high-finesse optical cavity realizing an efficient single atom detector

to the direction of the atomic beams. The two released atomic beams interfere with each other, resulting in a longitudinal interference pattern falling downwards. We detect the interference pattern in time with single atom resolution using a high finesse optical cavity, placed 36 mm below the center of the magnetic trap [33, 34]. An atom entering the cavity mode decreases the transmission of a probe beam resonant with the cavity. From the arrival times of the atoms we find the visibility of the interference pattern. From repeated measurements with different pairs of microwave frequencies we measure the visibility $\mathcal{V}(r)$ with r ranging from 0 to 4 λ_{dB}. The ability to continuously monitor the correlations also opens intriguing possibilities in the study of dynamics of phase transitions and quenches [34].

The temperature of the atomic sample can be controlled precisely by holding the atoms for several seconds in the trap where due to resonant stray light, fluctuations of the trap potential and background gas collisions energy is transferred to the atoms [35]. From absorption images we determine the heating rate to be $4.4{\pm}0.8\,\mathrm{nK/s}$. Using this technique we continuously cover a range of temperatures from $0.98\,T_c < T < 1.07\,T_c$ over a time scale of seconds during which we determine the visibility of the interference pattern. Segmentation of the acquired data into time bins of $\Delta t = 72\,\mathrm{ms}$ length allows for a temperature resolution of $0.3\,\mathrm{nK}$ which corresponds to $0.002\,T_c$. The time bin length was chosen to optimize between shot-noise limited determination of the visibility from the finite number of atom arrivals and sufficiently good temperature resolution. For the analysis we have chosen time bins overlapping by 50%. Using this analysis we determine $\mathcal{V}(r)$ for temperatures from T_c to up to $1.07\,T_c$.

4 Results

Figure 3 shows the measured visibility as a function of the separation r very close to the critical temperature T_c. We observe that the visibility decays on a much longer length scale than predicted by the thermal de Broglie wavelength λ_{dB}. We fit the long distance tail $r > \lambda_{dB}$ with equation (1) with $\eta = 0$ (dashed lines) and determine the correlation length ξ. The strong temperature dependence of the correlation function is directly visible. As T approaches T_c the visibility curves become more long ranged and similarly the correlation length ξ increases. The observation of long-range correlations shows how the size of the fluctuation regions strongly increases as the temperature is varied only minimally in vicinity of the phase transition temperature.

Figure 4 shows how the measured correlation length ξ diverges as the system approaches the critical temperature. Generally, an algebraic divergence of the correlation length is predicted. We fit our data with the power law according to (2) leaving the value of T_c as a free fit parameter. Therefore our analysis is independent of an exact calibration of both temperature and heating rate provided that the sample is thermalized and the heating rate is constant. The value of the critical exponent is averaged over various temporal offsets $0 < t_0 < \Delta t$ of the analyzing time bin window and the resulting value for the critical exponent is $\nu = 0.67 \pm 0.13$, where the error is the reduced χ^2 error. Systematic errors on the value of ν could be introduced by the detector response function. We find the visibility for a pure Bose–Einstein condensate to be 100% with a statistical error of 2% over the range of r investigated. This uncertainty of the visibility would amount to a systematic error of the critical exponent of 0.01 and is negligible compared to the statistical error. A non-linearity in the heating process could be introduced by a change of the heat capacity near the phase transition [21]. We estimate the corresponding error of ν to be less than 0.01.

The finite size and the inhomogeneity of our trapped gas pose interesting questions regarding the concept of phase transitions and universality for systems not in the thermodynamic limit [23]. Finite size effects are expected when the correlation length is large and they may lead to a slight underestimation of ν for our conditions. Moreover, the harmonic confining potential

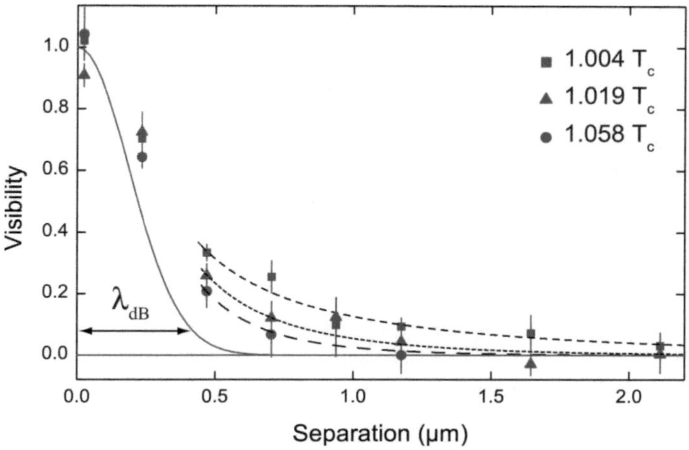

Fig. 3. Spatial correlation function of a trapped Bose gas close to the critical temperature. Shown is the visibility of a matter wave interference pattern originating from two regions separated by r in an atomic cloud shortly above the transition temperature. The *solid line* is a gaussian with a width given by the thermal de Broglie wavelength λ_{dB}. The experimental data show a long range phase coherence due to critical fluctuations which extends far beyond the scale set by λ_{dB}. The *dashed lines* are fits proportional to $\frac{1}{r}e^{-r/\xi}$. The error bars indicate the statistical error. Data adapted from [16]

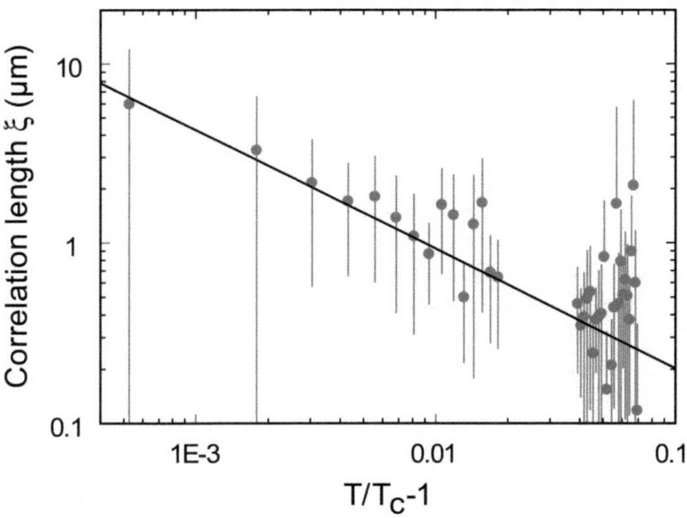

Fig. 4. The correlation length ξ as a function of temperature. The *line* is a fit of (2) to the data (with ν and T_c as free parameters). From the fit we find the critical exponent to be $\nu = 0.67 \pm 0.13$. The error bars indicate the statistical errors. Data adapted from [16]

introduces a spatially varying density. The phase transition takes place at the center of the trap and non-perturbative fluctuations are thus expected within a finite radius R [36]. Using the Ginzburg criterion as given in [22] we find $R \approx 10\,\mu\mathrm{m}$. The largest separation of output coupling regions we probe in our experiment is $2\,\mu\mathrm{m}$ which is well below this radius R.

So far, in interacting systems the critical exponent ν has been determined for the homogeneous case. The normal-to-superfluid phase transition in liquid Helium is among the most accurately investigated systems at criticality. However, in liquid Helium the critical exponent ν has not been measured directly. Instead, the critical exponent of the specific heat α has been measured to great accuracy in a spaceborne experiment [37]. Through the scaling relation $\alpha = 2 - 3\nu$ the value of the critical exponent $\nu = 0.67056 \pm 0.0006$ is inferred being in excellent agreement with theoretical predictions [38–40] as opposed to $\nu = 1/2$ predicted by mean-field theory. Alternatively, the exponent $\zeta \simeq 0.67$ (which is related to the superfluid density $\rho_s = |\Psi|^2$ instead of the order parameter Ψ) can be measured directly in second sound experiments in liquid Helium [41, 42]. Due to an argument by Josephson [43] it is believed that $\nu = \zeta$ for $\eta = 0$, however, a direct measurement of ν has still been missing.

5 Conclusion

In conclusion, we observe the long-range behaviour of the correlation function of a trapped Bose gas in the critical regime. The measurement reveals the behaviour of a phase transition in a finite-size system. Our measured value for the critical exponent does not coincide with that obtained by a simple mean-field model or with that of an ideal non-interacting Bose gas. The value is in good agreement with the expectations of renormalization group theory applied to a homogeneous gas of bosons and with measurements using strongly interacting superfluid helium.

We would like to thank G. Blatter, A. Kuklov, G. Shlyapnikov, M. Troyer, W. Zwerger for insightful discussions, SEP Information Sciences, OLAQUI Grant No. (EU FP6-511057), Qudedis (ESF), and QSIT for funding. T. Bourdel acknowledges funding by an EU Marie Curie fellowship under contract MEIF-CT-2005-023612.

References

[1] S. N. Bose: Plancks Gesetz und Lichtquantenhypothese, Z. Phys. **26**, 178 (1924)

[2] A. Einstein: Quantentheorie des einatomigen idealen Gases. Zweite Abhandlung, Sitzungsberichte der Preussischen Akademie der Wissenschaften **1**, 3 (1925)

[3] F. London: On the Bose-Einstein condensation, Phys. Rev. **54**, 947 (1938)
[4] M. H. Anderson, J. R. Ensher, M. R. Matthews, C. E. Wieman, E. A. Cornell: Observation of Bose-Einstein Condensation in a Dilute Atomic Vapor, Science **269**, 198 (1995)
[5] C. Raman, M. Köhl, R. Onofrio, D. S. Durfee, C. E. Kuklewicz, Z. Hadzibabic, W. Ketterle: Evidence for a Critical Velocity in a Bose-Einstein Condensed Gas, Phys. Rev. Lett. **83**, 2502 (1999)
[6] M. R. Matthews, B. P. Anderson, P. C. Haljan, D. S. Hall, C. E. Wieman, E. A. Cornell: Vortices in a Bose-Einstein Condensate, Phys. Rev. Lett. **83**, 2498 (1999)
[7] K. W. Madison, F. Chevy, W. Wohlleben, J. Dalibard: Vortex Formation in a Stirred Bose-Einstein Condensate, Phys. Rev. Lett. **84**, 806 (2000)
[8] M. P. A. Fisher, P. B. Weichman, G. Grinstein, D. S. Fisher: Boson Localization and the Superfluid-Insulator Transition, Phys. Rev. B **40**, 546 (1989)
[9] D. Jaksch, C. Bruder, J. I. Cirac, C. W. Gardiner, P. Zoller: Cold Bosonic Atoms in Optical Lattices, Physical Review Letters **81**, 3108 (1998)
[10] T. Stöferle, H. Moritz, C. Schori, M. Köhl, T. Esslinger: Transition from a Strongly Interacting 1D Superfluid to a Mott Insulator, Phys. Rev. Lett. **92**, 130403 (2004)
[11] B. Paredes, A. Widera, V. Murg, O. Mandel, S. Fölling, I. Cirac, G. Shlyapnikov, T. W. Hänsch, I. Bloch: Tonks–Girardeau Gas of Ultracold Atoms in an Optical Lattice, Nature (London) **429**, 277 (2004)
[12] T. Kinoshita, T. Wenger, D. S. Weiss: Observation of a One-Dimensional Tonks–Girardeau Gas, Science **305**, 1125 (2004)
[13] M. Köhl, H. Moritz, T. Stöferle, C. Schori, T. Esslinger: Superfluid to Mott Insulator Transition in One, Two, and Three Dimensions, Journal of Low Temperature Physics **138**, 635 (2005)
[14] Z. Hadzibabic, P. Krüger, M. Cheneau, B. Battelier, J. Dalibard: Berezinskii–Kosterlitz–Thouless Crossover in a Trapped Atomic Gas, Nature (London) **441**, 1118 (2006)
[15] I. B. Spielman, W. D. Phillips, J. V. Porto: Mott-insulator Transition in a Two-Dimensional Atomic Bose Gas, Physical Review Letters **98**, 080404 (2007)
[16] T. Donner, S. Ritter, T. Bourdel, A. Öttl, M. Köhl, T. Esslinger: Critical Behavior of a Trapped Interacting Bose Gas, Science **315**, 1556 (2007)
[17] Q. Niu, I. Carusotto, A. B. Kuklov: Imaging of Critical Correlations in Optical Lattices and Atomic Traps, Phys. Rev. A **73**, 053604 (2006)
[18] O. Penrose, L. Onsager: Bose–Einstein Condensation and Liquid Helium, Phys. Rev. **104**, 576 (1956)
[19] K. Huang: *Statistical Mechanics* (Wiley, New York 1987)
[20] L. Onsager: Crystal Statistics. I. A Two-Dimensional Model with an Order-Disorder Transition, Phys. Rev. **65**, 117 (1944)
[21] J. Zinn Justin: *Quantum Field Theory and Critical Phenomena* (Oxford University Press 1996)
[22] S. Giorgini, L. P. Pitaevskii, S. Stringari: Condensate Fraction and Critical Temperature of a Trapped Interacting Bose Gas, Phys. Rev. A **54**, R4633 (1996)
[23] K. Damle, T. Senthil, S. N. Majumdar, S. Sachdev: Phase Transition of a Bose Gas in a Harmonic Potential, Europhys. Lett **36**, 7 (1996)
[24] G. Baym, J.-P. Blaizot, M. Holzmann, F. Laloë, D. Vautherin: The Transition Temperature of the Dilute Interacting Bose Gas, Phys. Rev. Lett. **83**, 1703 (1999)

[25] G. Baym, J.-P. Blaizot, J. Zinn-Justin: The Transition Temperature of the Dilute Interacting Bose Gas for n Internal States, Europhysics Letters (EPL) **49**, 150 (2000)
[26] P. Arnold, G. Moore: BEC Transition Temperature of a Dilute Homogeneous Imperfect Bose Gas, Phys. Rev. Lett. **87**, 120401 (2001)
[27] V. A. Kashurnikov, N. V. Prokof'ev, B. V. Svistunov: Critical Temperature Shift in Weakly Interacting Bose Gas, Phys. Rev. Lett. **87**, 120402 (2001)
[28] N. Prokof'ev, O. Ruebenacker, B. Svistunov: Weakly Interacting Bose Gas in the Vicinity of the Normal-fluid–Superfluid Transition, Phys. Rev. A **69**, 053625 (2004)
[29] F. Gerbier, J. H. Thywissen, S. Richard, M. Hugbart, P. Bouyer, A. Aspect: Critical Temperature of a Trapped, Weakly Interacting Bose Gas, Phys. Rev. Lett. **92**, 030405 (2004)
[30] T. Esslinger, I. Bloch, T. W. Hänsch: Bose–Einstein Condensation in a Quadrupole-Ioffe-Configuration Trap, Phys. Rev. A **58**, 2664(R) (1998)
[31] A. Öttl, S. Ritter, M. Köhl, T. Esslinger: Hybrid apparatus for Bose–Einstein Condensation and Cavity Quantum Electrodynamics: Single Atom Detection in Quantum Degenerate Gases, Rev. Sci. Instrum. **77**, 063118 (2006)
[32] I. Bloch, T. W. Hänsch, T. Esslinger: Measurement of the Spatial Coherence of a Trapped Bose Gas at the Phasetransition, Nature (London) **403**, 166 (2000)
[33] T. Bourdel, T. Donner, S. Ritter, A. Öttl, M. Köhl, T. Esslinger: Cavity QED Detection of Interfering Matter Waves, Phys. Rev. A **73**, 043602 (2006)
[34] S. Ritter, A. Öttl, T. Donner, T. Bourdel, M. Köhl, T. Esslinger: Observing the Formation of Long-Range Order during Bose–Einstein Condensation, Phys. Rev. Lett. **98**, 090402 (2007)
[35] R. Gati, B. Hemmerling, J. Fölling, M. Albiez, M. K. Oberthaler: Noise Thermometry with Two Weakly Coupled Bose–Einstein Condensates, Phys. Rev. Lett. **96**, 130404 (2006)
[36] P. Arnold, B. Tomasik: T_c for Trapped Dilute Bose Gases: A Second-order Result, Phys. Rev. A **64**, 053609 (2001)
[37] J. A. Lipa, J. A. Nissen, D. A. Stricker, D. R. Swanson, T. C. P. Chui: Specific Heat of Liquid Helium in Zero Gravity Very Near the Lambda Point, Phys. Rev. B **68**, 174518 (2003)
[38] H. Kleinert: Critical Exponents from Seven-loop Strong-coupling φ^4 Theory in Three Dimensions, Phys. Rev. D **60**, 085001 (1999)
[39] M. Campostrini, M. Hasenbusch, A. Pelissetto, P. Rossi, E. Vicari: Critical Behavior of the Three-dimensional XY Universality Class, Phys. Rev. B **63**, 214503 (2001)
[40] E. Burovski, J. Machta, N. Prokof'ev, B. Svistunov: High-precision measurement of the Thermal Exponent for the Three-dimensional XY Universality Class, Phys. Rev. B **74**, 132502 (2006)
[41] L. S. Goldner, N. Mulders, G. Ahlers: Second Sound Very Near T_λ, J. Low Temp. Phys. **93**, 131 (1993)
[42] M. J. Adriaans, D. R. Swanson, J. A. Lipa: The velocity of Second Sound Near the Lambda Transition in Superfluid 4He, Physica B **194**, 733 (1993)
[43] B. D. Josephson: Relation between the Superfluid Density and Order Parameter for Superfluid He Near Tc, Phys. Lett. **21**, 608 (1966)

Part III

Ferromagnetic Films and Particles

Magnetic Anisotropy and Magnetization Switching in Ferromagnetic GaMnAs

W. Limmer, J. Daeubler, M. Glunk, T. Hummel, W. Schoch, S. Schwaiger, M. Tabor, and R. Sauer

Institut für Halbleiterphysik, Universität Ulm, 89069 Ulm, Germany
`wolfgang.limmer@uni-ulm.de`

Abstract. Characteristic features of diluted ferromagnetic semiconductors such as the anisotropic magnetoresistance or the spin polarization of charge carriers are intimately connected with a macroscopic magnetization. Since the orientation of the magnetization is controlled by magnetic anisotropy (MA), a detailed knowledge of this anisotropy is indispensable for the design of novel spintronic devices. In this article, angle-dependent magnetotransport is demonstrated to be an excellent tool for probing MA as an alternative to the standard ferromagnetic-resonance method. Moreover, its ability to trace the motion of the magnetization vector in a variable external magnetic field makes it ideally suitable for studying magnetization switching, a potential basic effect in future logical devices. The MA of a series of differently strained GaMnAs samples is analyzed by means of model calculations in a single-domain picture based on a series expansion of the resistivity tensor and a numerical minimization of the free enthalpy.

1 Introduction

In recent years, ferromagnetism has been implemented in several standard III–V semiconductors by incorporating high concentrations of magnetic elements into the group-III sublattice [1,2]. The resulting diluted ferromagnetic semiconductors are expected to play a key role in future spintronics technology where both the electrical charge and the spin of the carriers are utilized for information processing and storage. Today, one of the the most prominent representatives of this kind of semiconductor is (Ga,Mn)As, which is considered a potential candidate or at least an ideal test system for spintronic applications due to its compatibility with conventional semiconductor technology. In (Ga,Mn)As, magnetic Mn atoms with atomic structure [Ar]$3d^5 4s^2$ are predominantly incorporated on cation sites as d^5 divalent acceptors. For typical Mn concentrations between 2% and 8% this results in extremely high hole densities in the range $p = 10^{20}$–$10^{21}\,\mathrm{cm}^{-3}$. The permanent magnetic moments of the Mn^{2+} ions arise from the half-filled 3d-shell with a total spin of $S = 5/2$. (Ga,Mn)As is paramagnetic at room temperature and undergoes a transition to the ferromagnetic phase at the Curie temperature T_C, where values of up to 173 K have been reported so far. The ferromagnetic coupling between the Mn spins, successfully explained within a Zener kinetic-exchange mean-field model, is mediated by delocalized holes and originates from an an-

R. Haug (Ed.): Advances in Solid State Physics,
Adv. in Solid State Phys. **47**, 91–103 (2008)

tiferromagnetic exchange interaction between the holes and the Mn ions [2,3]. In order to prevent the formation of the MnAs phase, (Ga,Mn)As is grown by low-temperature molecular-beam epitaxy (LT-MBE) at typical temperatures of $T_S \approx 250\,^\circ$C. As a consequence, high concentrations of compensating point defects are incorporated, with As antisites (As_{Ga}) and Mn interstitials (Mn_I) being the major species. The density of Mn_I can be reduced by post-growth annealing at or below the growth temperature whereas As_{Ga} remains stable up to 450 °C. Since the incorporation of Mn_{Ga}, Mn_I, and As_{Ga} results in an increase of the lattice constant, (Ga,Mn)As layers grown on GaAs substrates are compressively strained.

During the last decade, considerable progress has been made in understanding the structural, electronic, and magnetic properties of (Ga,Mn)As [2]. In particular, anisotropic magnetoresistance (AMR), anomalous Hall effect (AHE), and magnetic anisotropy (MA) have been established as characteristic features, making (Ga,Mn)As potentially suitable for field-sensitive devices and non-volatile memories. Recently, we have shown that angle-dependent magnetotransport measurements are an excellent tool for probing the MA in compressively strained (Ga,Mn)As layers [4]. For this purpose, analytical expressions for the longitudinal and transverse resistivies as a function of magnetization orientation have been derived from a series expansion of the resistivity tensor up to second order with respect to the direction cosines of the magnetization.

In the present article, this model is first extended to an expansion up to fourth order and then used to analyze the strain-dependent MA of a series of (Ga,Mn)As layers grown on different (In,Ga)As templates. The model holds not only for (Ga,Mn)As but also for single-crystalline cubic and tetragonal ferromagnets in general. Moreover, in comparison to ferromagnetic-resonance (FMR) measurements usually performed to derive MA parameters, it additionally allows for tracing the motion of the magnetization vector in a variable external magnetic field.

2 Experimental Details

Differently strained (Ga,Mn)As layers with thicknesses of ∼180 nm and Mn concentrations of ∼5% were grown by LT-MBE on (In,Ga)As templates in the following way: After thermal deoxidation a 30-nm-thick GaAs buffer layer was grown at a temperature of $T_S \approx 580\,^\circ$C on semi-insulating GaAs(001) substrate. Then the growth was interrupted, the substrate temperature T_S was lowered to ∼430 °C, and a graded (In,Ga)As layer with a thickness between 0 μm and 5 μm was deposited following the method described in [5]. In order to minimize the number of threading dislocations and to end up with different lateral lattice constants in the (In,Ga)As templates, the In content was monotonously increased from 2% up to a maximum value of 10%. Prior to the epitaxy of (Ga,Mn)As, the growth was again interrupted and T_S was lowered to $\sim 250\,^\circ$C. High-resolution x-ray diffraction (HRXRD) reciprocal space mapping (RSM) of the (115) reflex was used to determine the vertical strain $\varepsilon_{zz} = (a_\perp - a_{\rm rel})/a_{\rm rel}$ of the (Ga,Mn)As layers, where the relaxed lattice constants $a_{\rm rel}$ were derived from the lateral and vertical lattice constants $a_\parallel$ and $a_\perp$, respectively. The values of ε_{zz} were found to gradually vary from +0.24% for the sample without (In,Ga)As template to −0.35 % for the sample with 10% In. Moreover, RSM showed that the (In,Ga)As layers were almost completely relaxed whereas the (Ga,Mn)As layers were fully strained. In addition to the series of (Ga,Mn)As layers described above, a 40-nm-thick (Ga,Mn)As film, used for a comparative study between FMR and magnetotransport, was grown on GaAs(001) substrate. Further details of the growth procedure can be found in [4].

For the magnetotransport studies Hall bars with Ti-AuPt-Au contacts were prepared on several pieces of the cleaved samples with the current direction along [100] and/or [110]. The width of the Hall bars was 0.3 mm and the longitudinal voltage probes were separated by 1 mm. Hole densities of the as-grown samples between $3 \times 10^{20}\,\mathrm{cm}^{-3}$ and $4 \times 10^{20}\,\mathrm{cm}^{-3}$ were determined by means of high-field magnetotransport measurements (up to 14.5 T) at 4.2 K. Least squares fits were performed to separate the contributions of the normal and anomalous Hall effect. Curie temperatures between 61 K and 83 K were estimated from the peak positions of the temperature-dependent sheet resistivities at 10 mT. For the angle-dependent magnetotransport measurements, carried out at 4.2 K, the Hall bars were mounted on the sample holder of a liquid-He-bath cryostat, which was positioned between the poles of an electromagnet system providing a maximum field strength of 0.68 T. The sample holder possesses two perpendicular axes of rotation, allowing for an arbitrary alignment of the Hall bars with respect to the applied magnetic field $\boldsymbol{H}$.

3 Theoretical Overview

In our theoretical considerations the whole (Ga,Mn)As layer is treated as a single homogeneous ferromagnetic domain. Although being a simplification, this model has been astoundingly successful in describing a large variety of magnetization-related phenomena in (Ga,Mn)As. We write the magnetization as a vector $\boldsymbol{M} = M\boldsymbol{m}$ where M denotes its magnitude and the unit vector $\boldsymbol{m}$ its direction. In terms of the polar and azimuth angles θ and φ, respectively, defined in Fig. 1, the components of $\boldsymbol{m}$ read as $m_x = \sin\theta\cos\varphi$, $m_y = \sin\theta\sin\varphi$, and $m_z = \cos\theta$. Throughout this paper, all vector components labelled by x, y, and z refer to the cubic coordinate system associated with the [100], [010], and [001] crystal directions, respectively.

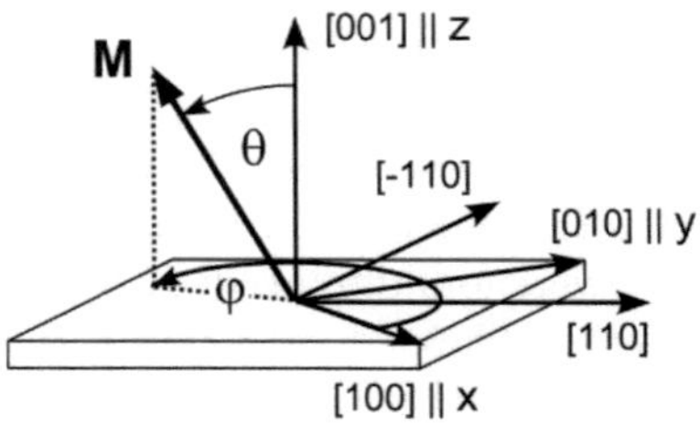

Fig. 1. Definition of the polar and azimuth angles θ and φ, respectively, used to describe the orientation of magnetization $\boldsymbol{M} = M\boldsymbol{m}$ with respect to the crystallographic axes

3.1 Magnetic Anisotropy

Magnetic anisotropy stands for the dependence of the free energy density F on the orientation of $\boldsymbol{M}$. In addition to the single-domain model, we assume that the magnitude M of the magnetization is nearly constant under the given experimental conditions. Instead of F we therefore consider the normalized quantity $F_M = F/M$, allowing for a more concise description of the MA. For a biaxially strained (Ga,Mn)As film with tetragonal distortion along [001] it can be written as [6]

$$F_M = B_{c4\|}\left(m_x^4 + m_y^4\right) + B_{c4\perp} m_z^4 + B_{c2\perp} m_z^2 + \frac{\mu_0 M}{2} m_z^2 + B_{\bar{1}10}(m_x - m_y)^2 \, . \tag{1}$$

The first three terms are intrinsic contributions arising from spin-orbit coupling in the valence band. The fourth and fifth terms are extrinsic contributions describing the demagnetization energy of an infinite plane and a uniaxial in-plane contribution whose origin is still under discussion [12–14], respectively. The two m_z^2 terms cannot be distinguished in our experiments and are therefore lumped into a single term $B_{001} m_z^2$. The anisotropy parameters B_i introduced above are in SI units. Expressed by the anisotropy fields H_i and $4\pi M_{\text{eff}}$ used in [6], they read as $B_{c4\|} = -\mu_0 H_{4\|}/4$, $B_{c4\perp} = -\mu_0 H_{4\perp}/4$, $B_{\bar{1}10} = -\mu_0 H_{2\|}/2$, and $B_{001} = \mu_0 2\pi M_{\text{eff}}$. Note that by using the trivial identity

$$|\boldsymbol{m}|^2 = m_x^2 + m_y^2 + m_z^2 = 1 \, , \tag{2}$$

Equation (1) can be easily converted to an equivalent expression where the in-plane contribution along $[\bar{1}10]$ is replaced by a contribution along [110]. In the presence of an external magnetic field $\boldsymbol{H}$ one has to additionally take into account the Zeeman energy and the total energy density is finally given by the free enthalpy

$$G_M = F_M - \mu_0 \boldsymbol{H} \cdot \boldsymbol{m} \, . \tag{3}$$

The various contributions to G_M are visualized in Fig. 2. The magnetization $\boldsymbol{M}$ aligns in such a way that G_M takes its minimum.

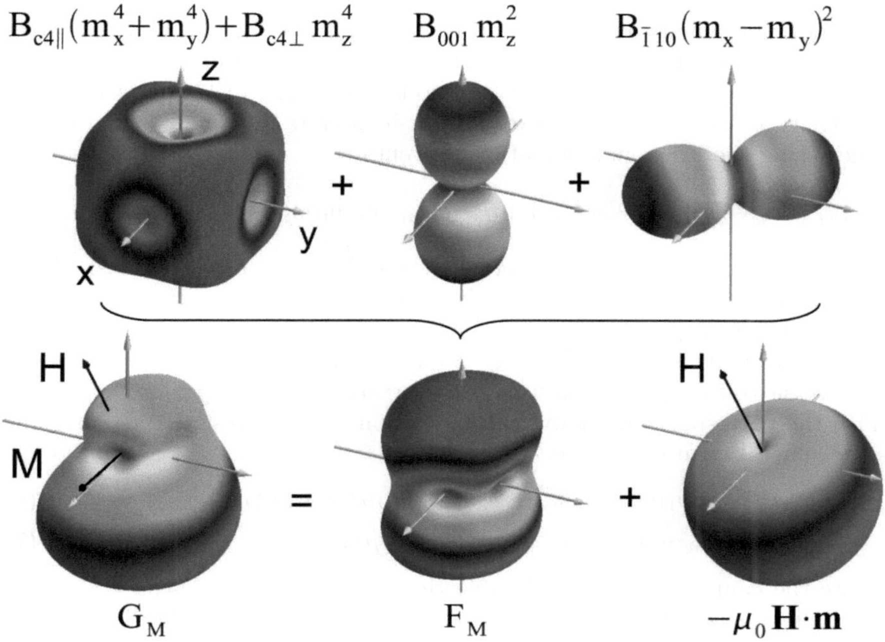

Fig. 2. 3D plots illustrating the different contributions to G_M

3.2 Longitudinal and Transverse Resistivities

The equations used in the discussion of the angle-dependent magnetotransport data can be written in a concise way by introducing the unit vectors $\boldsymbol{j}$, $\boldsymbol{n}$, and $\boldsymbol{t}$, which are adapted to the measurement configuration and specify the current direction, the surface normal, and an in-plane vector defined by $\boldsymbol{t} = \boldsymbol{n} \times \boldsymbol{j}$, respectively.

In standard magnetotransport measurements the longitudinal and transverse voltages, measured along and across the current direction, arise from the components $E_{\mathrm{long}} = \boldsymbol{j} \cdot \boldsymbol{E}$ and $E_{\mathrm{trans}} = \boldsymbol{t} \cdot \boldsymbol{E}$ of the electric field $\boldsymbol{E}$, respectively. Starting from Ohm's law $\boldsymbol{E} = \bar{\rho} \cdot \boldsymbol{J}$, where $\bar{\rho}$ represents the resistivity tensor and $\boldsymbol{J} = J\boldsymbol{j}$ the current density, the corresponding longitudinal resistivity ρ_{long} (sheet resistivity) and transverse resistivity ρ_{trans} (Hall resistivity) can be written as

$$\rho_{\mathrm{long}} = \frac{E_{\mathrm{long}}}{J} = \boldsymbol{j} \cdot \bar{\rho} \cdot \boldsymbol{j}\ , \quad \rho_{\mathrm{trans}} = \frac{E_{\mathrm{trans}}}{J} = \boldsymbol{t} \cdot \bar{\rho} \cdot \boldsymbol{j}\ . \tag{4}$$

In (Ga,Mn)As, as in other single crystalline ferromagnets, the resistivity tensor sensitively depends on the orientation of $\boldsymbol{M}$ with respect to the crystallographic axes [7]. Thus, in order to quantitatively model the measured resistivities in the general case of an arbitrarily oriented magnetization, a universal mathematical relationship between $\rho_{\rm long}$ and $\rho_{\rm trans}$ and the direction cosines m_i of $\boldsymbol{M}$ has to be derived. Following the ansatz of *Birss* [7] and *Muduli* et al. [8] we extend the model presented in [4] by writing the resistivity tensor $\bar{\rho}$ as a series expansion in powers of m_i up to the fourth order using the Einstein summation convention:

$$\rho_{ij} = a_{ij} + a_{kij}m_k + a_{klij}m_k m_l + a_{klmij}m_k m_l m_m + \ldots \,. \tag{5}$$

For cubic and tetragonal symmetry T_d and D_{2d}, respectively, most of the components a_{ij}, a_{kij}, ..., of the galvanomagnetic tensors vanish. Once the resistivity tensor $\bar{\rho}$ is known, (4) allows us to calculate $\rho_{\rm long}$ and $\rho_{\rm trans}$ for any direction of the current density $\boldsymbol{J}$ and for an arbitrary orientation of the transverse voltage probe relative to the crystal axes. For $\boldsymbol{j}$ along [100] and [110] the corresponding analytical expressions for the resistivities have the same general form reading as

$$\rho_{\rm long} = \rho_0 + \rho_1 m_j^2 + \rho_2 m_n^2 + \rho_3 m_j^4 + \rho_4 m_n^4 + \rho_5 m_j^2 m_n^2 \,, \tag{6}$$

$$\rho_{\rm trans} = \rho_6 m_n + \rho_7 m_j m_t + \rho_8 m_n^3 + \rho_9 m_j m_t m_n^2 \,, \tag{7}$$

where the components m_i of $\boldsymbol{M}$ referring to the crystal axes have been replaced by those referring to the coordinate system defined by $\boldsymbol{j}$, $\boldsymbol{n}$, and $\boldsymbol{t}$ according to

$$m_i = j_i m_j + t_i m_t + n_i m_n \,, \quad (i = x, y, z)\,. \tag{8}$$

The coefficients ρ_i are linear combinations of the expansion coefficients a_{ij}, a_{kij}, ..., and depend on the current direction. For unstrained layers (cubic symmetry) the relations $\rho_4 = \rho_5 = -\rho_2$ and $2(\rho_2 + \rho_4) = \rho_1 + \rho_3/2 - \rho_8$ hold for $\boldsymbol{j} \parallel [100]$ and $\boldsymbol{j} \parallel [110]$, respectively. The expressions for $\rho_{\rm long}$ and $\rho_{\rm trans}$ properly describe the AMR and the AHE, but (7) does not account for the ordinary Hall effect. For magnetic field strengths $\mu_0 H < 1\,\mathrm{T}$ and hole concentrations $p > 10^{20}\,\mathrm{cm}^{-3}$ as in our experiments, however, the maximum contribution of the ordinary Hall effect is $\mu_0 H/ep \approx 6 \times 10^{-6}\,\Omega\mathrm{cm}$ (e denotes the elementary charge), and is thus about two orders of magnitude smaller than the measured peak values of $\rho_{\rm trans}$ (see Sect. 4). Effects correlated with the magnitude B of the magnetic induction $\boldsymbol{B}$, such as the negative magnetoresistance, can be taken into account by considering B-dependent resistivity parameters.

4 Results and Discussion

The longitudinal and transverse resistivities of the (Ga,Mn)As layers were measured as a function of the magnetic field orientation at fixed field strengths of $\mu_0 H = 0.11$, 0.26, and 0.65 T. In order to probe the anisotropy in all three space directions, the applied magnetic field $\boldsymbol{H}$ was rotated within

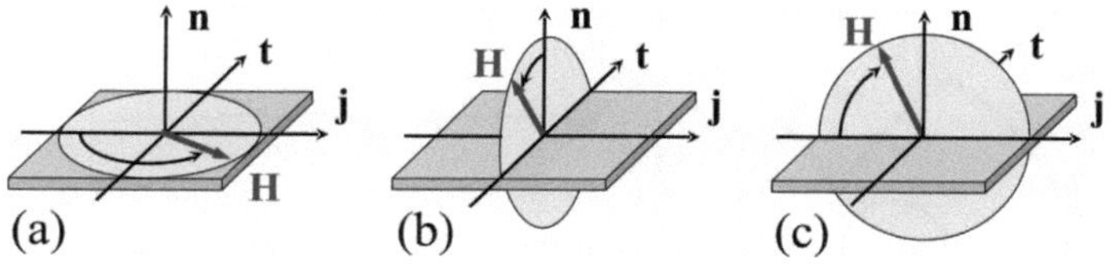

Fig. 3. The angle-dependent magnetotransport measurements were carried out for $\boldsymbol{H}$ rotated within (**a**) the layer plane, (**b**) a plane perpendicular to the current direction $\boldsymbol{j}$, and (**c**) a plane spanned by $\boldsymbol{j}$ and the normal vector $\boldsymbol{n}$

three different crystallographic planes perpendicular to $\boldsymbol{n}$, $\boldsymbol{j}$, and $\boldsymbol{t}$, respectively, as shown in Fig. 3. Prior to each angular scan, the magnetization $\boldsymbol{M}$ was put into a clearly defined initial state by raising the field to the maximum value where $\boldsymbol{M}$ is nearly saturated and aligned parallel to the external field. The field was then lowered to one of the above mentioned magnitudes and the scan was started.

4.1 Probing Magnetic Anisotropy by Magnetotransport

In this section, the determination of anisotropy parameters from angle-dependent magnetotransport data is exemplified by means of the 40-nm-thick (Ga,Mn)As layer. In the measurements $\boldsymbol{J}$ was applied along [110] and $\boldsymbol{H}$ was rotated within the (001), (110), and ($\bar{1}$10) planes. The angular dependences of ρ_{long} and ρ_{trans} for the first two configurations are shown in Fig. 4. At 0.65 T the Zeeman energy dominates the free enthalpy and the MA only plays a minor role. As a consequence, $\boldsymbol{M}$ nearly aligns with $\boldsymbol{H}$ and continuously follows its motion. In fact, the curves of ρ_{long} and ρ_{trans} at 0.65 T are smooth and largely reflect the anisotropy of the resistivity tensor. With decreasing magnetic field the influence of the MA increases and the orientation of $\boldsymbol{M}$ deviates more and more from the field direction. Accordingly, jumps and kinks occur in the curves at 0.26 and 0.11 T, arising from sudden movements of $\boldsymbol{M}$ caused by discontinuous orientation displacements of the minimum of G_M. The resistivity and anisotropy parameters from (6), (7), and (1), respectively, were determined by an iterative fit procedure. Starting with an initial guess for the anisotropy parameters B_i, the resistivity parameters ρ_i were obtained by fitting (6) and (7) to the experimental data recorded at 0.65 T. Then the anisotropy parameters were modified for an optimal agreement at 0.26 and 0.11 T, and the whole procedure was repeated until no further improvement of the fit could be achieved. The unit vector $\boldsymbol{m}$ at any given magnetic field $\boldsymbol{H}$ was calculated by numerically minimizing G_M with respect to the polar and azimuth angles of $\boldsymbol{M}$. With the exception of ρ_0, which decreases with increasing magnetic field and thus reflects the negative-magnetoresistance, the resistivity parameters turned out to be field independent within the accuracy of the fit. For the anisotropy parameters we obtained the values $B_{c4\|} = -0.02\,\text{T}$, $B_{c4\perp} = 0\,\text{T}$, $B_{001} = 0.20\,\text{T}$, and $B_{\bar{1}10} = 0.002\,\text{T}$. The theoretical

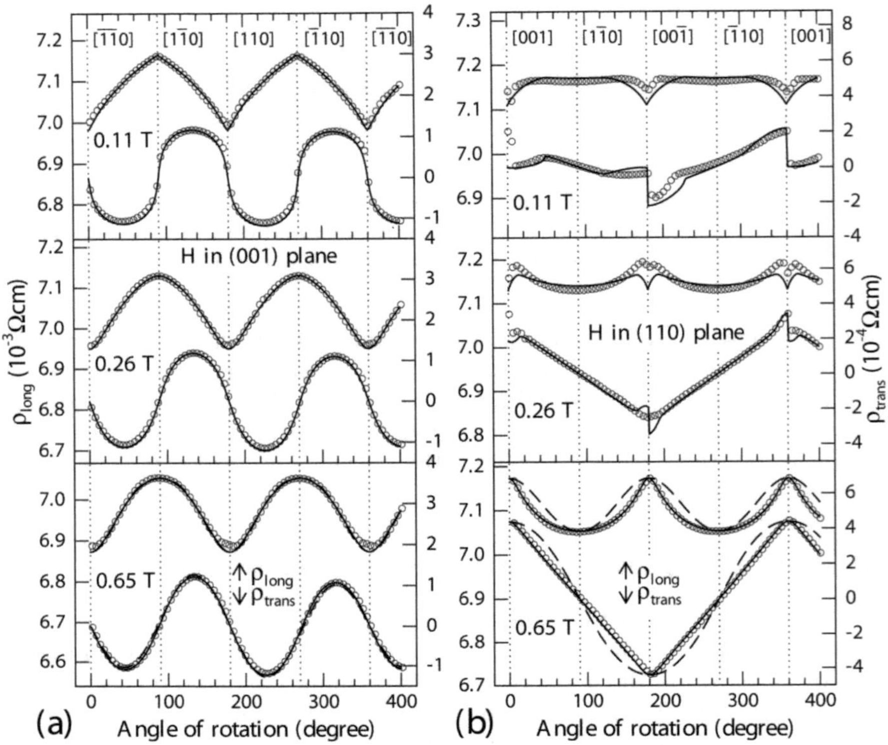

Fig. 4. Angle-dependent resistivities ρ_{long} (*left scales*) and ρ_{trans} (*right scales*) of the strained 40-nm-thick (Ga,Mn)As layer at 4.2 K. The measurements were carried out at fixed field strengths of $\mu_0 H = 0.11$, 0.26, and 0.65 T with $\boldsymbol{H}$ rotated in (**a**) the (001) and (**b**) the (110) plane. The *solid lines* represent fits to the experimental data using (6) and (7) and one set of resistivity and anisotropy parameters. The *dashed lines* at 0.65 T simulate the limiting case where $\boldsymbol{M}$ perfectly aligns with $\boldsymbol{H}$. In (**a**) the *dashed lines* completely coincide with the *solid lines*

curves calculated with these parameters and drawn as solid lines in Fig. 4 are in good agreement with the experimental data. The dashed lines at 0.65 T were calculated with $\boldsymbol{m}$ replaced by the vector $\boldsymbol{h} = \boldsymbol{H}/H$ in (6) and (7) and represent the limiting case where $\boldsymbol{M}$ perfectly aligns with $\boldsymbol{H}$. Once the anisotropy parameters are known, the orientations of the easy axes can be determined by minimizing F_M with respect to $\boldsymbol{m}$ at zero magnetic field. The easy axes are found to lie within the (001) layer plane ($\theta = 90°$) at the azimuth angles $\varphi_1 = 1.4°$ and $\varphi_2 = 88.6°$ (see Fig. 1). The slight deviation from the cubic [100] and [010] axes towards the [110] direction arises from the positive value of $B_{\bar{1}10}$.

Closer inspection of the calculated curves enables us to trace the motion of $\boldsymbol{M}$ in great detail. For $\boldsymbol{H}$ rotated within the (001) plane, the dashed curves

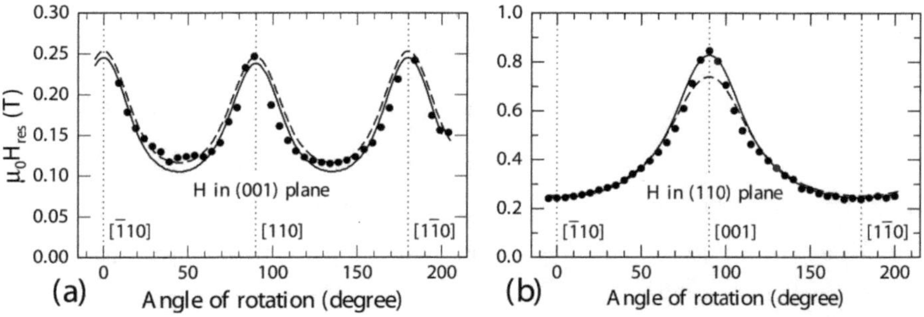

Fig. 5. Angle-dependent FMR fields of the 40-nm-thick (Ga,Mn)As layer at 5 K for $\boldsymbol{H}$ rotated (**a**) in the (001) and (**b**) in the (110) planes. The *solid lines* represent the result of a least squares fit, the *dashed lines* were calculated using the anisotropy parameters derived from the magnetotransport data

at 0.65 T coincide with the solid curves, meaning that the magnetization almost perfectly follows the motion of the magnetic field. For lower fields, $\boldsymbol{M}$ remains in the layer plane since [001] is a hard axis, but it increasingly deviates from $\boldsymbol{H}$ towards the easy [100] and [010] axes. At 0.11 T the direction of $\boldsymbol{M}$ abruptly switches whenever $\boldsymbol{H}$ approaches the slightly harder [110] and [$\bar{1}$10] axes, leading to the kinks observed for ρ_{long}. The rotation of $\boldsymbol{H}$ within a plane perpendicular to the layer is accompanied by significant differences in the orientations of $\boldsymbol{H}$ and $\boldsymbol{M}$, even for 0.65 T. This is clearly demonstrated in Fig. 4b, where the directions of $\boldsymbol{H}$ and $\boldsymbol{M}$ only coincide when $\boldsymbol{H}$ is oriented parallel or perpendicular to the layer plane. At lower fields this is no longer true and $\boldsymbol{M}$ avoids the perpendicular direction by tending towards the easy [100] and [010] axes. Accordingly, the differences between the minimum and maximum values of ρ_{long} and ρ_{trans} are drastically reduced at 0.26 and 0.11 T.

The anisotropy parameters derived from the magnetotransport measurements may be compared with those obtained from FMR measurements. Figure 5 shows the measured angular dependences of the resonance field H_{res} for $\boldsymbol{H}$ rotated within the (001) and (110) planes. A least squares fit, represented by the solid lines, yields a g factor of 1.9 and anisotropy parameters being almost identical to those obtained from magnetotransport, namely $B_{c4\|} = -0.02\,\mathrm{T}$, $B_{c4\perp} = 0\,\mathrm{T}$, $B_{001} = 0.24\,\mathrm{T}$, and $B_{\bar{1}10} = 0.002\,\mathrm{T}$. The dashed curves were calculated using the parameters from magnetotransport and demonstrate the good agreement between the two sets of anisotropy parameters. Details concerning FMR in (Ga,Mn)As can be found in [6, 15].

4.2 Strain Dependence of the Magnetic Anisotropy

The magnetotransport method presented in the preceeding section has been used to study the MA in the series of differently strained (Ga,Mn)As layers grown on (In,Ga)As templates. An overview of the results is given in Fig. 6.

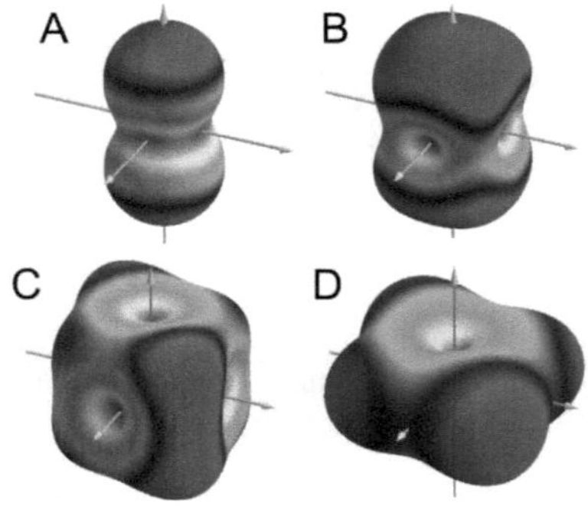

sample	ε_{zz}(%)	$B_{c4\parallel}$	$B_{c4\perp}$	B_{001}	$B_{\bar{1}10}$
A	0.24	-25	-30	190	5
B	0.01	-35	-35	35	-5
C	-0.20	-30	-20	-15	-50
D	-0.35	-35	10	-90	-10
			(10^{-3}T)		

Fig. 6. Anisotropy parameters of the differently strained (Ga,Mn)As layers and corresponding 3D plots of F_M

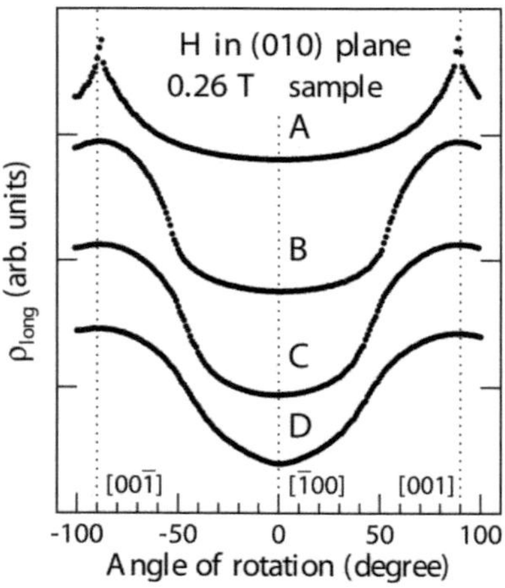

Fig. 7. The weakening of the out-of-plane axis with decreasing compressive and increasing tensile strain is accompanied by characteristic changes of the longitudinal resistivity, measured in the configuration of Fig. 3c

Whereas in the compressively strained and in the nearly unstrained layers (samples A and B, respectively) the easy axes are in-plane and nearly align with the [100] and [010] crystal axes, the easy axis in the layers with tensile strain (samples C and D) points along [001]. This is most clearly reflected by the anisotropy parameter B_{001}, which decreases from 0.19 T to -0.09 T for compressive to tensile strain. The weakening of the out-of-plane axis can also be seen by a mere qualitative inspection of resistivity curves. Figure 7 shows the longitudinal resistivities of the samples A–D recorded as a function of the magnetic field orientation at 0.26 T with $\boldsymbol{H}$ rotated in the (010) plane perpendicular to $\boldsymbol{j}$. The weakening of the [001] axis is accompanied by a

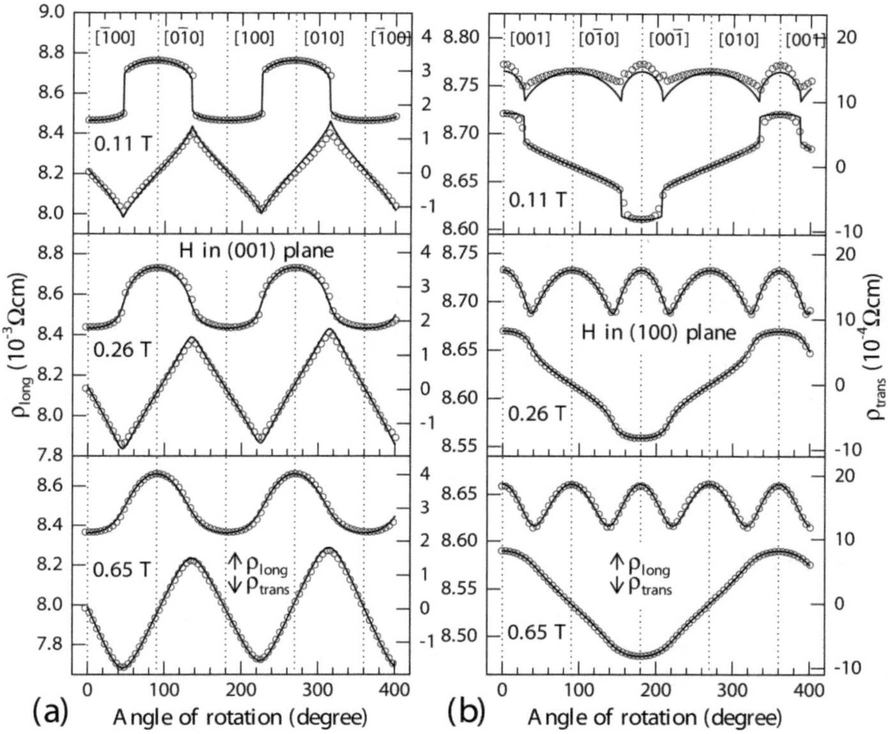

Fig. 8. Angle-dependent resistivities ρ_{long} and ρ_{trans} of the nearly unstrained (Ga,Mn)As layer ($\varepsilon_{zz} = 0.01\%$) at 4.2 K. The measurements were carried out at fixed field strengths of $\mu_0 H = 0.11$, 0.26, and 0.65 T with $\boldsymbol{H}$ rotated in (**a**) the (001) and (**b**) the (100) plane. The *solid lines* represent fits to the experimental data using (6) and (7) and one set of resistivity and anisotropy parameters

flattening of the curves in the range where $\boldsymbol{H}$ is nearly out-of-plane as well as by a narrowing when $\boldsymbol{H}$ approaches the $[\bar{1}00]$ in-plane axis.

Figure 8 shows as an example ρ_{long} and ρ_{trans} of the unstrained sample B, measured for $\boldsymbol{j}$ along [100] and for $\boldsymbol{H}$ rotated in the (001) and (100) planes. Two features are particularly important: Firstly, without fourth order terms in the series expansion of the resistivity tensor fundamental discrepancies arise between the simulations and the experimental traces of ρ_{long} in Fig. 8b. This differs from the results presented in [4], where terms up to second order were sufficient for a satisfactory agreement between theory and experiment. Secondly, the jumps and kinks in Fig. 8a at 0.11 T are extremely well pronounced, reflecting the occurence of a perfect magnetization switching where nearly all magnetic moments in the (Ga,Mn)As layer simultaneously change their orientation. In most other configurations significant differences between the calculated and the measured curves are observed at 0.11 T, giving evidence for a gradual breakdown of the single-domain model with decreasing magnetic field.

5 Summary

It is shown that angle-dependent magnetotransport measurements, performed at different magnetic field strengths, are an excellent tool for probing the magnetic anisotropy of differently strained (Ga,Mn)As layers grown on (In,Ga)As templates. For this purpose, analytical expressions for the longitudinal and transverse resistivities in single-crystalline ferromagnets with cubic and tetragonal symmetry were derived from a series expansion of the resistivity tensor with respect to the magnetization components up to fourth order. From compressive to tensile strain, the parameter B_{001} for the uniaxial out-of-plane anisotropy decreases and finally reverses sign, changing the [001] crystal axis from a magnetically hard to an easy axis.

Acknowledgements

This work was supported by the Deutsche Forschungsgemeinschaft (Li 988/4). The FMR data presented in this article were measured by the group of M. S. Brandt of the Walter-Schottky-Institut, TU München, Garching, Germany.

References

[1] H. Ohno, Science **281**, 951 (1998).
[2] T. Jungwirth, J. Sinova, J. Mašek, J. Kučera, and A. H. MacDonald, Rev. Mod. Phys. **78**, 809 (2006), and references therein.
[3] T. Dietl, H. Ohno, and F. Matsukura, Phys. Rev. B **63**, 195205 (2001).
[4] W. Limmer, M. Glunk, J. Daeubler, T. Hummel, W. Schoch, R. Sauer, C. Bihler, H. Huebl, M. S. Brandt, and S. T. B. Goennenwein, Phys. Rev. B **74**, 205205 (2006).
[5] J. C. Harmand, T. Matsuno, and K. Inoue, Jpn. J. Appl. Phys. **28**, 1101 (1989).
[6] X. Liu and J. K. Furdyna, J. Phys.: Condens. Matter **18**, R245 (2006).
[7] R. R. Birss, *Symmetry and Magnetism*, (North-Holland, Amsterdam, 1966).
[8] P. K. Muduli, K.-J. Friedland, J. Herfort, H.-P. Schönherr, and K. H. Ploog, Phys. Rev. B **72**, 104430 (2005).
[9] R. R. Birss, Proc. Phys. Soc. **75**, 8 (1960).
[10] T. McGuire and R. Potter, IEEE Trans. Magn. **11**, 1018 (1975).
[11] J. P. Jan, in *Solid State Physics*, edited by F. Seitz and D. Turnbull (Academic Press, New York, 1957), Vol. 5, pp. 1–96.
[12] M. Sawicki, K.-Y. Wang, K. W. Edmonds, R. P. Campion, C. R. Staddon, N. R. S. Farley, C. T. Foxon, E. Papis, E. Kamińska, A. Piotrowska, T. Dietl, and B. L. Gallagher, Phys. Rev. B **71**, 121302(R) (2005).

[13] U. Welp, V. K. Vlasko-Vlasov, A. Menzel, H. D. You, X. Liu, J. K. Furdyna, and T. Wojtowicz, Appl. Phys. Lett. **85**, 260 (2004).
[14] K. Hamaya, T. Watanabe, T. Taniyama, A. Oiwa, Y. Kitamoto, and Y. Yamazaki, Phys. Rev. B **74**, 045201 (2006).
[15] S. T. B. Goennenwein, T. Graf, T. Wassner, M. S. Brandt, M. Stutzmann, J. B. Philipp, R. Gross, M. Krieger, K. Zürn, P. Ziemann, A. Koeder, S. Frank, W. Schoch, and A. Waag, Appl. Phys. Lett. **82**, 730 (2003).

Highly Spin-Polarized Tunneling in Fully Epitaxial Magnetic Tunnel Junctions with a Co-Based Full-Heusler Alloy Thin Film and a MgO Barrier

Masafumi Yamamoto, Takao Marukame, Takayuki Ishikawa,
Ken-ichi Matsuda, and Tetsuya Uemura

Division of Electronics for Informatics, Graduate School of Information Science and Technology, Hokkaido University,
N14, W9, Kita-ku, 060-0814 Sapporo, Japan
yamamoto@nano.ist.hokudai.ac.jp

Abstract. Co-based full-Heusler alloy (Co_2YZ) thin films are highly preferable ferromagnetic materials in spintronic devices because of the half-metallic ferromagnetic nature at room temperature (RT) theoretically predicted for some of these alloys. We developed fully epitaxial magnetic tunnel junctions (MTJs) that have a Co_2YZ thin film of $Co_2Cr_{0.6}Fe_{0.4}Al$ (CCFA), Co_2MnSi (CMS), or Co_2MnGe (CMG) as a lower electrode, and a MgO tunnel barrier, and have demonstrated a relatively high tunnel magnetoresistance (TMR) ratio of 109% at RT (317% at 4.2 K) for CCFA/MgO/$Co_{50}Fe_{50}$ MTJs and a TMR ratio of 90% at RT (192% at 4.2 K) for CMS/MgO/$Co_{50}Fe_{50}$ MTJs. A high tunneling spin polarization of 0.88 at 4.2 K was estimated for epitaxial CCFA films with the *B*2 structure. The demonstrated high TMR ratios confirmed that fully epitaxial MTJs with a MgO tunnel barrier are promising as a key device structure for fully utilizing the high spin polarization of Co-based full-Heusler alloy thin films.

1 Introduction

In conventional semiconductor electronics, information is carried by the electron charge and the spin of the electron has not played any role. There has been growing interest, though, in an emerging technology called spintronics that utilize the spin degree of freedom of the electron [1]. Employing spin-polarized electrons is essential for spintronic devices. Half-metallic ferromagnets (HMFs) are characterized by an energy gap for one spin direction at the Fermi level (E_F), leading to a complete spin polarization at E_F [2]. The potentially high spin polarization of HMFs is widely applicable and advantageous for ferromagnetic electrodes used in spintronic devices in terms of achieving high tunnel magnetoresistance (TMR) ratios in magnetic tunnel junctions (MTJs), efficient spin injection from ferromagnetic electrodes into semiconductors, and current-induced magnetization switching in MTJs.

Co-based full-Heusler alloys (Co_2YZ) have attracted much interest as a promising ferromagnetic electrode material for spintronic devices, especially

R. Haug (Ed.): Advances in Solid State Physics,
Adv. in Solid State Phys. **47**, 105–116 (2008)

for MTJs [3–15]. This is because of the half-metallic ferromagnetic nature theoretically predicted for some of these alloys [16, 17], and because of their high Curie temperatures, which are well above room temperature (RT) [18].

Our purpose in the present study has been to demonstrate high TMR ratios at RT for MTJs using a Co-based Heusler alloy thin film. Our approach was to develop fully epitaxial MTJs with a Co-based full-Heusler alloy thin film and a MgO tunnel barrier [8, 9]. The lattice mismatch between some Co_2YZ – for example $Co_2Cr_{0.6}Fe_{0.4}Al$ (CCFA), Co_2MnGe (CMG), and Co_2MnSi (CMS) – and MgO for a 45° in-plane rotation is relatively small, so MgO is a good candidate for use as a tunnel barrier for epitaxial MTJs with a Co_2YZ thin film. We developed fully epitaxial MTJs that have a Co_2YZ thin film of CCFA, CMG or CMS as a lower electrode, and a MgO tunnel barrier [8–15], and have demonstrated a relatively high TMR ratio of 109% at RT (317% at 4.2 K) for CCFA/MgO/$Co_{50}Fe_{50}$ MTJs [8, 11, 13, 14] and a TMR ratio of 90% at RT (192% at 4.2 K) for CMS/MgO/$Co_{50}Fe_{50}$ MTJs [12]. A high tunneling spin polarization of 0.88 at 4.2 K was estimated for epitaxial CCFA films with the $B2$ structure [13].

2 Experimental Methods

Full-Heusler alloys are ternary inter-metallic compounds with the composition X_2YZ in the $L2_1$ cubic structure (space group:$Fm\overline{3}m$). A typical full-Heusler alloy consists of two different transition metals X and Y, and a non-magnetic element Z as shown in Fig. 1a. The lattice consists of four different fcc sub-lattices. Each has an atom basis as follows: X element at (1/4, 1/4, 1/4) and (3/4, 3/4, 3/4), Y at (0, 0, 0), and Z at (1/2, 1/2, 1/2). The lattice mismatch between CCFA (lattice constant, a = 0.5737 nm), CMG (a = 0.5743 nm) or CMS (a = 0.5654 nm) and MgO (a = 0.4212 nm) on a 45° in-plane rotation is relatively small (about −3.6% for CMG, −3.7% for CCFA and −5.1% for CMS). Therefore, it is reasonable to expect that CCFA, CMG and CMS films will grow epitaxially with their [100] direction rotated by 45° from the MgO [100] direction in the (001) plane (Fig. 1b). Similarly, MgO is a good candidate for use as a tunnel barrier for epitaxial MTJs with a thin film of CCFA, CMG or CMS.

We fabricated single-crystal, epitaxially grown Co_2YZ thin films of CCFA [8,19,20], Co_2MnGe [9,21], or Co_2MnSi [22] on MgO-buffered MgO(001) substrates. Given these epitaxial thin films, we developed fully epitaxial MTJs that had a Co_2YZ thin film of CCFA [8, 11, 13, 14], Co_2MnGe [9, 10, 15] or Co_2MnSi [12] as a lower electrode and a MgO tunnel barrier. We used an epitaxially grown $Co_{50}Fe_{50}$ thin film on a MgO tunnel barrier as an upper electrode throughout the present study. The lattice mismatch between $Co_{50}Fe_{50}$ (a = 0.2842 nm) and MgO is −4.1% on a 45° in-plane rotation.

Now, we will describe the fabrication procedure for Co_2YZ/MgO/$Co_{50}Fe_{50}$ MTJs with exchange biasing, where Co_2YZ represents CCFA, CMG or

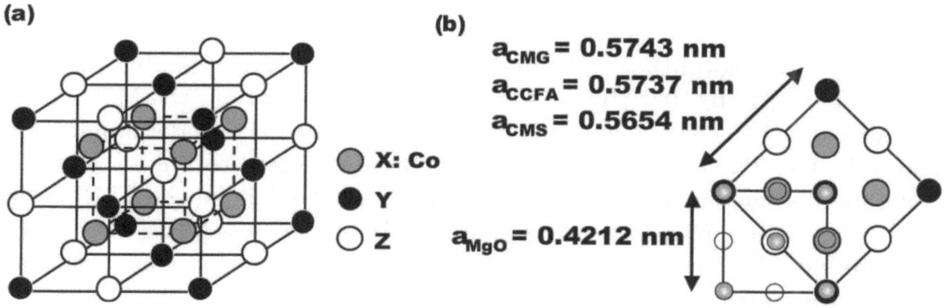

Fig. 1. (**a**) Schematic view of the $L2_1$ crystal structure with composition X_2YZ. The lattice consists of four different fcc sublattices. Each has an atom basis as follows: X element at (1/4, 1/4, 1/4) and (3/4, 3/4, 3/4), Y at (0, 0, 0), and Z at (1/2, 1/2, 1/2). (**b**) Top view of Heusler alloy crystal structure ($L2_1$ structure). MgO cubic structure is superimposed

CMS. Our approach for fabricating exchange-biased MTJs was to use an upper electrode of $Co_{50}Fe_{50}$ film in an antiferromagnetically coupled (i.e., synthetic ferrimagnetic) $Co_{50}Fe_{50}$/Ru/$Co_{90}Fe_{10}$ trilayer exchange-biased by an IrMn antiferromagnetic layer through the $Co_{90}Fe_{10}$/IrMn interface to obtain a high exchange-bias field value (H_{ex}) for epitaxial $Co_{50}Fe_{50}$ electrodes. The fabricated MTJ layer structure (from the substrate side) was as follows: MgO buffer (10 nm)/Co_2YZ (50 nm)/MgO tunnel barrier (typically 2.4 nm)/$Co_{50}Fe_{50}$ (typically 3.4 nm)/Ru (0.8 nm)/$Co_{90}Fe_{10}$ (2 nm)/ IrMn (10 nm)/Ru cap (5 nm). All layers in these MTJs were successively deposited on MgO(001) single-crystal substrates in an ultrahigh vacuum chamber (with a base pressure of about 8×10^{-8} Pa) through the combined use of magnetron sputtering and electron beam evaporation. The Co_2YZ lower electrode was deposited by rf magnetron sputtering at RT and subsequently annealed *in situ* at 500°C to 600°C. The Co_2YZ film composition was determined, with an accuracy of 2 to 3% for each element, through inductively coupled plasma analysis. The MgO tunnel barrier was deposited by electron beam evaporation at RT. The pressure during the deposition of the MgO tunnel barrier was about 6×10^{-7} Pa. The layers of $Co_{50}Fe_{50}$, Ru, $Co_{90}Fe_{10}$, and IrMn were all deposited by magnetron sputtering at RT. We carried out *in situ* reflection high-energy electron diffraction (RHEED) observations for each successive layer during fabrication. We investigated the structural properties of the fabricated Co_2YZ films through x-ray Bragg scans and x-ray pole figure measurements (Bruker AXS D8 DISCOVER Hybrid). Structural characterization of the fabricated MTJ layer structures was done by cross-sectional high-resolution transmission electron microscope (HRTEM) observation, and transmission electron diffraction. The surface morphologies of the fabricated Co_2YZ films were observed using atomic force microscopy (Digital Instruments).

We fabricated fully epitaxial MTJs with the layer structure described above by using photolithography and Ar ion milling. The fabricated junction size was $8 \times 8\,\mu m$ to $10 \times 10\,\mu m$. After the micro-fabrication, the MTJs were annealed at 175°C for one hour in a vacuum of 5×10^{-2} Pa under a magnetic field of 5 kOe. The magnetoresistance was then measured through a dc four-probe method at temperatures from RT to 4.2 K. We defined the TMR ratio as $(RA_{AP} - RA_P)/RA_P$, where RA_{AP} and RA_P are the respective resistance-area products for the antiparallel and parallel magnetization configurations between the upper and lower electrodes.

3 Results and Discussion

3.1 Structural Properties

To fabricate high quality MTJs, a lower ferromagnetic electrode with little surface roughness must be prepared. Our AFM measurements indicated that 100-nm-thick CCFA films grown on 10-nm-thick MgO buffer layers at RT and subsequently annealed at 500°C had sufficiently flat surface morphologies with roughness of about 0.23 nm rms [19]. Similarly, our AFM measurements showed that 50-nm-thick CMS films grown on 10-nm-thick MgO buffer layers at RT and subsequently annealed at 600°C had sufficiently flat surface morphologies with rms roughness of about 0.22 nm. In contrast, AFM measurements showed that as-deposited CMS films had surface morphologies with roughness of 0.34 nm rms. The improved surface flatness revealed by AFM measurement after post-deposition annealing at 600°C was consistent with the RHEED observations, which indicated the streak patterns of the annealed CMS film were sharper and more distinct than those of the as-deposited film [22].

In x-ray pole figure measurements of CMS films deposited at RT and subsequently annealed at 600°C, CMS 111 diffraction peaks with fourfold symmetry with respect to the sample rotation angle, ϕ, were clearly observed at a tilt angle χ of 54.7°. This provides direct evidence that the film annealed at 600°C was epitaxial and crystallized in the $L2_1$ structure. Because the ϕ values for the CMS 111 peaks were shifted by 45° with respect to those of the MgO 111 peaks, the crystallographic relationship was CMS (001)[100]||MgO (001)[110]. These results were commonly obtained for both the Co-rich CMS film (a film composition of $Co_2Mn_{0.84}Si_{0.80}$) and the Co-deficient CMS film (a film composition of $Co_2Mn_{1.23}Si_{1.19}$). Similarly, it was confirmed that both Co-rich CMG films (a film composition of $Co_2Mn_{0.74}Ge_{0.43}$) and Co-deficient CMG films (a film composition of $Co_2Mn_{1.23}Ge_{1.17}$), deposited in both cases at RT and subsequently annealed at 500°C to 600°C, grew epitaxially and crystallized in the $L2_1$ structure. X-ray diffraction measurement of the 50-nm-thick CCFA thin film annealed *in situ* at 500°C showed that the film grew epitaxially and crystallized in the $B2$ structure.

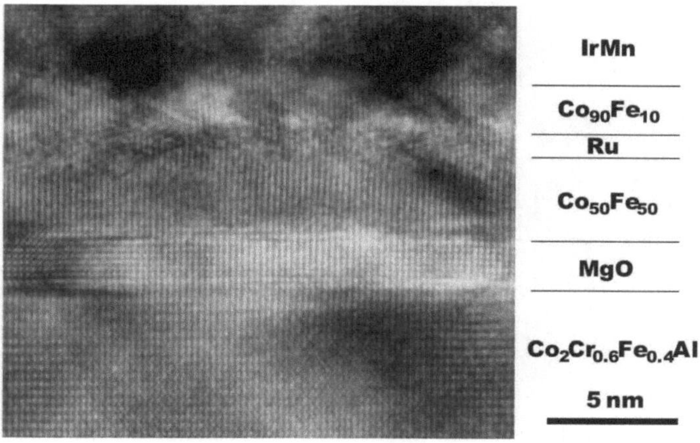

Fig. 2. Cross-sectional high-resolution transmission electron microscope image of a fully epitaxial MTJ layer structure consisting of $Co_2Cr_{0.6}Fe_{0.4}Al$ (CCFA) (50 nm)/ MgO (2 nm)/$Co_{50}Fe_{50}$ (3 nm)/Ru (0.8 nm)/$Co_{90}Fe_{10}$ (2 nm)/IrMn (10 nm)/Ru cap (5 nm), along the [1−10] direction of the CCFA [13]

We will now describe the structural characterization of the fabricated layer structures. First, we will describe the structural properties of the CCFA/ MgO/$Co_{50}Fe_{50}$ MTJ layer structure. RHEED patterns observed *in situ* for each layer during fabrication clearly indicated that the CCFA lower electrode, MgO tunnel barrier, and $Co_{50}Fe_{50}$ upper electrode grew epitaxially. Figure 2 shows a cross-sectional HRTEM image of an MTJ layer structure consisting of $Co_2Cr_{0.6}Fe_{0.4}Al$ (CCFA) (50 nm)/MgO (2 nm)/$Co_{50}Fe_{50}$ (3 nm)/Ru (0.8 nm)/$Co_{90}Fe_{10}$ (2 nm)/IrMn (10 nm)/Ru cap (5 nm), along the [1−10] direction of the CCFA film. This image clearly reveals that all the layers of the CCFA/MgO/$Co_{50}Fe_{50}$ basic tunnel junction trilayer – i.e., the CCFA lower electrode, MgO tunnel barrier, and $Co_{50}Fe_{50}$ upper electrode – were grown epitaxially and were single crystalline [13]. It also confirmed that extremely smooth and abrupt interfaces were formed. Next, we will describe the structural properties of the $Co_{50}Fe_{50}$/Ru/$Co_{90}Fe_{10}$/IrMn quadrilayer, which was part of the MTJ layer structure. We also observed streak patterns in RHEED patterns that were dependent on the incident direction of the electron beam for layers of Ru, $Co_{90}Fe_{10}$, and IrMn, indicating that the layers grew epitaxially on the single-crystal $Co_{50}Fe_{50}$ electrode. Furthermore, cross-sectional HRTEM lattice images (Fig. 2) clearly showed that all the layers of Ru, $Co_{90}Fe_{10}$, and IrMn were grown epitaxially on the single-crystal $Co_{50}Fe_{50}$ electrode and were single crystalline.

Figure 3a shows a cross-sectional HRTEM lattice image of an MTJ layer structure consisting of Co_2MnSi (CMS) (50 nm)/MgO (2.4 nm)/$Co_{50}Fe_{50}$ (3 nm)/Ru (0.8 nm)/$Co_{90}Fe_{10}$ (2 nm)/IrMn (10 nm)/Ru cap (5 nm), along the [1−10] direction of the CMS. This image clearly shows that all the layers

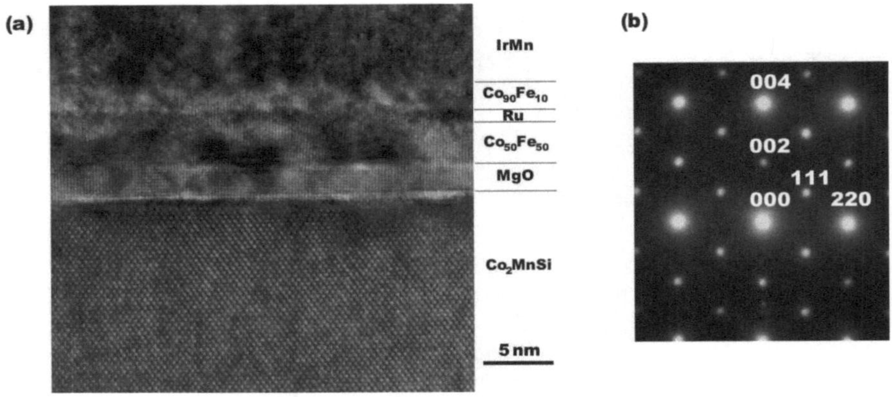

Fig. 3. (**a**) Cross-sectional high-resolution transmission electron microscope image of a fully epitaxial MTJ layer structure consisting of Co_2MnSi (CMS) (50 nm)/MgO (2.4 nm)/$Co_{50}Fe_{50}$ (3 nm)/Ru (0.8 nm)/$Co_{90}Fe_{10}$ (2 nm)/IrMn (10 nm)/Ru cap (5 nm), along the [1−10] direction of the CMS [12]. (**b**) Electron diffraction pattern for the CMS layer. The electron beam diameter was 10 nm

from the CMS lower electrode to the IrMn layer were also grown epitaxially and were single crystalline. Figure 3a also shows that extremely smooth and abrupt interfaces were formed in the CMS/MgO/$Co_{50}Fe_{50}$ trilayer. Figure 3b shows a microbeam electron diffraction pattern with a beam diameter of 10 nm for the CMS film. In the electron diffraction pattern, 111 spots were clearly observed, indicating that the 600°C-annealed CMS films grown on MgO buffer layers had the $L2_1$ structure, and so were consistent with the results obtained through x-ray pole figure measurements. Furthermore, we observed 111 spots, through micro-beam electron diffraction patterns with a beam diameter of 10 nm, for any spot region of either the Co-rich CMS film ($Co_2Mn_{0.84}Si_{0.80}$) or the Co-deficient CMS film ($Co_2Mn_{1.23}Si_{1.19}$); this indicated that these films consisted of the $L2_1$ structure, although the possibility of structural disorder within the $L2_1$ structure cannot be excluded. In contrast, micro-beam electron diffraction patterns indicated the Co-rich and Co-deficient CMG films both had the $L2_1$ structure with some residual regions of the $B2$, $A2$, and unknown structures. Micro-beam electron diffraction for the CCFA films indicated that these films crystallized in the $B2$ structure, which was consistent with the results of x-ray pole figure measurements.

3.2 Spin-Polarized Tunneling Characteristics

3.2.1 $Co_2Cr_{0.6}Fe_{0.4}Al$/MgO/$Co_{50}Fe_{50}$ MTJs

Figure 4a shows typical magnetoresistance curves at RT and 4.2 K for a fabricated fully epitaxial, exchange-biased CCFA/MgO/$Co_{50}Fe_{50}$ MTJ. The CCFA film composition was $Co_2Cr_{0.57}Fe_{0.39}Al_{1.12}$. The applied bias voltage

(V) was 5 mV. The MTJ exhibited clear exchange-biased TMR characteristics with high TMR ratios of 109% at RT and 317% at 4.2 K [13]. These values are significantly higher than our previously reported values of 90% at RT and 240% at 4.2 K for fully epitaxial CCFA/MgO/$Co_{50}Fe_{50}$ MTJs, in which we used the difference in the coercive forces to form the antiparallel magnetization configurations between the CCFA lower electrode and the $Co_{50}Fe_{50}$ upper electrode [11]. We obtained relatively high H_{ex} values of about 350 Oe at RT and about 1000 Oe at 4.2 K as shown in Fig. 4a. We can reasonably attribute the high H_{ex} values obtained for the fabricated MTJs to a lower net saturation magnetization of the synthetic ferrimagnetic trilayer compared with a saturation magnetization of the $Co_{50}Fe_{50}$ electrode [23].

Figure 4b shows the TMR ratio at $V = 5$ mV for a CCFA/MgO/$Co_{50}Fe_{50}$ MTJ (CCFA-MTJ) as a function of temperature (T) from 4.2 K to RT (this is the same MTJ as shown in Fig. 4a). For comparison, the TMR ratio as a function of T is also plotted for a fully epitaxial, exchange-biased $Co_{50}Fe_{50}$/MgO/$Co_{50}Fe_{50}$ MTJ (a reference $Co_{50}Fe_{50}$-MTJ) identically fabricated with the same layer structure as that of the exchange-biased CCFA-MTJ except that the lower electrode CCFA was replaced with $Co_{50}Fe_{50}$. The $Co_{50}Fe_{50}$-MTJs were post-fabrication annealed under the same annealing conditions as for the CCFA-MTJs (i.e., at 175°C under a magnetic field of 5 kOe). The layer structure (from the substrate side) was $Co_{50}Fe_{50}$ (50 nm)/MgO (2.2 nm)/$Co_{50}Fe_{50}$ (3 nm)/Ru (0.8 nm)/$Co_{90}Fe_{10}$ (2 nm)/IrMn (10 nm)/Ru cap (5 nm), and the structure was grown on a MgO-buffered MgO substrate. The $Co_{50}Fe_{50}$-MTJs showed TMR ratios of 185% at 4.2 K and 125% at RT. As shown in Fig. 4b, the TMR ratio of the CCFA-MTJ was definitely higher than that of the $Co_{50}Fe_{50}$-MTJ below about 220 K, and it reached 317% at 4.2 K for the CCFA-MTJ (although it was slightly lower at RT).

In other words, the CCFA-MTJ exhibited a strong T dependence of the TMR ratio from 4.2 K to RT, while the $Co_{50}Fe_{50}$-MTJ showed a more moderate T dependence. If we use parameter $\gamma = \alpha(4.2\text{K})/\alpha(\text{RT})$, where α is the TMR ratio, to represent the degree of T dependence of the TMR ratio, γ for the fabricated CCFA-MTJs was 2.9, which was much higher than $\gamma = 1.5$ for the $Co_{50}Fe_{50}$-MTJ.

Figure 4c shows the TMR ratio, as well as RA_{AP} and RA_P, as a function of T from 4.2 to 297 K. RA_{AP} also decreased with increasing T, while RA_P was almost independent of T for the CCFA-MTJs. (These T dependences of RA_{AP} and RA_P were similar to those observed for $Co_{70}Fe_{30}$/MgO/$Co_{84}Fe_{16}$ MTJs [24].)

We next estimated the tunneling spin polarization of the epitaxial CCFA electrode from the obtained TMR ratios. The TMR ratios for MTJs have been traditionally related to the spin polarizations at E_F, P_1, and P_2, of the ferromagnetic electrodes through Jullière's model [25]; i.e., TMR $= 2P_1P_2/(1 - P_1P_2)$. Jullière's model was derived by assuming a loss of coherence in tunneling (i.e., non-conservation of the electron's wave vector component parallel to the interface) [26]. However, a straightforward appli-

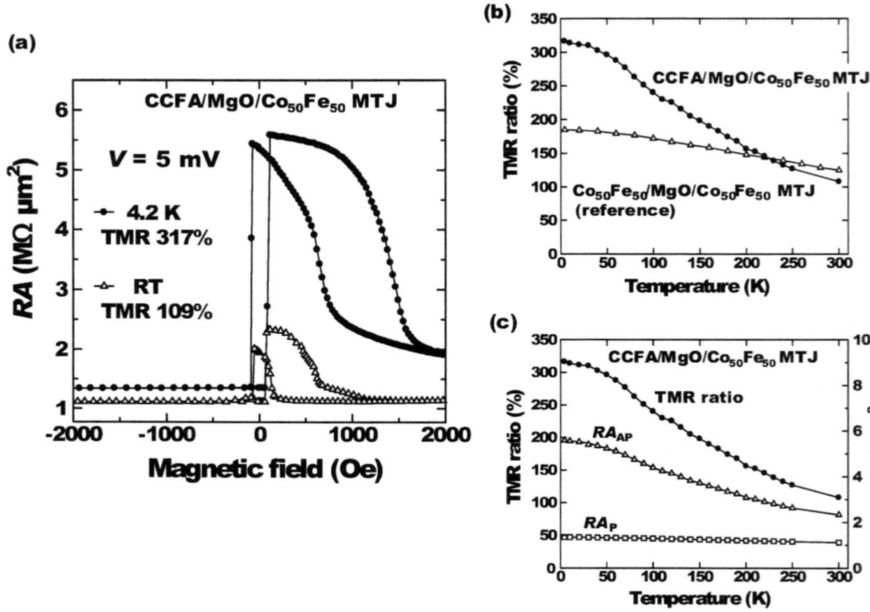

Fig. 4. Tunnel magnetoresistance characteristics of a fully epitaxial, exchange-biased $Co_2Cr_{0.6}Fe_{0.4}Al$/MgO (2.4 nm)/$Co_{50}Fe_{50}$ MTJ. The junction size was 10 × 10 μm. The bias voltage V was 5 mV. (**a**) Typical magnetoresistance curves at RT and 4.2 K. (**b**) TMR ratio as a function of temperature (T) from 4.2 K to RT. As a reference, the TMR ratio as a function of T for an identically fabricated fully epitaxial, exchange-biased $Co_{50}Fe_{50}$/MgO/$Co_{50}Fe_{50}$ MTJ is also shown. (**c**) TMR ratio, as well as RA_{AP} and RA_{P}, as a function of T from 4.2 K to RT [13, 14]

cation of Jullière's model for a TMR ratio of 317% at 4.2 K for fully epitaxial CCFA/MgO/$Co_{50}Fe_{50}$ MTJs with a $Co_{50}Fe_{50}$ electrode spin polarization of 0.50, derived from dI/dV curves of superconductor/AlO_x/$Co_{50}Fe_{50}$ tunnel structures [27], corresponding to the originally defined spin polarization using majority- and minority-spin band density of states at E_F, results in an unreallistically high P value exceeding 1.0 for the epitaxial CCFA electrode. This result indicates the TMR ratio was enhanced by a coherent tunneling contribution for fully epitaxial CCFA/MgO/$Co_{50}Fe_{50}$ MTJs. Furthermore, the obtained TMR ratios of 185% at 4.2 K and 125% at RT clearly indicate enhancement of the TMR ratio by a coherent tunneling contribution for the reference $Co_{50}Fe_{50}$-MTJs [28, 29]. Therefore, we estimated the tunneling spin polarization for the epitaxial $Co_{50}Fe_{50}$ electrode, P_{CoFe}, by applying Jullière's model for the TMR ratio of 185% at 4.2 K (125% at RT) of the reference $Co_{50}Fe_{50}$-MTJs. We obtained a P_{CoFe} value of 0.69 at 4.2 K (0.62 at RT), which was higher than the P value of 0.50 derived from superconductor/AlO_x/$Co_{50}Fe_{50}$ tunnel structures [27]. Similarly, we estimated the tunneling spin polarization for an epitaxial CCFA electrode in fully epitaxial CCFA/MgO/$Co_{50}Fe_{50}$ MTJs, P_{CCFA}, by applying Jullière's model for the TMR ratio of 317% at 4.2 K (109% at RT) of the CCFA-MTJs, along with a P_{CoFe} value of 0.69 at 4.2 K (0.62 at RT) derived from the TMR ratio for the reference $Co_{50}Fe_{50}$-MTJs; in this case, we obtained a high tunneling spin polarization of 0.88 at 4.2 K (0.57 at RT) for the epitaxial CCFA thin film with the $B2$ structure. Although a rigorous comparison is not justified, the thus obtained P_{CCFA} value of 0.88 is larger than the theoretically predicted P_{CCFA} value of 0.78 [30] even though we assumed a tunneling spin polarization of 0.69 at 4.2 K (0.62 at RT) for the epitaxial $Co_{50}Fe_{50}$ electrode in the estimation rather than 0.50 at 4.2 K as was derived from superconductor/AlO_x/$Co_{50}Fe_{50}$ tunnel structures. This result also indicates a coherent tunneling contribution for fully epitaxial CCFA/MgO/$Co_{50}Fe_{50}$ MTJs. It is thus suggested that fully epitaxial MTJs with a Co_2YZ film or Co_2YZ films and a MgO tunnel barrier could provide high TMR ratios due to both the half-metallicity and coherent tunneling [31].

3.2.2 Co_2MnSi/MgO/$Co_{50}Fe_{50}$ MTJs and Co_2MnGe/MgO/$Co_{50}Fe_{50}$ MTJs

In this section, we describe the TMR characteristics for exchange-biased Co_2MnSi/MgO/$Co_{50}Fe_{50}$ and Co_2MnGe/MgO/$Co_{50}Fe_{50}$ MTJs (CMS-MTJs and CMG-MTJs, respectively).

Figure 5a shows typical magnetoresistance curves at V of 5 mV at RT and 4.2 K for a fabricated CMS-MTJ where we used a Co-rich CMS film with the film composition of $Co_2Mn_{0.84}Si_{0.80}$. Clear exchange-biased TMR characteristics were obtained with relatively high TMR ratios of 90% at RT and 192% at 4.2 K. Figure 5b shows the TMR ratio at $V = 5$ mV, as well as RA_{AP} and RA_P, as a function of T from 4.2 to RT. As T decreased from RT to 4.2 K, the TMR ratio increased by a factor of 2.1 (i.e., $\gamma = 2.1$). As shown in Fig. 5b, RA_{AP} also increased with decreasing T, while RA_P was almost independent of T. These behaviors were similar to what was observed for the CCFA-MTJs.

For CMG-MTJs we also demonstrated a relatively high TMR ratio of 83% at RT (185% at 4.2 K), where we used a Co-rich CMG film with the film composition of $Co_2Mn_{0.74}Ge_{0.43}$. (i.e., Co-rich CMG) [15]. The TMR characteristics of the CMG/MgO/$Co_{50}Fe_{50}$ MTJ showed a similar T dependence; i.e., RA_{AP} increased with decreasing T, while RA_P was almost independent of T [15].

To clarify a guideline for obtaining high TMR ratios for MTJs with a Co_2MnZ thin film, where Z stands for Si and Ge, we systematically fabricated Co_2MnZ/MgO/$Co_{50}Fe_{50}$ MTJs with a Co-rich CMS film and a Co-deficient CMS film for CMS-MTJs and with a Co-rich CMG film and a Co-deficient CMS film for CMG-MTJs. Figure 6 plots the TMR ratio as a

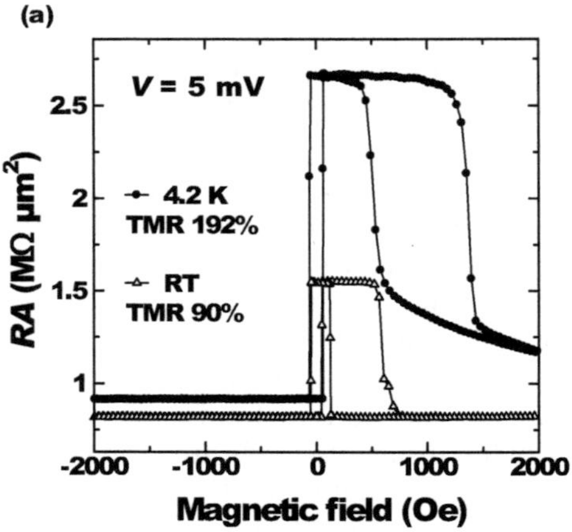

Fig. 5. Tunnel magnetoresistance characteristics of a fully epitaxial, exchange-biased Co_2MnSi/MgO (2.4 nm)/$Co_{50}Fe_{50}$ MTJ. The junction size was $10 \times 10\,\mu m$. The bias voltage V was 5 mV. (**a**) Typical magnetoresistance curves at RT and 4.2 K. (**b**) TMR ratio, as well as RA_{AP} and RA_P, as a function of T from 4.2 K to RT. The CMS film composition was $Co_2Mn_{0.84}Si_{0.80}$ (Co-rich CMS) [12]

function of T for the fabricated MTJs; Fig. 6a plots that for the CMS/MgO/$Co_{50}Fe_{50}$ MTJs with a Co-rich CMS film ($Co_2Mn_{0.84}Si_{0.80}$) and a Co-deficient one ($Co_2Mn_{1.23}Si_{1.19}$), and Fig. 6b plots that for the CMG/MgO/$Co_{50}Fe_{50}$ MTJs with a Co-rich CMG film ($Co_2Mn_{0.74}Ge_{0.43}$) and a Co-deficient one ($Co_2Mn_{1.05}Ge_{1.17}$). As shown in Fig. 6a the TMR ratios of the CMS-MTJs with the Co-rich CMS film (192% at RT 4.2 K and 90% at RT, $\gamma = 2.1$) [12] were much higher than those of the MTJs with the Co-deficient CMS film (84% at 4.2 K and 20% at RT, $\gamma = 4.2$). Furthermore, the TMR ratios of the CMS-MTJs with the Co-rich CMS film were less dependent on T. These characteristic features were commonly observed for the CMG-MTJs as shown in Fig. 6b; i.e., (1) the TMR ratios of the CMG-MTJs with the Co-rich CMG film (185% at RT 4.2 K and 83% at RT, $\gamma = 2.2$) [15] were much higher than those of the MTJs with the Co-deficient CMG film (70% at 4.2 K and 14% at RT, $\gamma = 4.2$) [10], and (2) the TMR ratios of the CMG-MTJs with the Co-rich CMG film were less dependent on T. These results indicate that a promising approach for obtaining Co_2MnZ (Z=Si, Ge) films, for Co_2MnZ/MgO-MTJs, is to have a film composition close to a stoichiometric one of 2:1:1 from the Co-rich film side.

4 Conclusion

We developed fully epitaxial magnetic tunnel junctions (MTJs) with a Co-based full-Heusler alloy thin film and a MgO tunnel barrier, and demonstrated a high tunnel magnetoresistance (TMR) ratio of over 100% at room

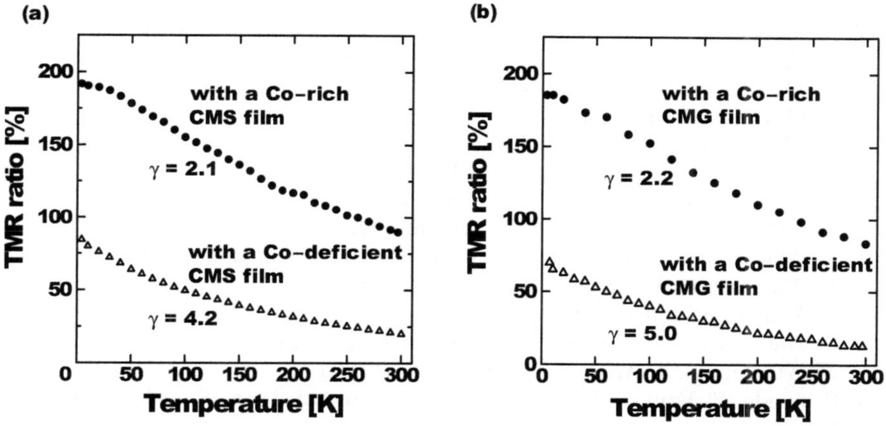

Fig. 6. (**a**) TMR ratios as a function of T for CMS/MgO/$Co_{50}Fe_{50}$ MTJs with a Co-rich CMS film ($Co_2Mn_{0.84}Si_{0.80}$) and a Co-deficient CMS film ($Co_2Mn_{1.23}Si_{1.19}$). (**b**) TMR ratios as a function of T for CMG/MgO/$Co_{50}Fe_{50}$ MTJs with a Co-rich CMG film ($Co_2Mn_{0.74}Ge_{0.43}$) and a Co-deficient CMS film ($Co_2Mn_{1.05}Ge_{1.17}$)

temperature (RT). A high tunneling spin polarization of 0.88 at 4.2 K was estimated for epitaxial $Co_2Cr_{0.6}Fe_{0.4}Al$ (CCFA) films in the fully epitaxial CCFA/MgO/$Co_{50}Fe_{50}$ MTJ layer structure. The demonstrated high TMR ratios at RT suggest that an epitaxial Co-based full-Heusler alloy thin film in combination with an epitaxial MgO tunnel barrier is highly promising for ferromagnetic electrodes used in spintronic devices.

References

[1] S. A. Wolf, D. D. Awschalom, R. A.Buhrman, J. M. Daunghton, S. von Molnar, M. L. Roukes, A. Y. Chtchelkanova, and D. M. Treger, Science **294**, 1488 (2001).
[2] R. A. de Groot, F. M. Mueller, P. G. van Engen, and K. H. J. Buschow, Phys. Rev. Lett. **50**, 2024 (1983).
[3] C. Felser, G. H. Fecher, and B. Balke, Angew. Chem., Int. Ed. **46**, 668 (2007).
[4] K. Inomata, S. Okamura, R. Goto, and N. Tezuka, Jpn. J. Appl. Phys., Part 2 **42**, L419 (2003).
[5] N. Tezuka, N. Ikeda, S. Sugimoto, and K. Inomata, Appl. Phys. Lett. **89**, 252508 (2006).
[6] S. Kämmerer, A. Thomas, A. Hütten, and G. Reiss, Appl. Phys. Lett. **85**, 79 (2004).
[7] Y. Sakuraba, M. Hattori, M. Oogane, Y. Ando, H. Kato, A. Sakuma, T. Miyazaki, and H. Kubota, Appl. Phys. Lett. **88**, 192508 (2006).
[8] T. Marukame, T. Kasahara, K.-i. Matsuda, T. Uemura, and M. Yamamoto, Jpn. J. Appl. Phys., Part 2 **44**, L521 (2005).

[9] M. Yamamoto, T. Marukame, T. Ishikawa, K Matsuda, T. Uemura, and M. Arita, J. Phys. D: Appl. Phys. **39**, 824 (2006).
[10] T. Marukame, T. Ishikawa, K.-i. Matsuda, T. Uemura, and M. Yamamoto, J. Appl. Phys. **99**, 08A904 (2006).
[11] T. Marukame, T. Ishikawa, K.-i. Matsuda, T. Uemura, and M. Yamamoto, Appl. Phys. Lett. **88**, 262503 (2006).
[12] T. Ishikawa, T. Marukame, H. Kijima, K.-i. Matsuda, T. Uemura, M. Arita, and M. Yamamoto, Appl. Phys. Lett. **89**, 192505 (2006).
[13] T. Marukame, T. Ishikawa, S. Hakamata, K.-i. Matsuda, T. Uemura, and M. Yamamoto, Appl. Phys. Lett. **90**, 012508 (2007).
[14] T. Marukame and M. Yamamoto, J. Appl. Phys. **101**, 083906 (2007).
[15] S. Hakamata, T. Ishikawa, T. Marukame, K.-i. Matsuda, T. Uemura, and M. Yamamoto, J. Appl. Phys. **101**, 09J513 (2007).
[16] S. Ishida, S. Fujii, S. Kashiwagi, and S. Asano, J. Phys. Soc. Jpn. **64**, 2152 (1995).
[17] S. Picozzi, A. Continenza, and A. J. Freeman, Phys. Rev. B **66**, 094421 (2002).
[18] P. J. Webster, J. Phys. Chem. Solids **32**, 1221 (1971).
[19] T. Marukame, T. Kasahara, K.-i Matsuda, T. Uemura, and M. Yamamoto, IEEE Trans. Magn. **41**, 2603 (2005).
[20] K.-i. Matsuda, T. Kasahara, T. Marukame, T. Uemura, and M. Yamamoto, J. Cryst. Growth **286**, 389 (2006).
[21] T. Ishikawa, T. Marukame, K.-i. Matsuda, T. Uemura, M. Arita, and M. Yamamoto, J. Appl. Phys. **99**, 08J110 (2006).
[22] H. Kijima, T. Ishikawa, T. Marukame, H. Koyama, K. Matsuda, T. Uemura, and M. Yamamoto, IEEE Trans. Magn. **42**, 2688 (2006).
[23] T. Ishikawa, T. Marukame, K.-i. Matsuda, T. Uemura, and M. Yamamoto, IEEE Trans. Magn. **42**, 3002 (2006).
[24] S. S. P. Parkin et al., Nature Mater. **3**, 862 (2004).
[25] M. Jullière, Phys. Lett. A **54**, 225 (1975).
[26] J. Mathon and A. Umerski, Phys. Rev. B **60**, 1117 (1999).
[27] D. J. Monsma and S. S. P. Parkin, Appl. Phys. Lett. **77**, 720 (2000).
[28] W. H. Butler, X.-G. Zhang, T. C. Schulthess, and J. M. Maclaren, Phys. Rev. B **63**, 054416 (2001).
[29] J. Mathon and A. Umerski, Phys. Rev. B **63**, 220403R (2001).
[30] Y. Miura, K. Nagao, and M. Shirai, Phys. Rev. B **69**, 144413 (2004).
[31] Y. Miura, H. Uchida, Y. Oba, K. Nagao, and M. Shirai, to be published in J. Phys.: Condens. Matter.

Structural and Magnetic Properties of Transition Metal Nanoparticles from First Principles

Markus Ernst Gruner, Georg Rollmann, Alfred Hucht, and Peter Entel

Department of Physics, University of Duisburg-Essen, Campus Duisburg, Lotharstr. 1, 47048 Duisburg, Germany
Markus.Gruner@uni-duisburg-essen.de

Abstract. Until recently, the simulation of transition metal particles in the nanometer range was only feasible with semi-empirical approaches and classical molecular dynamics simulations. However, the close interrelation of electronic and structural properties often leaves no alternative to a fully quantum mechanical treatment. The evolution of modern supercomputer technology nowadays allows the simulation of nanometer-sized objects from first principles in the framework of density functional theory (DFT). A technologically relevant example is the search for ultra-high density magnetic recording media where the decrease of the magnetic grain size competes with the onset of superparamagnetism. Here, Fe-Pt nanoparticles are discussed as a promising solution to the problem due to their large magnetocrystalline anisotropy in the ordered $L1_0$ phase. However, in experiment also other, less favorable, structures are observed. Therefore, a systematic *ab initio* investigation of the morphologies of transition metal nanoparticles with respect to their energetics and magnetism appears highly desirable. Within this contribution, we discuss the results of recent DFT calculations of Fe and Fe-Pt clusters with up to 561 atoms including full geometric optimization.

1 Introduction

In nanotechnology, it is naturally size that matters. While on the experimental side the struggle goes on to find the bottom by fabrication and characterization of increasingly finer particles, e. g., by transmission electron imaging with sub-Ångström resolution [1], the tendency for theoretical atomistic investigations is to extend the system sizes to the relevant range for practical applications.

Typical approaches for the atomistic simulation of materials properties at realistic length scales are the use of semi-empirical potentials to model the interaction between the atoms, or hybrid schemes like QM/MM, in which only a core region is described by a quantum-mechanical approach. The former is hampered by accuracy problems, since frequently no adequate parameterization can be found for the material under consideration. The latter is only a solution if quantum mechanics does not affect the physics considerably outside a confined region of the simulated system. Because of the increased influence of the surfaces, with its distinct electronic properties due to the

R. Haug (Ed.): Advances in Solid State Physics,
Adv. in Solid State Phys. **47**, 117–128 (2008)

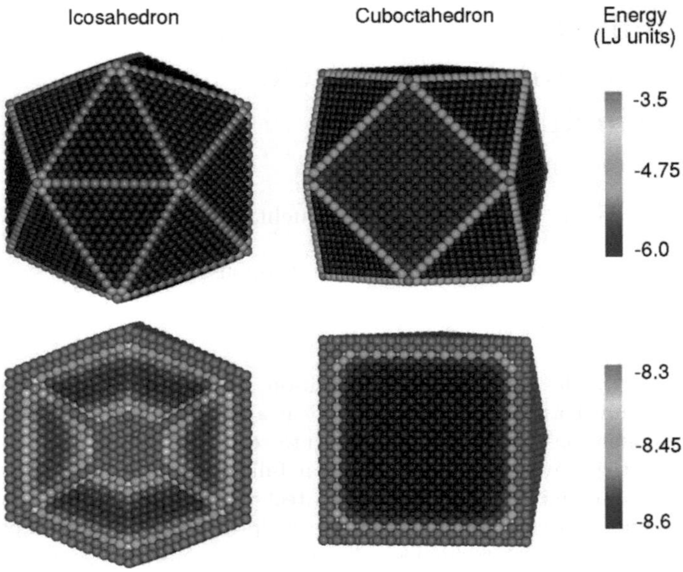

Fig. 1. Surface vs. volume energy contributions of geometrically optimized 12431 atom Lennard–Jones icosahedra and cuboctahedra (15 closed geometric shells). The color scale refers to the potential energy of the specific atom in Lennard–Jones units

decreased coordination of the atoms, there is often no alternative to a fully quantum mechanical treatment for the investigation of the morphologies of ultra-fine particles. This is frequently done in the framework of density functional theory (DFT), which has proven to be a very effective tool in the field of materials science (for a recent overview, see, e.g., [2]). Currently, several thousands of atoms can be treated in the case of passivated semiconductors with a well defined band gap and no relaxation due to surface reconstruction. For transition metals, the feasible system sizes are much smaller. Spin polarization further increases the computational effort.

Transition metal clusters with a few nanometers in diameter are receiving growing attention from the fundamental as well as from the technological point of view [3–7]. Depending on the material under consideration, the surface-to-volume ratio reaches a critical magnitude typically in the range of a few hundred to several thousands of atoms, leading to crossover effects between different geometries. This has been studied in depth for Lennard–Jones systems [8–10] and also by using empirical model potentials [11, 12]. The reason for this crossover can often be traced back to a competition between surface and volume energy, which is visualized in Fig. 1 for the example of a Lennard–Jones cluster with 15 closed geometric shells (12431 atoms). While the icosahedron possesses 20 energetically favorable (111)-surfaces, it is hampered by the internal strain caused by its 20-fold twinned structure. The cuboctahedron has no internal twinning, which at a certain size overweighs the energy penalty of the six unfavorable (100)-surfaces.

One major technological driving force for this development is certainly the increasing need for miniaturization of functional units which is especially evident in the field of ultra-high density magnetic recording, where the long-lasting exponential increase in storage density over time still seems unbroken: While eight years ago, densities of 35 GBit/in^2 were state-of-the-art [13], current lab demos reach values around 400 GBit/in^2 and the expectations of the manufacturers go up to 10 to 50 TBit/in^2 in the future. However, to reach this goal it is necessary to switch to new technologies like self-assembled patterned media and new materials like Fe-Pt and Co-Pt, which might allow grain sizes of 3–4 nm in diameter, without compromising long-term information stability through the superparamagnetic effect [14–16]. This is due to their extremely large uniaxial magnetocrystalline anisotropy, being more than one order of magnitude larger than that provided by current recording media [13]. In the following, we give two examples how state-of-the-art DFT calculations can contribute to this development.

2 Methodological Details

The calculations were carried out using the Vienna Ab-initio Simulation Package (VASP) [17], where the electronic wavefunctions of the valence electrons are expanded in a plane wave basis set. The plane wave energy cutoff was set to 268 eV and the interaction with the nuclei and the core electrons was described within the projector augmented wave (PAW) approach [18]. For the exchange-correlation functional, the generalized gradient approximation (GGA) was used in the formulation of *Perdew* and *Wang* [19, 20] in connection with the spin interpolation formula of *Vosko*, *Wilk* and *Nusair* [21]. Since the system is non-periodic, we restricted the integration in k-space to the Γ-point only. All clusters were placed in a cubic supercell, the largest with a fixed edge length of 33 Å.

The calculations include geometrical optimizations, which were carried out on the Born–Oppenheimer surface using the conjugate gradient method. The structural relaxations were stopped when the energy difference between two consecutive relaxations was less than 0.1 meV.

A few numbers may give an impression of the computational demand: Working on a single processor only (IBM PPC440d, as used in the IBM Blue Gene/L), the average time needed for an unconstrained geometric optimization of a cluster with $N = 147$ atoms (diameter $d \approx 1.5$ nm) adds up to around 177 days, while it reaches 6 years for clusters with $N = 561$ ($d \approx 2.5$ nm). This yields a scaling of the overall computation time as $N^{2.5}$. To fulfill the computational requirements we used up to 2048 processors on the IBM Blue Gene/L installation at Forschungszentrum Jülich. The efficiency of the VASP code on massively parallel architectures has been documented in [22].

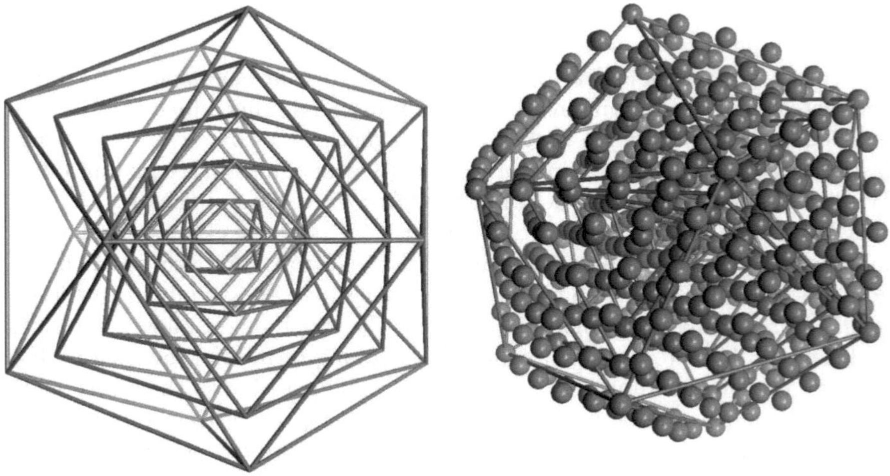

Fig. 2. Shell-wise Mackay-transformation (SMT) of an icosahedral Fe_{561} cluster. *Left*: Edge model of the shape of each shell, by connecting the respective corner atoms. The outermost shell (*red*) is icosahedral, while the innermost shell (*blue*) has a cuboctahedral character. *(online color) Right*: Full picture of the Fe_{561} cluster including all atomic positions from a different perspective. The atoms are colored according to their shell affiliation corresponding to the left panel

3 Structural Transitions in Iron Nanoparticles

Iron is one of the most important bulk materials of our time. Many of its applications arise from its unusual allotropy. The body centered cubic (bcc) α-phase, which is stabilized by magnetism at low temperatures, changes to the face centered cubic (fcc) γ-phase at 1185 K, which again disappears in favor of the paramagnetic bcc δ-phase at 1667 K. In the form of nanoparticles, iron provides a lot of interesting applications, especially in the field of biomedicine (for an overview, see, e.g., [23]). However, until recently, no conclusive information on the preferred constitution of iron particles in the range of one up to several nanometers in diameter was available. For example, for the 13-atom cluster, the lowest energy has been found for a Jahn–Teller distorted icosahedron [24, 25] from *ab initio* calculations, while transmission electron micrographs (TEM) of 6 nm particles [26] rather suggest a bcc lattice structure.

However, *ab initio* calculations performed by the present authors [27] have revealed that iron nanoclusters containing more than around 100 to 150 atoms should possess a bcc ground-state structure. This critical size is much smaller than the value predicted by previous calculations on the basis of empirical model potentials [28]. Another important result of this study was that perfect icosahedra and cuboctahedra, which are frequently used as reference structures, proved to be unstable against a partial transformation

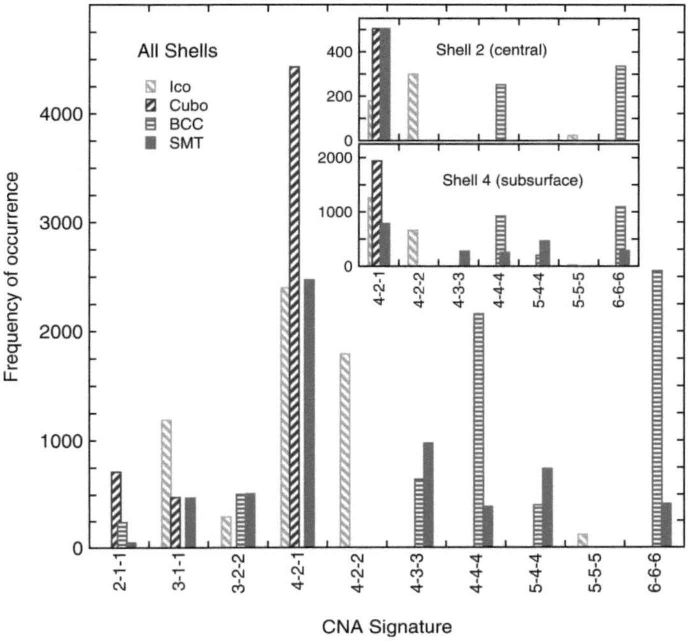

Fig. 3. Comparison of the results of a common neighbor analysis (CNA) of various geometrically optimized morphologies of Fe_{561}. These are the icosahedron (Ico), cuboctahedron (Cubo), bcc and SMT isomers. The inset shows the partial CNA performed only for atoms of shell 2 and 4, corresponding to the second shell from the center and the subsurface shell, respectively

along the Mackay-path [29], which connects both structures by a diffusionless transformation mechanism. In the course of the Mackay transformation, twelve (out of 20) pairwise adjacent triangular (111)-faces of the icosahedron stretch along their common edge and turn into the same plane to form the six square (100)-faces of the cuboctahedron. The geometries resulting from the structural optimization process can be understood as partially transformed ones between both endpoints. Each complete geometric shell converts to a different, continuously varying degree (Fig. 2). The surface shell maintains a nearly icosahedral shape which is obtaining an increasingly cuboctahedral character towards the inside. For Fe_{561}, the shell-wise Mackay-transformed (SMT) structure lies 42 meV/atom lower in energy than the icosahedron, but only 25 meV/atom higher than the bcc isomer, which is a temperature equivalent of 290 K.

A more detailed structural analysis leads to the origin of this unusual structural transformation. The common neighbor analysis (CNA) [31, 32] allows a comprehensive comparison of the SMT structure with other morphologies. If we restrict ourselves to the first (for bcc up to the second)

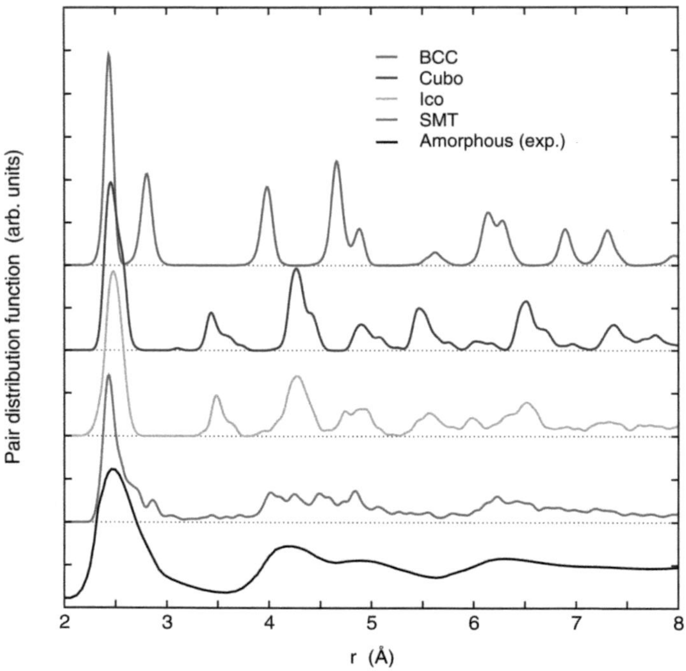

Fig. 4. Pair distribution functions (PDF) of the Fe_{561} isomers discussed in Fig. 3. The experimental PDF for an amorphous Fe-film [30] is shown for comparison

neighbor shell, a set of three-index-signatures provides a classification of the local structural environments of the atoms in the cluster. The indices refer – in the respective order – to the number of nearest neighbors both atoms have in common, the number of bonds between those neighbors, and the longest chain of bonds connecting them. The ideal bcc structure is then characterized by 4-4-4 and 6-6-6 signatures, referring to nearest and next nearest neighbors, the fcc structure solely by the 4-2-1 signature. The icosahedron contains 4-2-1, 4-2-2 and 5-5-5 signatures, the latter referring to the icosahedral spine. For surface and subsurface atoms other signatures apply, especially those with a smaller first index. A comparison of the CNA of the SMT with the other isomers is shown in Fig. 3. The surprising result is that this morphology with icosahedral outer shape and cuboctahedral core possesses features of the bcc and fcc structure but completely misses 4-2-2 and 5-5-5 signatures, which are typical for the icosahedron. A closer look to the individual geometric shells reveals that there is nearly perfect coincidence with the fcc cuboctahedron in the core part of the cluster, while particularly the subsurface shell exhibits correspondence with the bcc isomer (inset of Fig. 3). So, the dramatic energy gain of the SMT isomers with respect to the cuboctahedra and icosahedra

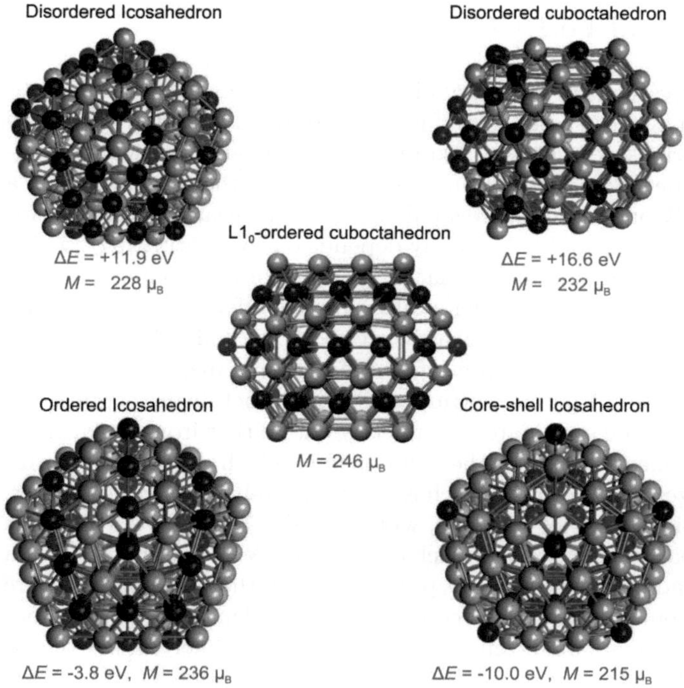

Fig. 5. Different morphologies of $Fe_{67}Pt_{80}$ nanoclusters, their relative energies and magnetic moments. Fe atoms are depicted in *blue*, Pt in *purple (online color)*. All energies are relative to the $L1_0$ ordered cuboctahedron (*center*)

can be traced back to the manifestation of a local bcc-like environment for the subsurface atoms.

The inhomogeneous structural environment of the SMT results in a pair distribution function (PDF) which is completely different from the other morphologies (Fig. 4) but exhibits features which are rather observed for amorphous alloys. Thus, the SMT may be interpreted as an amorphization mechanism similar to the rosette-like structural excitations which were recently discovered for Pt and Au nanoclusters [33]. Amorphous iron has been produced by low-temperature condensation of evaporated iron on thin films [30] or by sonochemical methods [34] and has raised considerable attention concerning its structural and magnetic properties in the past [35–37]. Recently, an intermediate amorphous phase was reported from empirical molecular dynamics simulations of 2.5 nm $Fe_{80}Ni_{20}$ clusters [38], which shows striking similarities concerning its PDF and other structural motifs as the wave-like bending of atomic columns and planes, which is visible in Fig. 2. This can be seen as a further indication that SMT morphologies may in fact be realized in the experiment under certain environmental conditions.

4 Morphologies of Fe-Pt Clusters

The systematic investigation of binary alloy clusters raises many more technical difficulties due to the compositional degree of freedom. Apart from the competition of various geometries, also order-disorder and surface segregation effects must be taken into account. Furthermore, the possibility for a perfectly chemically ordered or segregated core-shell structure exists only for particular compositions – and this usually depends on the morphology, which hampers a direct comparison of the relative stability of the isomers.

It is therefore helpful, if one isomer stands out as a reference configuration because of its technological relevance, as it is the case in Fe-Pt. Here, it is the stoichiometric $L1_0$ phase which is preferred due the large magnetocrystalline anisotropy. In the case of perfect, symmetric $L1_0$ cuboctahedra, two of the (001)-surfaces have to be terminated completely with either iron or platinum. This results in an excess of one of the components, which decreases with increasing cluster size. For $N = 147$, which is approximately 1.5 nm in diameter, this species occupies 56.3 % of the sites, while for $N = 561$ ($d \approx 2.5$ nm) it is still 52.8 %. Comparisons between Fe-rich and Pt-rich $N = 147$ cuboctahedra reveal that it is energetically strongly preferred, that the (001)-surfaces are occupied by Pt-atoms, even if this requires a partial depletion of the inner platinum layers, which are maintaining the $L1_0$ structure. Therefore, we chose to focus on clusters with an excess of Pt atoms, since it appears to stabilize the perfectly ordered $L1_0$ morphology.

A qualitative picture of the energetic tendencies can already be gained at rather small system sizes. In Fig. 5, we compare different optimized morphologies of $Fe_{67}Pt_{80}$ nanoparticles. The disordered isomers are in general significantly higher in energy than the $L1_0$ isomers, with the disordered cuboctahedra being the least favorable morphology. Here, considerable reconstructions especially along the edges of the six (001) surfaces occur. A straightforward way to construct an ordered icosahedral structure is to apply a full transformation along the Mackay-path starting from the cuboctahedron with $L1_0$ order. This isomer is by 24 meV/atom lower in energy than the $L1_0$ cuboctahedron. The tilting which is involved in the transformation of the Pt-covered (001)-surfaces to triangular (111)-facets leads to a distortion of the $L1_0$-typical parallel layering and, thus, a considerable reduction of the magnetocrystalline anisotropy must be expected. Segregation of Fe and Pt leads to even lower magnetocrystalline anisotropy energies. With 68 meV/atom, core-shell isomers with iron mainly in the core and Pt exclusively on the shell are found to be the most favorable conformation at this size. The preference of multiply twinned icosahedra over single-domain cuboctahedra as shown in our study is in direct contradiction with previous semi-empirical structure optimizations of 147 atom Fe-Pt clusters, where a general preference of cuboctahedra was reported except for the case of pure iron [39]. This underlines the importance of the corroboration of calculations using model potentials or semi-empirical methods by expensive but accurate *ab initio* approaches.

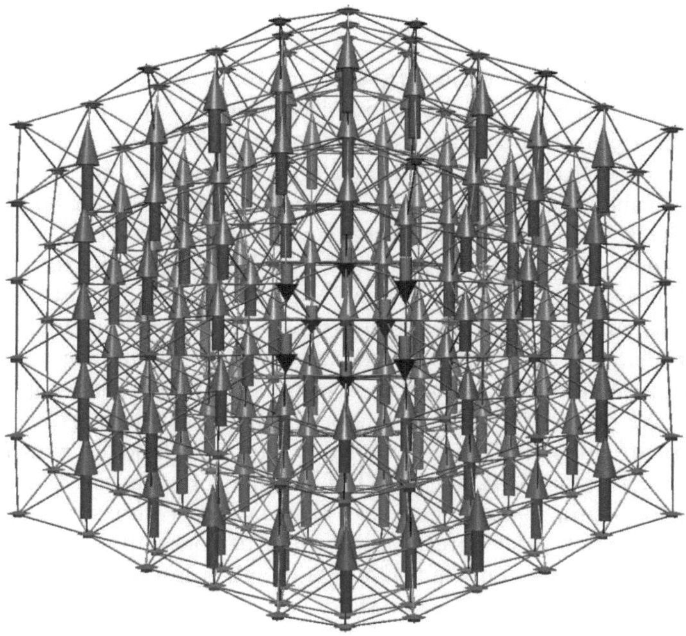

Fig. 6. Ferrimagnetic structure of an icosahedral $Fe_{265}Pt_{296}$ core-shell nanoparticle (cut). Fe spins are depicted by *blue* and *green arrows* for spin up and spin down, respectively. Pt spins are represented by *purple arrows* (only one direction). The length of the arrows corresponds to the magnitude of the atomic magnetic moments *(online color)*

Calculations of larger clusters show that the energetic trends favoring the multiply twinned structures do in fact decrease with increasing system size. However, they persist up to the technologically interesting range of particle sizes, which is beginning at a diameter of about 3 nm. A detailed discussion of the size-dependent changes in the energetic order will be given elsewhere.

The average moments per atom of all studied isomers lie within $0.2\,\mu_B$ per atom together and close to the expected bulk value, which is in agreement with previous DFT studies of unrelaxed Fe-Pt clusters with up to 135 atoms [40]. The maximum separation of Fe and Pt atoms leads to smallest magnetization for the core-shell cluster, while the largest is given by the $L1_0$ cuboctahedron. However, in larger core-shell clusters (cf. Fig. 6) ferrimagnetic tendencies in the Fe-core may play a decisive role due to the affinity of fcc iron towards antiferromagnetism [41].

5 Conclusion

We have presented two examples of how large-scale *ab initio* calculations help to understand genuine properties of nanometer-sized objects. The first is the determination of the ground-state structure of nanometer-sized iron clusters, which cannot be decided from the experimental data obtained so far. Here, *ab initio* calculations show that for diameters around 1.5 nm and larger body centered cubic isomers have the lowest energies. A new conformation was found within the range of thermal energies that provides, while maintaining an icosahedral outer shell, bcc- and fcc-like environments for the iron atoms resulting in a pair distribution function which is typical for amorphous materials. The second topic, which is closely related to technological applications, is the quest for ultra-high-density recording materials. Here, our investigations show that in the range of a few nanometers in diameter, the perfectly ordered $L1_0$ isomers are in strong competition with other morphologies which are of limited use for this purpose. Future DFT calculations may certainly give decisive hints for a proper optimization of the material composition or the experimental production procedures. Considering the contemporary evolution of supercomputing power and the advances in sub-Ångstrom imaging methods, the hope, that first principles methods and experimental characterization and preparation methods can work jointly on the atomistic design of nanoscale systems for technological applications, appears not to be out of reach.

Acknowledgements

We thank J. R. Chelikowsky and M. L. Tiago (ICES, University of Texas, Austin) for the fruitful exchange on iron nanoparticles. Our thank goes also to M. Farle, O. Dmitrieva, D. Sudfeld, C. Antoniak and M. Spasova (University of Duisburg-Essen) for many stimulating discussions concerning the Fe-Pt system. Large parts of the calculations have been performed on the IBM Blue Gene/L supercomputer of the John von Neumann Institute for Computing (NIC) in Jülich. We thank the staff for their continuous support and also Dr. P. Vezolle of IBM for his help in optimizing the VASP code for the Blue Gene/L system.

References

[1] M. A. O'Keefe, C. Hetherington, Y. Wanga, E. Nelson, J. Turner, C. Kisielowski, J.-O. Malm, R. Mueller, J. Ringnald, M. Pan, A. Thust: Sub-Angstrom high-resolution transmission electron microscopy at 300 keV, Ultramicroscopy **89**, 215 (2001)

[2] Density functional theory in materials research, MRS Bulletin **31**, No. 9, pp. 659–692 (2006)

[3] I. M. L. Billas, A. Châtelain, W. A. de Heer: Magnetism from the atom to the bulk in iron, cobalt, and nickel clusters, Science **265**, 1682 (1994)
[4] J. A. Alonso: Electronic and atomic structure, and magnetism of transition metal clusters, Chem. Rev. **100**, 637 (2000)
[5] J. Bansmann, S. H. Baker, C. Binns, J. A. Blackman, J.-P. Bucher, J. Dorantes-Dávila, V. Dupuis, L. Favre, D. Kechrakos, A. Kleibert, K.-H. Meiwes-Broer, G. M. Pastor, A. Perez, O. Toulemonde, K. N. Trohidou, J. Tuaillon, J. Xie: Magnetic and structural properties of isolated and assembled clusters, Surf. Sci. Rep. **56**, 189 (2005)
[6] M. L. Tiago, Y. Zhou, M. M. G. Alemany, Y. Saad, J. R. Chelikowsky: Evolution of magnetism in iron from the atom to the bulk, Phys. Rev. Lett. **97**, 147201 (2006)
[7] F. W. Payne, W. Jiang, L. A. Bloomfield: Magnetism and magnetic isomers in free chromium clusters, Phys. Rev. Lett. **97**, 193401 (2006)
[8] B. W. van de Waal: Stability of face-centered cubic and icosahedral Lennard–Jones clusters, J. Chem. Phys. **90**, 3407–3408 (1988)
[9] B. Raoult, J. Farges, M.-F. de Feraudy, G. Torchet: Comparison between icosahedral, decahedral and crystalline Lennard–Jones models containing 500 to 6000 atoms, Phil. Mag. B **60**, 881–906 (1989)
[10] J. P. K. Doye, F. Calvo: Entropic effects on the size dependence of cluster structure, Phys. Rev. Lett. **86**, 3570 (2001)
[11] C. L. Cleveland, U. Landmann: The energetics and structure of nickel clusters: Size dependence, J. Chem. Phys. **94**, 7376 (1991)
[12] F. Baletto, R. Ferrando: Structural properties of nanoclusters: Energetic, thermodynamic, and kinetic effects, Rev. Mod. Phys. **77**, 371 (2005)
[13] D. Weller, A. Moser: Thermal effect limits in ultrahigh-density magnetic recording, IEEE Trans. Magn. **35**, 4423 (1999)
[14] S. Sun, C. B. Murray, D. Weller, L. Folks, A. Moser: Monodisperse FePt nanoparticles and ferromagnetic FePt nanocrystal superlattices, Science **287**, 1989 (2000)
[15] M. L. Plumer, J. van Ek, D. Weller (Eds.): *The Physics of Ultra-High-Density Magnetic Recording* (Springer, Berlin 2001)
[16] S. Sun: Recent advances in chemical synthesis, self-assembly, and applications of FePt nanoparticles, Adv. Mater. **18**, 393 (2006)
[17] G. Kresse, J. Furthmüller: Efficient iterative schemes for ab initio total-energy calculations using a plane-wave basis set, Phys. Rev. B **54**, 11169 (1996)
[18] G. Kresse, J. Furthmüller: From ultrasoft pseudopotentials to the projector augmented-wave method, Phys. Rev. B **59**, 1758 (1999)
[19] J. P. Perdew: in P. Ziesche, H. Eschrig (Eds.): *Electronic Structure of Solids '91* (Akademie Verlag, Berlin 1991)
[20] J. P. Perdew, K. Burke, Y. Wang: Generalized gradient approximation for the exchange-correlation hole of a many-electron system, Phys. Rev. B **54**, 16533 (1996)
[21] S. H. Vosko, L. Wilk, M. Nusair: Accurate spin-dependent electron liquid correlation energies for local spin density calculations: A critical analysis, Can. J. Phys. **58**, 1200 (1980)
[22] M. E. Gruner, G. Rollmann, A. Hucht, P. Entel: Massively parallel density functional theory calculations of large transition metal clusters, Lecture Series on Computer and Computational Sciences **7**, 173 (2006)

[23] D. L. Huber: Synthesis, properties, and applications of iron nanoparticles, Small **1**, 482 (2005)
[24] P. Bobadova-Parvanova, K. A. Jackson, S. Srinivas, M. Horoi: Density-functional investigations of the spin ordering in Fe_{13} clusters, Phys. Rev. B **66**, 195402 (2002)
[25] G. Rollmann, P. Entel, S. Sahoo: Competing structural and magnetic effects in small iron clusters, Comput. Mater. Sci. **35**, 275 (2005)
[26] T. Vystavel, G. Palasantzas, S. A. Koch, J. T. M. De Hosson: Nanosized iron clusters investigated with in situ transmission electron microscopy, Appl. Phys. Lett. **82**, 197 (2003)
[27] G. Rollmann, M. E. Gruner, A. Hucht, P. Entel, M. L. Tiago, J. R. Chelikowsky: Shell-wise mackay transformation in iron nano-clusters, submitted for publication
[28] N. A. Besley, R. L. Johnston, A. J. Stace, J. Uppenbrinck: Theoretical study of the structures and stabilities of iron clusters, J. Mol. Struct. (Theochem) **341**, 75 (1995)
[29] A. L. Mackay: A dense non-crystallographic packing of equal spheres, Acta Cryst. **15**, 916 (1962)
[30] T. Ichikawa: Electron diffraction study of the local atomic arrangement in amorphous iron and nickel films, Phys. Stat. Sol. (a) **19**, 707 (1973)
[31] H. Jónsson, H. C. Andersen: Icosahedral ordering in the Lennard-Jones liquid and glass, Phys. Rev. Lett. **60**, 2295 (1988)
[32] D. Faken, H. Jónsson: Systematic analysis of local atomic structure combined with 3D computer graphics, Comput. Mater. Sci. **2**, 279 (1994)
[33] E. Aprà, F. Baletto, R. Ferrando, A. Fortunelli: Amorphization mechanism of icosahedral metal nanoclusters, Phys. Rev. Lett. **93**, 065502 (2004)
[34] K. S. Suslick, S.-B. Choe, A. A. Cichowlas, M. W. Grinstaff: Sonochemical synthesis of amorphous iron, Nature **353**, 414 (1991)
[35] U. Krauss, U. Krey: Local magneto-volume effect in amorphous iron, J. Magn. Magn. Mater **98**, L1 (1991)
[36] R. F. Sabiryanov, S. K. Bose, O. N. Mryasov: Effect of topological disorder on the itinerant magnetism of Fe and Co, Phys. Rev. B **51**, 8958 (1995)
[37] N. P. Kovalenko, Y. P. Krasny, U. Krey: *Physics of Amorphous Metals* (Wiley-VCH, Berlin 2001)
[38] L. E. Kar'kina, I. N. Kar'kin, Y. N. Gornostyrev: Structural transformations in Fe-Ni-alloy nanoclusters: Results of molecular-dynamic-simulation, Phys. Met. Met. **101**, 130 (2006)
[39] A. Fortunelli, A. M. Velasco: Structural and electronic properties of Pt/Fe nanoclusters from EHT calculations, J. Mol. Struct. (Theochem) **487**, 251 (1999)
[40] H. Ebert, S. Bornemann, J. Minár, P. H. Dederichs, R. Zeller, I. Cabria: Magnetic properties of Co- and FePt-clusters, Comput. Mater. Sci. **35**, 279 (2006)
[41] M. E. Gruner, G. Rollmann, S. Sahoo, P. Entel: Magnetism of close packed Fe_{147} clusters, Phase Transitions **79**, 701 (2006)

First-Principles Study of Ferromagnetic Heusler Alloys: An Overview

Silvia Picozzi

Consiglio Nazionale delle Ricerche – Istituto Nazionale di Fisica della Materia (CNR-INFM), CASTI Regional Lab,
Via Vetoio 10, 67010 Coppito (L'Aquila), Italy
silvia.picozzi@aquila.infn.it

Abstract. A particularly attractive class of materials for spintronic applications is represented by half-metals (i.e., materials showing carriers at the Fermi level 100% spin-polarized), among which ferromagnetic Heusler alloys occupy a prominent position. Here, we present a review of first-principles calculations based on the density functional theory, focusing in particular on *i*) several mechanisms that are predicted to destroy half-metallicity and *ii*) magneto-optical properties such as the Kerr rotation. We show that defect-induced states at the Fermi level in Co-based Heusler-alloys such as Co_2MnSican drastically reduce the degree of spin-polarization, pointing to the necessity of keeping point-defects under control during experimental growth. In the context of spin-injection into mainstream semiconductors, interface states are also predicted to be induced by the junction between a full-Heusler and GaAs. A careful analysis of the magneto-optical Kerr effect is performed for several Co-based full-Heusler alloys. Although the size of the Kerr rotation is found to be of the order of 0.5° and therefore not technologically appealing, we show that first-principles calculations can be efficiently used as a tool to predict novel half-metals with large Kerr angles.

1 Introduction

In the last ten years, i.e., ever since "Magneto-electronics" was first proposed in the ninties [1], the spin degree of freedom in conventional electronics has turned from a revolutionary concept to a standard ingredient in our tool kit for materials- and device-design: the so-called spintronics is considered nowadays as one of the most promising and rich branches of future technology. Within this area, Heusler alloys represent one of the most interesting class of materials. In fact, these intermetallic alloys are characterized by several appealing properties: *i*) they are generally ferromagnetic with remarkably high Curie temperatures (cfr Co_2MnSiand Co_2MnGewith ordering temperatures $T_C > 900\,\mathrm{K}$); *ii*) their density of states shows the celebrated half-metallicity (HM), i.e., the majority and minority spin channels show a metallic and semiconducting character, respectively, with the technologically-attractive consequence that carriers at the Fermi level (E_F) are 100% spin-polarized; *iii*) their crystal structure (cubic $L2_1$) along with their lattice constants are very well compatible with corresponding quantities of mainstream semiconductors (such as Si or GaAs). As a result, the

R. Haug (Ed.): Advances in Solid State Physics,
Adv. in Solid State Phys. **47**, 129–141 (2008)

interest (both from the theoretical as well as from experimental point of view) towards these compounds has steeply increased in the last decade. Remarkably, prototype devices have been realized using Heusler alloys (such as Co_2MnSi) as electrodes for magnetic tunnel junctions (MTJs) Fully epitaxial, exchange-biased MTJs were fabricated with a Co-based full-Heusler alloy Co_2MnSithin film as a lower electrode, a MgO tunnel barrier, and a $Co_{50}Fe_{50}$ upper electrode. The tunnel magnetoresistance ratios were 90% at room temperature and 192% at 4.2 K. The bias voltage dependence of differential conductance (dI/dV) for the parallel and antiparallel magnetization configurations suggested the existence of a basic energy gap structure for the minority-spin band of the Co_2MnSielectrode with an energy difference of about 0.4 eV between the bottom of the empty minority-spin conduction band and the Fermi level [2]. Epitaxially grown spin-valve-type magnetic tunnel junctions with $Co_2FeAl_{0.5}Si_{0.5}$ full-Heusler alloys for top and bottom electrodes and a MgO barrier with various thicknesses (t_{MgO}). A TMR ratio up to 175% at room temperature was obtained for $t_{MgO} = 2.0$ nm, suggesting a large tunneling spin polarization as well as high thermal stability of $L2_1$-$Co_2FeAl_{0.5}Si_{0.5}$ [3]. Still from the technological point of view, progresses have been made in the context of spin-injection into semiconductors: *Dong* et al. [4] demonstrated electrical spin injection from the Heusler alloy Co_2MnGe into a p-i-n $Al_{0.1}Ga_{0.9}As/GaAs$ light emitting diode, with a maximum steady-state spin polarization of approximately 13% at 2 K measured in two types of heterostructures. The injected spin polarization at 2 K was calculated to be 27 % based on a calibration of the spin detector using Hanle effect measurements.

In parallel, many theoretical studies, based on the density-functional theory, have been conducted, addressing basic physics issues. For example, *Galanakis* et al. [5] showed that the Heusler alloys satisfy of a sort of "Slater–Pauling" rule, well-known to describe the behaviour of intermetallic binary alloys [6]: the total spin-moment M_t was found to scale linearly with the number of valence electrons Z_t, such that $M_t = Z_t - 24$ for the full-Heuslers and $M_t = Z_t - 18$. More recently, *J. Kübler* showed that the Curie temperature of several Heusler compounds follows an almost linear dependence on the magnetic moment [7] The increasing technological interest towards this class of half-metals stimulated *ab-initio* studies focused all on Heuslers but branching into different directions, ranging from interfaces of Heuslers with semiconductors or with oxides (ferroelectric or not [8]), surfaces or point-defects in full- or half-Heuslers, effects of correlations and nature of magnetism, *etc.* Within this context, we offer, in this contribution, an overview of two relevant aspects in the materials science related to Heusler alloys: *i)* the loss of HM, as induced by several different mechanisms and *ii)* magneto-optical properties of different full-Heusler compounds. The chapter is therefore organized as follows: after briefly exposing our first-principles approach in Sect. 2, we move in Sect. 3 to the discussion of *ab-initio* calculations of defect-induced and interface-induced states in $Co_2MnSi$and Co_2MnGe, along with an overview on other examples of loss of HM as predicted via first-principles methods. In

Sect. 4 we briefly recall our theoretical approach to evaluate magneto-optical (MO) properties in ferromagnets and then we discuss mainly the Kerr rotation in several Heuslers, in terms of their underlying electronic structure. Finally, we draw our conclusions in Sect. 6.

2 Computational and Structural Details

First-principles calculations within the density-functional-theory have been performed using both the generalized gradient approximation (GGA) in the Perdew–Burke–Erzenhof implementation [9] and the standard local-spin density approximation (LSDA) [10]. The calculations have been mainly performed using the full-potential linearized augmented plane-wave code (FLAPW) [11], which represents one of the most accurate density functional methods: it is an all-electron approach where a fully-relativistic and scalar-relativistic treatments are performed for core- and valence electrons, respectively. For the simulations of the magneto-optical effects, we included spin-orbit coupling on the valence electrons as well (see below Sect. 4). Muffin tin radii were chosen to be R_{MT} = 2.1 a.u. for all the atoms; angular momenta (wave vector) up to $l_{max} = 8$ (K_{max} = 3.6 a.u.) were used for the expansions of both wave functions and charge density. The Brillouin zone was sampled using special **k** points according to the Monkhorst–Pack scheme [12].

3 Loss of Half-Metallicity

Ab-initio calculations predict full-HM for many Heusler materials; [5, 13] experiments on the degree of spin-polarization, however, do not always support this picture. For example, in the case of Co_2MnSi, the degree of measured spin-polarization was found to be about 50–60% [14, 15]. In order to explain the discrepancies between theory and experiments, several different mechanisms were proposed to reduce the full spin-polarization. In Subsect. 3.1 we focus on the effects of point defects and of interfaces with standard semiconductors as mechanisms for the loss of HM, whereas in Subsect. 3.2 we discuss other possible sources of reduction of P as predicted from ab-initio calculations.

3.1 Point Defects and Interfaces with Semiconductors

We focus here on the electronic structure of *i*) non-stoichiometric Co-antisite and Mn-antisite as well as stoichiometric Co-Mn swaps in Co_2MnSi and *ii*) [001]-ordered Co_2MnGe/Ge and Co_2MnGe/GaAs interfaces.

Let's start with point-defects [16]. Our results show that the Mn-antisite (recall that in this case, one Mn "impurity" replaces one Co atom) has a low formation energy (i.e., the energy cost to form this kind of defect is about 0.3 eV), consistently with experimental observations of a high-level of disorder in the Mn-Co sublattice. This kind of defects, however, preserves HM, with small charge rearrangements occurring in the majority spin-states just above E_F (density of states not shown). On the other hand, in the case of the Co-antisite, the formation energy is higher (about 0.8 eV), indicating that these impurities occur less frequently than Mn-antisites, but still in quite high percentage. As for the Co-Mn swaps, this "double-defect" can be considered as the superposition of the Co and Mn antisites (both in terms of energetics and electronic structure) and will therefore not be further discussed [16]. From the electronic structure point-of-view the introduction of Co-antisites has dramatic consequences on the HM character. In fact, as shown in Fig. 1c, the Co antisite (mainly with *d*-character) introduces a sharp peak in the total density of states (TDOS), consistently a the rather undispersed band just at E_F which destroys the full-spin-polarization of the ideal bulk Co_2MnSi(the resulting spin-polarization is as low as 6 %). We point out that, when Co replaces Mn, we have two electrons more to be placed; being the up-spin component of the TDOS very similar to the unperturbed case (cfr bold red and thin blue line in Fig. 1c), the total difference in terms of electrons is taken over by the peak in the TDOS located in proximity to the Fermi level. This is consistent with the change in magnetic moments: we go from an ideal situation in which the total moment per unit-cell $M_t = 40\ \mu_B$ (recall we have considered 8 formula-units per supercell) and the local Mn moment is about $3\ \mu_B$ to a situation in which $M_t \sim 38\ \mu_B$ and the Co antisite shows a local moment of about 1 Bohr magneton. The character of the peak is extremely localized in energy as well as in space: a charge density plot of this same peak (not shown) reveals that the perturbation is efficiently screened by the surrounding Co atoms in about one atomic shells.

The second important question we want to address here is: is HM preserved when a Heusler materials is interfaced with a standard semiconductor? Our results are shown in Fig. 1a: half-metallicity is lost on both interfacial Mn and Ge atoms. Furthermore, HM is lost also on the semiconducting side (not shown). This situation is representative also for other atomic junction-terminations (such as Co-terminated case) as well as for different semiconductors (Ge or GaAs) and semiconducting termination (anion- vs cation-terminated) [17,18]. Therefore, interface states kills spin-polarization in much the same way as defect-induced states do. There is, however, an important difference with the point-defect case, previously discussed: the perturbation extends for a few layers (from 3 to 5) in both the semiconducting and Heusler side, at variance with the extremely localized character of the defect-states which is fully screened in basically one atomic shell.

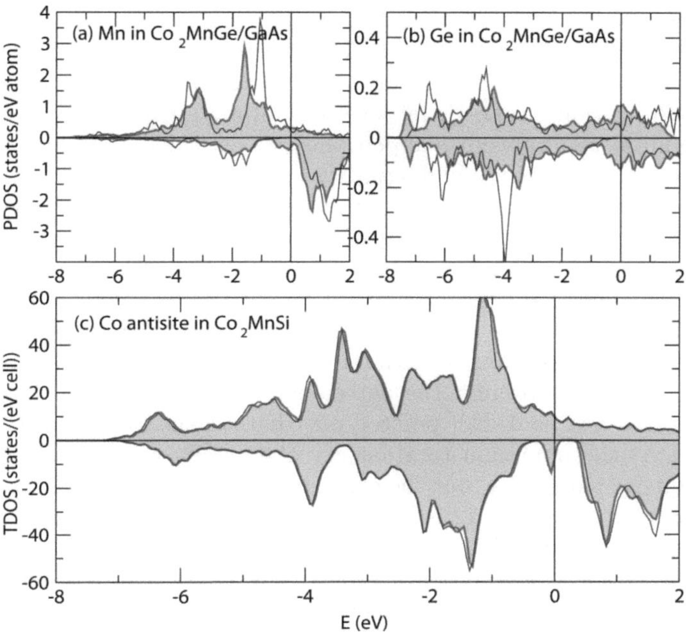

Fig. 1. PDOS for (**a**) Mn and (**b**) Ge atoms: *bold red* (*blue thin*) *lines* represent atoms in [001]-ordered $Co_2MnGe/GaAs$ junction (single bulk Co_2MnGe). TDOS for the Co-antisite case in Co_2MnSi: *bold red* (*blue thin*) *lines* represent atoms in the antisite case (single bulk Co_2MnSi)

3.2 Other Mechanisms: Predictions from Density-Functional-Theory

Surface-effects were first suggested as mechanisms for the loss of half-metallicity. *Ishida* et al. [19] performed pioneering simulations on the (001) and (111) $Co_2MnGe$and Co_2MnSifilms, showing that the film thickness and the surface termination affects the degree of spin-polarization; for example, the (111) film with the Si or Ge atom surface was predicted to be half-metallic. Recently, the stability, as well as the electronic and magnetic structure of $Co_2MnSi(001)$ thin films for many different atomic terminations were studied by *Hashemifar* et al. [20]. Surface states appeared in many termination. On the other hand, the pure Mn termination, due to its strong surface-subsurface coupling, was predicted to be half-metallic. *Galanakis* [21] focused on the electronic and magnetic properties of the (001) surfaces of several full- and half-Heusler alloys. The MnSb-terminated surfaces of the half-Heusler alloys (such as NiMnSb) showed properties similar to those of the bulk and a high spin polarization at the Fermi level was obtained (although slightly reduced from 100% in the bulk). On the other hand, $Co_2MnGe$and Co_2MnSisurfaces of full-Heuslers were predicted to lose their bulk half-metallic character, with a negligible degree of spin polarization at the Fermi level.

Lezaic et al. focused on the temperature dependence of both magnetization and spin polarization at the Fermi level in half-metallic half-Heusler NiMnSb [22]. Ab initio results were mapped onto an extended Heisenberg model (where the effect of longitudinal fluctuations was explicitely included), solved by a Monte Carlo method. In [22], it was suggested that, due to the hybridization of states forming the spin gap (in turn depending on thermal spin fluctuations), the polarization is expected to flop at temperatures much lower than the ferromagnetic Curie temperature.

Correlation effects beyond the standard treatment within density functional theory were also suggested as sources for changing the predicted spin-polarization, improving in particular the agreement with experiments. For example, *Chandra Kandpal* et al. [23] pointed out that, in the local density approximation (LDA) the minimum total energy of Co_2FeSi was found for the experimental lattice parameter, but the calculated magnetic moment was lower than experiments by about 12%. Moreover, HM, along with a magnetic moment equal to the experimental value of 6 μ_B were predicted only after increasing the lattice parameter by more than 6%. In order to reconcile the discrepancy between theory and experiments, the LDA+U approach was used to statically treat on-site electron correlations. For Co_2FeSi, an effective Coulomb exchange interaction $U_{eff}=U-J$ in the range of about 2–5 eV was found to result in half-metallic ferromagnetism and to the measured integer magnetic moment at the experimental lattice parameter. Finally, it was also shown that, if correlation effects become too strong (>2 eV for Co_2MnSi and >5 eV for Co_2FeSi), then a loss of half-metallic behavior is expected.

4 Magneto-Optical Properties

4.1 Introduction and Methodology to Calculate the Complex Kerr Angle

Linearly polarized light reflected from a ferromagnetic compound becomes elliptically polarized with its major axis rotated with respect to the incident polarization plane: this effect is known as the magneto-optical Kerr-effect (MOKE) [24]. Magnetooptics is nowadays a very important branch in magnetism, mainly for two reasons: first, one can exploit the Kerr rotation to read suitably magnetically stored information in an optical way in modern high density data storage technology (for these applications a large value of the Kerr rotation is required); second, the MOKE is currently considered as a powerful probe in many fields of research, such as surface magnetism, magnetic interlayer coupling in multilayers, plasma resonance effects in thin layers, structural and magnetic anisotropies. [25]. The polar Kerr-effect, where both magnetization and incident light are normal to the surface, is technologically relevant since it gives the biggest rotation angle compared to others, i.e., the longitudinal or transverse Kerr-effects (which both have in-plane magnetization while the polarization of light is parallel and perpendicular to the incident plane, respectively).

As for the theoretical approach to evaluate magneto-optical properties within the density functional theory, we recall [24] that in the random-phase approximation (RPA), the optical conductivity tensor in the long wavelength limit can be written as a function of the photon energy, ω, as

$$\sigma_{\alpha\beta}(\omega) = \frac{i}{\omega V}\int \frac{d\mathbf{k}}{(2\pi)^3}\sum_{i,j} f(\epsilon_i)(1-f(\epsilon_j))\left(\frac{\Pi^{\alpha}_{ji}\Pi^{\beta}_{ij}}{\omega + i\delta - \epsilon_{ij}} - \frac{\Pi^{\alpha}_{ij}\Pi^{\beta}_{ji}}{\omega + i\delta + C}\right), \tag{1}$$

neglecting the local field effects (V is the unit-cell volume, f is the occupation factor and δ is an interband broadening factor, ϵ_{ij} is the energy difference between final and initiale states). Here we recall [24] that the momentum operator is defined as

$$\mathbf{\Pi}_{ij} = \int \phi^*_{i,\mathbf{k}}(\mathbf{r})\left[\nabla + \left(\frac{1}{4m}\right)(\sigma \times \nabla V(\mathbf{r}))\right]\phi_{j,\mathbf{k}}(\mathbf{r})d\mathbf{r}\,. \tag{2}$$

Although (1) provide the formula for the optical conductivity tensor, it is instructive to rewrite these equations introducing $\Pi^{\pm} = \Pi^x \pm \Pi^y i$ for the polar geometry where $\alpha, \beta = x$ or y, as

$$\begin{aligned}\sigma_{xy}(\omega) &= \frac{i}{V}\int \frac{d\mathbf{k}}{(2\pi)^3}\sum_{i,j} f(\epsilon_i)(1-f(\epsilon_j))\frac{|\Pi^+_{ij}|^2 - |\Pi^-_{ij}|^2}{4\epsilon_{ij}} \\ &\quad \left(\frac{1}{\omega + i\delta - \epsilon_{ij}} - \frac{1}{\omega + i\delta + \epsilon_{ij}}\right), \\ \sigma_{xx}(\omega) &= \frac{i}{V}\int \frac{d\mathbf{k}}{(2\pi)^3}\sum_{i,j} f(\epsilon_i)(1-f(\epsilon_j))\frac{|\Pi^+_{ij}|^2 + |\Pi^-_{ij}|^2}{4\epsilon_{ij}} \\ &\quad \left(\frac{1}{\omega + i\delta - \epsilon_{ij}} + \frac{1}{\omega + i\delta + \epsilon_{ij}}\right).\end{aligned} \tag{3}$$

Π^+_{ij} and Π^-_{ij} correspond to the optical matrix for the left and right circularly polarized light, respectively; it is therefore obvious that σ_{xy} is proportional to the *difference* of the optical transition between the left and right circularly polarized light mediated by an electron transition between an occupied initial state i and an unoccupied final state f and that σ_{xx} measures the corresponding *average*.

Therefore for the incident light of frequency ω, the reflected light is characterized by the complex Kerr angles, $\theta + i\varepsilon$, which can be written as

$$\theta + i\varepsilon = -\frac{\sigma_{xy}}{\sigma_{xx}(1 + i(4\pi/\omega)\sigma_{xx})^{(1/2)}} = -\frac{\sigma_{xy}(\omega)}{D(\omega)} \tag{4}$$

in the limit of small θ (Kerr rotation), ε (Kerr ellipticity) and σ_{xy}. $D(\omega)$ denotes the denominator of the Kerr angle, containing diagonal terms of the optical conductivity only.

In addition to *interband* transitions, the *intraband* transitions are described with the empirical Drude term (τ_D being the intraband relaxation time):

$$\sigma = \frac{\omega_p^2}{4\pi(1 - i\omega\tau_D)} \tag{5}$$

where the plasma frequency is expressed as:

$$\omega_p = \frac{4\pi e^2}{m^2 V} \sum_{i\mathbf{k}} \delta(\varepsilon_{i\mathbf{k}} - E_F)|\Pi_{ii}^{\alpha}|^2 \tag{6}$$

We performed the calculations of the optical and magneto-optical properties evaluating matrix-elements on about 1000 special **k**-points [12]. In order to consider the effect of finite lifetimes, as well as the experimental resolution, a Lorentzian broadening of 0.4 eV was applied for both the inter- and intra-band contributions.

5 Electronic and Magneto-Optical Properties of Co-Based Full-Heusler Alloys

In this section, we discuss the Kerr rotation in several full-Heuslers: Co_2MnSi, Co_2MnGe, Co_2MnSn, Co_2MnGa and Co_2MnIn. According to the Slater-Pauling rule as applied to the Heusler compounds, when the anion is a group IV (III) element, the magnetic moment is equal to 5 (4) μ_B. In particular, as shown in Fig. 2 where we report the total density of states for all the alloys of interest, full HM is present in $Co_2MnSi$and Co_2MnGewhereas it is lost in Co_2MnSn, Co_2MnGa and, Co_2MnIn (due to states tailing from the minority conduction band below E_F). Figure 2 also shows that most of the weight of majority states is in the occupied part, whereas in the minority channel both the occupied and unoccupied branches contribute. This is important when magneto-optics is concerned: since this latter involves optical transitions between occupied and unoccupied states, we expect mainly the minority spin-channel to contribute to the main spectral shapes.

Let's first discuss the Kerr angle in Co_2MnSi(cfr Fig. 3a). [26] As shown in (4), the Kerr angle is given by a magneto-optical numerator (and therefore due to the simultaneous presence of exchange and spin-orbit coupling) and by an optical denominator (therefore arising from the diagonal-only part of the conductivity). For this reason, we show separately the numerator (real and imaginary parts, Fig. 3b and inverse denominator (real and imaginary parts, Fig. 3b, whose product roughly gives the main contribution to the Kerr

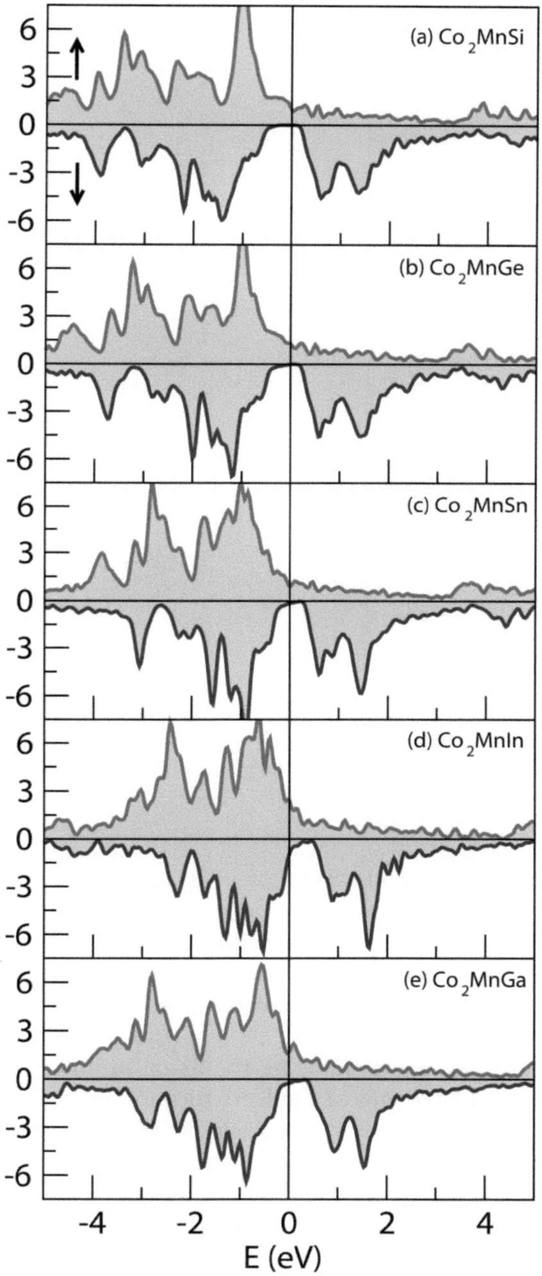

Fig. 2. Total density of states (in states/eV cell) for (**a**) Co_2MnSi, (**b**) Co_2MnGe, (**c**) Co_2MnSn, (**d**) Co_2MnGa, and (**e**) Co_2MnInvs. energy. Positive (negative) lines show the majority (minority) spin channels. The zero of the energy scale marks the Fermi level

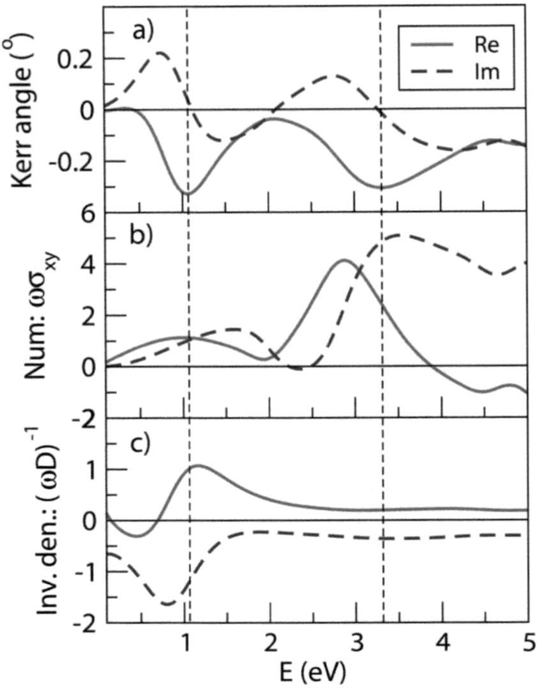

Fig. 3. (**a**) Kerr angle, (**b**) numerator of the expression for the Kerr angle (conductivity times frequency) and (**c**) inverse denominator in the expression of the Kerr angle. *Solid red* (*dashed blue*) *lines* mark the real (imaginary) part. *Vertical dashed lines* are guide-to-the-eye for main peaks in the Kerr angle

rotation. In this way, we are able to attribute the first peak in θ mainly to optical contributions, whereas the second valley has a prevailing magneto-optical origin (for clarity, vertical lines corresponding to the main peaks have been drawn as a guide to the eye). This scenario is valid also for other full-Heusler compounds (not shown) and also for other half-Heuslers – consistent with [27].

Let's now compare the Kerr angles when changing the anion in different Heusler alloys (see Fig. 4). The calculated size of the Kerr rotation is around 0.5 degrees: this is a standard value for 3*d* based materials, which are generally not very active from the magneto-optical point of view. An exception is PtMnSb where a giant Kerr rotation was observed both experimentally [28] and theoretically [27]. This suggests that in order to achieve a high Kerr angle, one needs, in addition to a large magnetic moment – in this case provided by Mn –, a heavy element (such as Pt, Au) that provides a large spin-orbit coupling, possibly strongly hybridized with Mn *d* states. Our series always shows another transition metal (i.e., Co) as other "cation", whose spin–orbit coupling is rather small. On the other hand, we have different *p* elements,

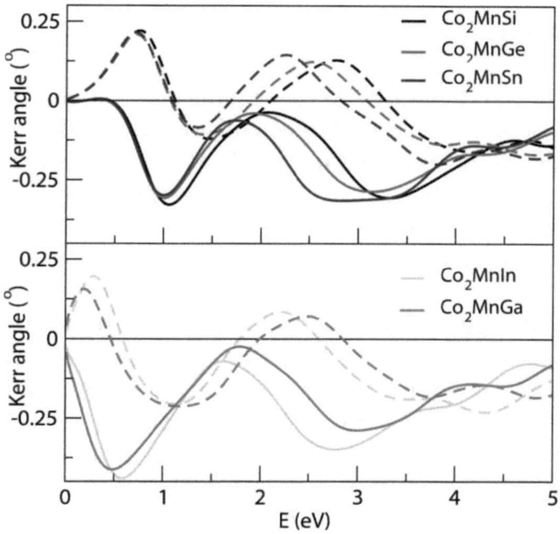

Fig. 4. *Top panel*: *black*, *red* and *blue lines* mark the Kerr angles for Co_2MnSi, Co_2MnGe, and Co_2MnSn, respectively. *Bottom panel*: *green* and *pink lines* show the Kerr angles for $Co_2MnGa$and Co_2MnIn, respectively *(online color)*. In all panels, the *solid* (*dashed*) *lines* show the real (imaginary) contribution as a function of the photon frequency

ranging from light (i.e., Si) to heavy (i.e., Sn); however, the effect of an increased spin-orbit coupling with the atomic number of the *p*-source elements is negligible. Among the compounds studied, $Co_2MnGa$and Co_2MnIn show the highest rotation (cfr. with Co_2MnSi, Co_2MnGe, and Co_2MnSn); at the same time, the energy position of the peak is shifted towards lower energies when replacing a IV-like element (such as Si, Ge, Sn) with a III-like element (like Ga or In). We point out that, upon reduction of the broadening values in a physically meaningful range (i.e., > 0.1 eV, compatible with achievable experimental resolution) the Kerr angle increases up to 0.7^o in $Co_2MnGa$and Co_2MnIn [26]. Our generally small Kerr angles show that half-metallicity alone (such as that typical of $Co_2MnSi$and Co_2MnGe) is not sufficient to guarantee strong MO effects. Moreover, irrespective of the anion, the Kerr angles look pretty similar all over the series: the Kerr rotation is generally negative and shows two valleys, whereas the Kerr ellipticity exhibits some zero crossings and two main positive peaks. The similarity in the magneto-optical properties are consistent with an overall similar underlying electronic structure.

6 Conclusions and Outlook

We have presented an overview on first-principles calculations based on Heusler alloys, focusing in particular on several mechanisms that can reduce the full spin-polarization predicted for ideal materials at 0 K. Antisites, for example, were suggested as a possible source of defect-induced states at the Fermi level and our findings suggest therefore that disorder in the cationic sublattices should be kept minimal during growth. Magneto-optical properties were also discussed for several full-Heuslers and some guidelines to achieve a high Kerr effect were suggested on the basis of a careful discussion of magneto-optical properties in terms of the underlying band-structure. The potential of Heusler alloys in the spintronics framework is undoubtful and it is clearly witnessed by the increasing interests from both theoretical and experimental communities. If materials issues (such as the still rather low quality of the samples) will be solved, it is not irrealistic to expect a technological breakthrough coming from this class of materials.

References

[1] G. Prinz: Science **282**, 1660 (1998)
[2] T. Ishikawa, T. Marukame, H. Kijima, K.-I. Matsuda, T. Uemura, M. Arita, and M. Yamamoto: Appl. Phys. Lett. **89**, 192505 (2006)
[3] N. Tezuka, N. Ikeda, S. Sugimoto, and K. Inomata: Appl. Phys. Lett. **89**, 252508 (2006)
[4] X. Y. Dong, C. Adelmann, J. Q. Xie, and C. J. Palmstrom, X. Lou, J. Strand, P. A. Crowell, J.-P. Barnes, and A. K. Petford-Long: Appl. Phys. Lett. **86**, 102107 (2005)
[5] I. Galanakis, P. H. Dederichs, and N. Papanikolau, Phys. Rev. B **66**, 134428 (2002).
[6] J. Kübler: Physica B **127**, 257 (1984).
[7] J. Kübler, unpublished.
[8] K. Yamauchi and S. Picozzi, in preparation.
[9] J. P. Perdew, K. Burke, and M. Ernzerhof: Phys. Rev. Lett. **77**, 3865 (1996)
[10] U. Von Barth and L. Hedin: J. Phys C **5**, 1629 (1972).
[11] E. Wimmer, H. Krakauer, M. Weinert, and A. J. Freeman, Phys. Rev. B **24**, 864 (1981).
[12] H. J. Monkhorst and D. J. Pack, Phys. Rev. B **13**, 5188 (1976).
[13] S. Picozzi, A. Continenza , and A. J. Freeman: Phys. Rev. B **66**, 094421 (2002).
[14] B. P. Raphael et al.: Appl. Phys. Lett. **79**, 4396 (2001)
[15] A. Kohn, V. K. Lazarov, L. J. Singh, Z. H. Barber, and A. K. Petford-Long: J. Appl. Phys **101** 023915 (2007).
[16] S. Picozzi, A. Continenza, and A. J. Freeman, Phys. Rev. B **69**, 094423 (2002).
[17] S.Picozzi, A.Continenza, and A. J. Freeman: J. Appl. Phys. **94**, 4723 (2003).
[18] S.Picozzi, A.Continenza, and A. J. Freeman: J. Phys. Chem. Sol. **64**, 1697 (2003).

[19] S. Ishida, T. Masaki, S. Fujii, and S. Asano: Physica B- Cond. Mat. **245**, 1 (1998).
[20] S. J. Hashemifar, P. Kratzer, and M. Scheffler: Phys. Rev. Lett. **94**, 096402 (2006).
[21] I. Galanakis: J. Phys. Cond. Matt. **14**, 6329 (2002).
[22] M. Lezaic, Ph. Mavropoulos, J. Enkovaara, G. Bihlmayer, and S. Blügel: Phys. Rev. Lett. **97** 026404 (2006).
[23] H. C. Kandpal, G. H. Fecher, C. Felser, and G. Schönhense: Phys. Rev. B **73**, 094422 (2006).
[24] H. Ebert: Rep. Prog. Phys. **59**, 1665 (1996).
[25] K. H. J. Buschow, *Ferromagnetic Materials*, edited by E. P. Wohlfarth and K. H. J. Buschow (North-Holland, Amsterdam, 1988), Vol. 4, p. 493.
[26] S. Picozzi, A. Continenza, and A. J. Freeman, J. Phys. D: Appl. Phys. **39**, 851 (2006).
[27] V. N. Antonov, A. N. Yaresko, A. Ya. Perlov, V. V. Nemoshkalenko, P. M. Oppeneer, and H. Eschrig: Low Temp. Phys. **25**, 387 (1999).
[28] P. G. Van Engen, K. H. J. Buschow, R. Jonegreur, and M. Erman: Appl. Phys. Lett. **42**, 202 (1983).

Part IV

Graphene

Dirac Particles in Epitaxial Graphene Films Grown on SiC

Claire Berger[1,2], Xiaosong Wu[1], Phillip N. First[1], Edward H. Conrad[1], Xuebin Li[1], Michael Sprinkle[1], Joanna Hass[1], François Varchon[2], Laurence Magaud[2], Marcin L. Sadowski[2], Marek Potemski[2], Gérard Martinez[2], and Walt A. de Heer[1]

[1] School of physics, Georgia Institute of Technology, Ga-30328 Atlanta, USA
claire.berger@physics.gatech.edu
[2] CNRS-Institut Néel, 38042 Grenoble Cedex 9, France

Abstract. We report on the transport and structural properties of graphene layers grown epitaxially on hexagonal SiC. Experimentally, the charge carriers in epitaxial graphene are found to be chiral and the band structure is clearly related to the Dirac cone. To lowest order, epitaxial graphene appears to consist of stacked graphene sheets; the first layer is highly charged with the others carrying much lower charge.

1 Introduction

The potential of carbon nanotubes for nanoelectronics has been amply demonstrated: transport is ballistic even at room temperature [1], carbon nanotubes have extremely weak electron-phonon coupling, excellent FET characteristics [2], and the material itself is very robust. Because these properties directly stem from a single sheet of graphite (graphene), it was realized that 2D graphene should in principle retain these essential properties [3].

Ando [4] indeed showed that the absence of backward scattering in carbon nanotubes can be traced back to an anomalous Berry's phase, which corresponds to a sign change of the wave function under a rotation in the wave vector space of two-dimensional graphene. This is related to the Dirac–Weyl equation for the Hamiltonian $H = v_F \cdot \sigma \cdot p$, where σ are the Pauli matrices, p the momentum and v_F the velocity. The essential characteristics are a particle-hole symmetry, a conical dispersion relation (the energy $E = \pm v_F p$ is linear with momentum) and an additional quantum number label, the pseudo-spin. The pseudo-spin characterizes the component of the wavefunction from the two equivalent crystalline sublattices (A and B) that compose the graphene honeycomb lattice. Electrons have opposite chirality in the two valleys K and K' of the Brillouin zone. On a specific cone the pseudospins of oppositely directed electrons are also reversed. Hence, backscattering from intravalley events is suppressed because it does not conserve pseudospin, and resistance is reduced. This is at the origin of ballistic

R. Haug (Ed.): Advances in Solid State Physics,
Adv. in Solid State Phys. **47**, 145–157 (2008)

properties of 1D graphene based structures. In a magnetic field B the energy is quantized in Landau levels $E_n = \pm\sqrt{2e\hbar v_F^2 Bn}$ [4,5]. Note the square-root dependence on magnetic field (it is linear in case of a quadratic dispersion relation), the particle-hole symmetry, and the presence of a level at $n = 0$ independent of magnetic field.

The realization of a single sheet of graphene was until recently quite elusive. Single graphene layers were grown on various metallic surfaces [6], and on semiconducting SiC by thermal decomposition [7]. Multiple exfoliation of graphite has been perfected [8] to ultimately one layer graphene [9, 10]. This very fascinating material clearly demonstrates the chiral nature of the charge carriers, as is manifested in several properties, of which the anomalous phase observed in a room temperature quantum Hall effect [11] is the most striking. The thermal growth on SiC provides a road map towards large scale graphene production; fast, reproducible and routine processes are developed and surface tools are available for characterization [3].

Here we show that multilayered films grown on SiC have characteristics of graphene: transport properties show evidence for a Dirac-chiral particle with nontrivial Berry's phase and weak anti-localization; magneto infrared spectroscopy reveals the Landau level structure of graphene. This surprising result can been understood considering that the transport is dominated by the charged interface layer. Above the interface layer, the other layers are quasi-neutral, and do not present a graphitic character but a non regular Bernal stacking and a unique azimuthal orientational disorder.

2 Interface States

Hexagonal SiC decomposes at high temperature in vacuum by sublimation of Si. Depending on temperature, the SiC surface undergoes a series of carbonrich reconstructions, the last of which is a graphene layer [3, 7]. The 4H (6H)-SiC unit cell consists of a stack of 4 (6) hexagonal Si-C bilayers, so that the $(000\bar{1})$face is C-terminated and the (0001)face is Si-terminated. N-layer epitaxial graphene (EGn) films were grown on the Si-face of SiC by heating to about 1300–1400°C in ultra-high vacuum. Films have also been prepared on both faces in moderate vacuum conditions using ovens with controlled background gas. Graphene grows fast on the C-face, so that EGn on the C-face have a minimum of 4–5 layers. A slow layer by layer growth can be achieved on the Si-face up to 3–4 layers [7, 12].

Because the ultra-thin films lie on a surface, substrate interaction has to be considered. The best example (from which graphite actually gets its name) is drawing a graph with a pencil: the transfer of graphite from the rod to the paper shows that substrate/graphene interaction is stronger than the interlayer graphene interaction.

We expect the SiC/graphene interface to be charged due to the built-in electric field necessary to equilibrate the work-function difference δW between

SiC and graphene. A rough estimate gives a charge of a few $10^{12}\,\mathrm{cm}^{-2}$ from $\delta\mathrm{W} = 0.3\,\mathrm{eV}$ [14]. A charge density of about $4 \times 10^{12}\,\mathrm{cm}^{-2}$ is consistently found in transport experiments (see below). Because of screening, we also expect the charge transfer to be restricted to 1–3 layers, so that the top layers should be almost neutral [10, 12, 16, 17].

In fact the first carbon layer (EG0) grown above the SiC substrate has no graphitic electronic properties and acts as a buffer layer between the substrate and subsequent graphene layers. X-ray reflectivity data on the C-face [18], show that atoms in this first plane form strong covalent bonds with the SiC substrate. The SiC/first layer distance $\gamma = 0.16\,\mathrm{nm}$ is close to that of diamond and much smaller than the graphite interlayer distance ($\gamma = 0.3354\,\mathrm{nm}$).

This short bond is also found in Density Functional Theory (DFT) calculations on both the Si-and the C-terminated interfaces [13]. As a first approximation to the actual structure we used a $\sqrt{3}\times\sqrt{3}\mathrm{R}30$ SiC surface reconstruction. This cell is commensurate with a slightly expanded 2×2 graphene cell. DFT calculations show that the first C-layer acts as a buffer layer. It exhibits a large gap and a Fermi level pinned by a state with a small dispersion. These states are related to the dangling bonds and consequently depend on the interface geometry used in the calculation. They remain unchanged when further graphene layers are added on top of the buffer layer. Details of the interface states for the EG0 buffer layer and of its conduction properties are under investigation by Scanning Tunneling Microscopy (STM) and Scanning Tunneling Spectroscopy [19, 20]. First results seem to indicate that at least on the Si-face this EG0 layer is poorly conducting.

On top of this buffer layer, the next graphene layer (EG1) has an electronic dispersion of graphene with two linear bands intersecting at the K or K' point. Calculations confirm an electron charge transfer from SiC to the graphene layer, in agreement with ARPES and XPS measurements [21, 22]. X-ray diffraction shows the EG0-EG1 separation is 0.341 nm, slightly larger than that of Bernal graphite [18]. The exact n-type doping depends on the interface states. For instance, calculations on the C-face for a C-deficient interface layer, as suggested by X-ray scattering data, indicate a larger charge transfer than for a bulk truncated geometry. It is significant that although details of the charge transfer (dangling bonds) are different for the Si- and C-faces and sensitive to the geometry, the first C-layer on top of a SiC surface acts as a buffer layer and allows the next graphene layer to behave electronically like an isolated graphene sheet.

3 Non-Bernal Stacking of Graphene Layers

The EGn layers grown on the Si-face or the C-face have different characteristics. X-ray diffraction shows that the vertical coherence of the EG is limited to lateral dimensions $\lambda_c \sim 29\,\mathrm{nm}$ on the Si-face grown in UHV [23], and much larger on the C-face grown in moderate vacuum ($\lambda_c > 300\,\mathrm{nm}$) [18]. On the

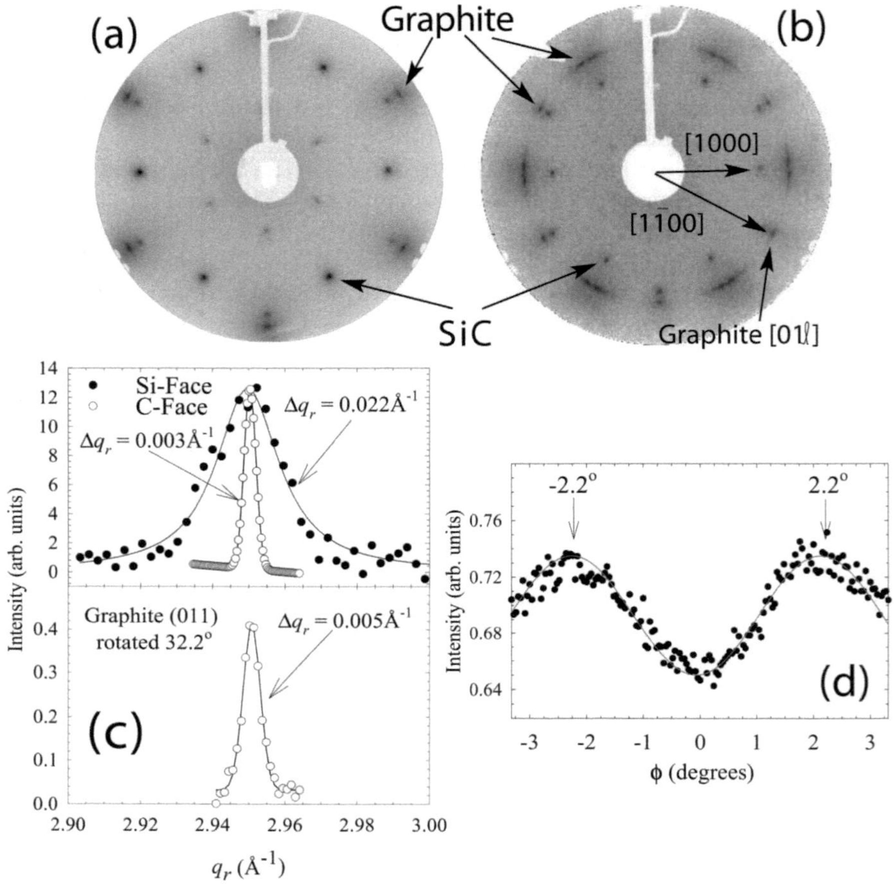

Fig. 1. LEED and x-ray diffraction from multilayer graphene grown on 4H-SiC(000$\bar{1}$) substrates. (**a**) LEED pattern (71 eV) for ~ 3 ML graphene, (**b**) LEED pattern (103 eV) for ~ 4 ML graphene (unlabeled sets of 6-fold spots in (a) and (b) are from a $\sqrt{3} \times \sqrt{3}R30°$ SiC interface reconstruction). (**c**) Radial x-ray scans through (*top*) the (10ℓ) graphite rod, and (*bottom*) across the diffuse arcs seen in (b). (**d**) Azimuthal x-ray scans across the graphite diffuse rods seen in (b)

C-face the measured diffraction peak width could be primarily limited by the SiC terrace widths of typically 300 nm. We note that there is ample evidence that the graphene layers drape continuously over the SiC steps [14, 20] therefore the domain size relevant for electrical transport may be much larger than the domains measured by X-rays.

EGn layers grown on the Si-face of 4H- and 6H-SiC were observed by STM. Images show large flat regions with a weak hexagonal corrugation of 0.03 nm on a 1.9-nm period [3, 19]. This is attributed to the quasimatching of 6 unit

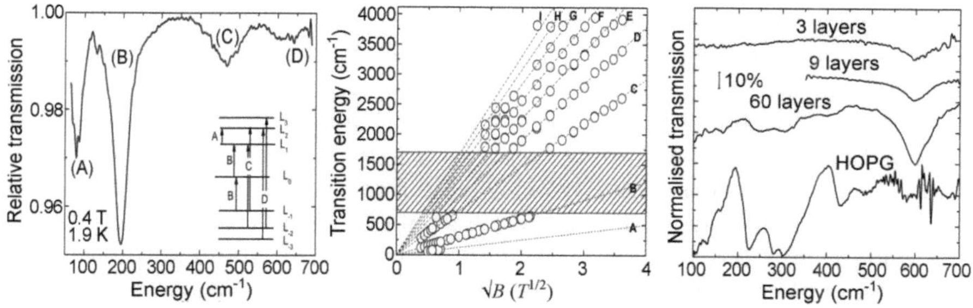

Fig. 2. Infrared spectroscopy of a 10 layer EGn. (**a**) IR transmission spectrum at $B = 0.4\,\mathrm{T}$ and $T = 1.9\,\mathrm{K}$. (Inset) Schematic diagram of the Landau levels. The *arrows* indicate the observed transitions (**b**) The absorption maxima positions as a function of field showing the $\sqrt{B}$ dependence. The hatched strip corresponds to the opacity range of the SiC substrate (**c**) spectra for various EGn thicknesses, as indicated, and comparison with graphite (HOPG)

cells of SiC on 13 graphene cells rotated by 30 degrees, and accompanying interface reconstruction. Low energy electron diffraction (LEED) confirms the orientational relation between SiC and graphene (Fig. 1), indicating that graphene grows epitaxially on the Si face. For an EG1 layer, imaging by STM at low bias voltage shows the honeycomb pattern of graphene [20].

By contrast the layers are extremely flat on the C-face with an rms roughness $< 0.005\,\mathrm{nm}$, as shown by X-ray reflectometry [18]. This is confirmed by cross-sectional transmission electron microscopy [24]. Reflectivity data are consistent with stacking faults occurring every other layer on average. The interlayer distance ($\gamma = 0.3370 \pm 0.0005\,\mathrm{nm}$) is larger than in graphite, and close to turbostratic graphite [18]. Also the graphene layers grown on top of the C-face present an azimuthal rotational disorder, revealed by LEED and X-ray diffraction [25]. Three rotated phases ($30°$, $+2.2°$ and $-2.2°$) are stacked with nearly equal concentrations, giving rise to films with a high concentrations of rotated boundaries (i.e., stacking faults with a density similar to that of the X-ray data). Recently, calculations of rotated graphene layers have shown that the electronic structure presents two linearly dispersing bands, just like graphene, and that contrary to the case of a Bernal bilayer, the bands are linear in k down to the charge neutrality point [26].

Therefore the picture on the C-face is that of flat, highly-ordered and weakly interacting graphene layers, not in a graphitic Bernal stacking. The first charged graphene layer EG1 is separated from the interface by a buffer layer, and protected from the environment by the top graphene multilayers.

4 Landau Level Spectroscopy of Epitaxial Graphene

In contrast with the interface charged layer, we now present evidence for the presence of quasi-neutral graphene layers (top layers). A representative infrared magneto-spectroscopy spectrum is presented in Fig. 2 for a 10 layer EGn in a weak magnetic field of 0.4 Tesla [17]. All the observed absorption lines are shown. Their energy shift with increasing magnetic field accurately follows a $\sqrt{B}$ dependence. All the lines are identified as transitions between various Landau levels, according to $E_n = v_0\sqrt{2ne\hbar B}$, with as single parameter, the effective renormalized velocity $v_0 = 1.03 \times 10^6$ m/s. It is found to be close to the Fermi velocity for exfoliated graphene. Note that v_0 is largerthat the value from band structure calculations, what might be due to electron-electronand electron-phonon renormalization effects [27]. Measurements for layers of various thicknesses show that the intensity of the signal scales with the thickness of the film. Moreover the spectra is very different from that of graphite even for thick films (about 60 layers). The $\sqrt{B}$ dispersing lines are the only one observed, in sharp contrast with measurements on graphite.

The exact $\sqrt{B}$ dependence is the hallmark of a linear dispersion relation; quadratic $E(k)$ gives a linear B dependence. The absorption transition involving the $n = 0$ Landau level is observed at a magnetic field as low as 0.16 T, which indicates that already at this field the Fermi energy is pinned to the $n = 0$ level. Hence, at zero magnetic field : $-15\,\text{meV} \leq E_F \leq 15\,\text{meV}$ and $n \leq 1.5 \times 10^{10}\,\text{cm}^{-2}$ and the Fermi wavelength is ≥ 300 nm. These experiments demonstrate that epitaxial graphene consists of stacked graphene layers, whose electronic band structure is characterized by a Dirac cone. Remarkably, there is no evidence for a gap nor for a deviation of the linear density of states: undistorted Dirac cone properties are directly observed as close as 15 meV to the Dirac point.

5 Dirac Particle Transport

Transport measurements of patterned Hall bars for EGn on both Si and C-faces show evidence for graphene properties. As shown below these include an anomalous Berry's phase (from Landau plots), the correct graphene-type Landau level energy spacing (as determined from the temperature dependence of the Shubnikov-de Haas oscillations) and weak anti-localization.

Figure 3 shows the magnetoresistance (MR) of a wide Hall bar for a thin EG1-EG2 film grown in UHV on the Si-face similar to the one of [3]. Shubnikov-de Haas (SdH) oscillations are observed with a 1/B frequency B_1=14 T. Resistance maxima are expected at fields B_n when the Fermi energy E_F intercepts the Landau levels E_n, hence $B_n = (E_F/v_F)^2/2ne\hbar = B_1/n$. For normal electrons, maxima are found when $E_F = (n+1/2)eB_n\hbar/m$, hence a plot of n versus $1/B_n$(Landau plot) intercepts the origin for a Dirac particle, and the y axis at one half for a normal electron. The phase shift for graphene is related to the Berry's phase acquired by Dirac particles in a magnetic field. The Landau plot in Fig. 3 passes through the origin indicating that the Berry's phase is anomalous. From B_1 and $v_F = 10^6$ m/s, we deduce E_F=1680 K, thus a carrier density $n_s = 1.9 \times 10^{12}\,\text{cm}^{-2}$, that is the

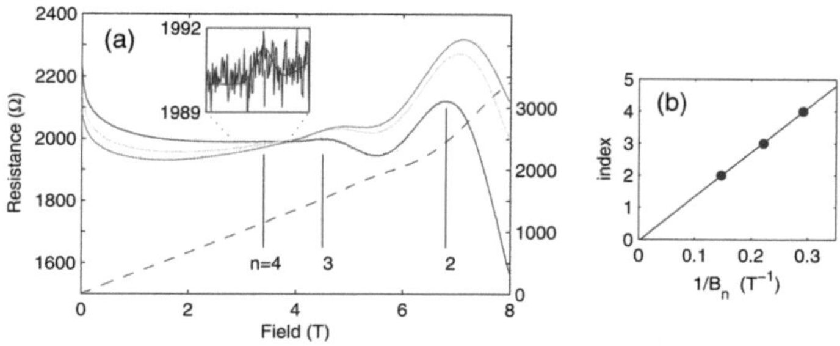

Fig. 3. 2D transport measured in a 400 μm×600 μm Hall bar for a EG1-EG2 on the Si face. (**a**) Magnetoresistance at T =0.3, 2, and 4 K showing well developed SdH peaks, indicated with their Landau indices n; the Hall resistance at 0.3 K (*dashed line*) shows a weak feature at the expected Hall plateau position. (**b**) Landau plot; the linear extrapolation passes through the origin demonstrating the anomalous Berry's phase characteristic of graphene

value found with the Hall coefficient R_H=330 Ω/T. This charge density is in agreement with photoemission data for an EG1-EG2 film grown on the Si face [21].

Thermal population of the Landau levels reduces the amplitude of the SdH oscillations following the Lifshitz–Kosevich (LK) equation : $A_n(T) \sim A_0 u/\sinh(u)$ where $u = 2\pi^2 k_B T/\Delta E(B)$ [28]. This provides a measurement of the Landau level energy spacings for the highly charged layers. From this fit we find in Fig. 3 that at $B = 7\,\mathrm{T}$, $(E_3(B) - E_2(B))/k_B = 250\,\mathrm{K}$ (compared with 340 K predicted for graphene at this carrier density) and that the Dirac point is about 1290 K below E_F.

This sample shows evidence for Dirac electrons although there is significant point-defect scattering as indicated by the intense weak localization peak near $B = 0$ and a relatively low mobility (1100 cm^2/Vs). The mean free path estimated from the weak localization peak [3] is comparable to the average scattering length $\zeta_c \sim 100\,\mathrm{nm}$ estimated from STM on similar samples [20].

Graphene films grown on the C-face have much higher mobilities [12]. This is in agreement with the longer structural coherence length (see above). The films are also considerably thicker so that the high carrier-density layer at the interface is more protected [18]. For the dozen or so samples showing SdH oscillations each shows only one frequency B_1. As seen in Fig. 4, high field MR measurements up to 23 Tesla and at 100 mK on a patterned ribbon (width=1 μm) confirms a single frequency up to the second Landau level with a charge carrier of $3.7 \times 10^{12}\,\mathrm{cm}^{-2}$ [31]. Another example is shown in Fig. 4 in the 3 T–9 T range. The resistance fluctuations corresponds to Universal Conductance Fluctuations (UCF) for highly coherent samples (see below).

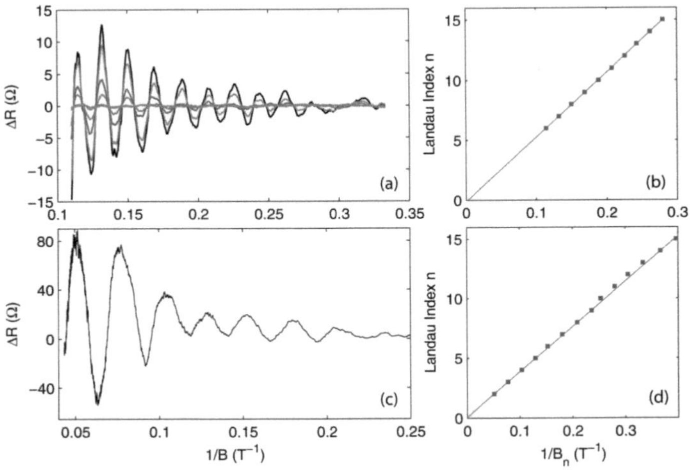

Fig. 4. Two EGn (C-face) patterned Hall bars of moderate width: (**a**) magnetoresistance (3 T–9 T) of a 1 μm × 5 μm Hall bar plotted as a function of 1/B after subtracting a smooth background at several temperatures (4, 10, 20, 30, 50, 70 K); The zero field resistance is R(H=0)=502 Ω. (**b**) Landau plot for data in (a): $B_1 = 53\,\mathrm{T}$, y-axis intercept 0.13±0.02 (**c**) same as (a) for a 1 μm × 6.5 μm Hall bar at high field (up to 23 T) and low temperature (100 mK) (**d**) Landau plot for data in (c): $B_1 = 36\,\mathrm{T}$, intercept 0.08±0.1

6 Weak Anti-Localization

Magnetoresistance measurements of a wide Hall bar (100 μm × 1000 μm) at several temperatures [29] present similar Dirac characteristic features. A Landau plot reveals the anomalous Berry's phase (Fig. 5d and the LK analysis of the peak amplitude (Fig. 5e) agrees with the expected Landau level spacing for a Dirac particle. Furthermore, the charge density $3.8 \times 10^{12}\,\mathrm{cm}^{-2}$ is comparable to the value from the Hall effect $n_s = 4.6 \times 10^{12}\,\mathrm{cm}^{-2}$. This indicates that transport is dominated by the highly charged layer at the interface. The other layers are expected to also contribute, and more so in 2D structures than in quasi 1D structures, which could explain the carrier density discrepancy as well the low field kink observed in the Hall effect.

Significantly, the weak localization peak is very small, $\Delta G \sim 0.07e^2/h$, compared with $\Delta G \sim 1e^2/h$ in the low mobility sample of Fig. 3. This indicates that point defect density in this sample is low in accordance with the higher mobility $\mu = 11600\,\mathrm{cm}^2/\mathrm{V.s}$. The MR shows a marked positive upturn brought out by subtracting a temperature independent background (i.e., the data at 50 K). This positive field dependence suggests weak antilocalization (WAL) that is expected due to the chiral nature of electrons in graphene.

Because of chirality, pseudospin conserving scattering processes that do not distinguish between A and B sublattices (i.e., long-range scatterers) can-

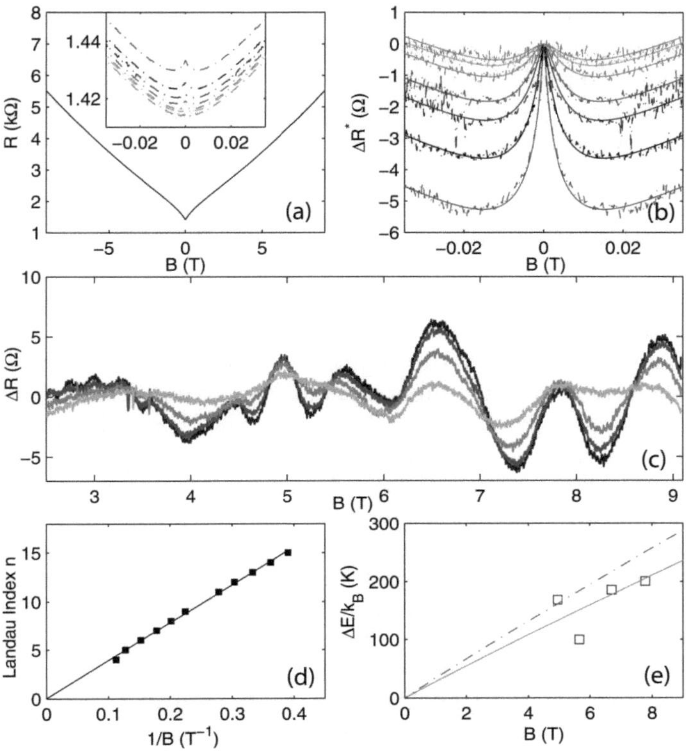

Fig. 5. 2D transport in a 100 µm × 1000 µm Hall bar on a ∼10 layer EGn on the C-face. (**a**) Resistance as a function of the magnetic field. Inset, *dash-dot lines*, low field MR at various temperatures (1.4, 4.2, 7, 10, 15, 20, 30, 50 K). (**b**) Low field MR after subtracting 50 K data as a background. *Dash-dot lines*, experimental data. *Solid lines*, fits to the theory by McCann *et al.* (**c**) High field MR after subtracting a parabolic background at several temperatures (4, 7, 15, 30 K) (**d**) Landau plot for SdH maxima oscillations. (**e**) Landau level spacing obtained by Lifshitz–Kosevich analysis. *Squares*: experiment. *Solid line*, theoretical prediction for ΔE assuming $v_F = 0.82 \times 10^8$ cm/s, *dash-dot line*: $v_F = 10^8$ cm/s

not backscatter [4]. The magnetic field introduces an additional phase shift to the wave function which relaxes this condition, resulting in a positive magnetoresistance (WAL). In contrast, point defects that locally break the sublattice degeneracy, thereby causing intervalley scattering, restore the usual weak localization. Note that due to the Bernal stacking in graphite, the sublattice degeneracy is lifted so that WAL is neither expected nor observed. In fact the amplitude, field and temperature dependence of the magnetoresistance of Fig. 5 match predictions [30] of WAL very well. This shows that transport is dominated by chiral electrons which are scattered by long-range potentials. These could be due to the localized counterions in the SiC substrate.

Note that this high mobility 2D samples shows a featureless (except for extremely weak ripples) Hall resistance and small amplitude of the SdH oscillations (Fig. 5c) superimposed on a large positive magnetoresistance. This is in contrast with narrower patterned ribbons such as shown in Fig. 4, for which the amplitude of the SdH is stronger.

7 Coherence and Confinement

The high structural quality of EGn on the C-face is reflected in long electronic phase coherence lengths l_ϕ. This is clearly manifested in smaller ribbons, when l_ϕ becomes comparable to the sample size. In mesoscopic systems, coherent scattering on random static impurities build up an interference pattern which is affected by a magnetic field. The conductance hence fluctuates with field, with an amplitude that reflects the degree of coherence [33]. Reproducible conductance fluctuations of the order of $1e^2/h$ are observed in patterned ribbons, indicative of micrometer long l_ϕs. An example is given in Fig. 6-inset for a $0.5\,\mu\text{m} \times 6\,\mu\text{m}$ ribbon. In narrow ribbons, weak anti-localization is not observed. Instead a well defined sharp magnetoresistance peak centered at zero field can be analyzed by weak localization in the 1D limit [12], with an elastic scattering length limited by the width of the patterned sample. In the sample of Fig. 6, the temperature dependence of both the weak localization peak and the conductance fluctuations gives L_ϕ=1.1 µm at 4 K. The Casimir–Onsager relation (symmetry of the current leads and voltage probes by inversion of the magnetic field) is also well verified, attesting the good reproducibility of the fluctuations.

Narrow ribbons also show evidence for quantum confinement. From SdH oscillations, a carrier density of $4 \times 10^{12}\,\text{cm}^{-2}$ is deduced. Note that this corresponds to a Fermi wavelength $\lambda_F \approx 20\,\text{nm}$, whereas the low-density layers have $\lambda_F \geq 400\,\text{nm}$. These layers are therefore expected to contribute little to the transport in ribbons that are narrower than 500 nm. For very narrow ribbons ($\leq 100\,\text{nm}$) with rough edges, the low-density layers are expected to be insulating, since there are no propagating modes (channels).

Figure 6 shows the magnetoresistance of a narrow ribbon (see [12] for details). The Landau levels for a graphene ribbon of width W are approximately given by $E_n(B,W) \approx [E_n(W)^4 + E_n(B)^4]^{1/4}$ where $E_B(n)$ are the 2D graphene Landau levels and $E_W(n) = n\pi\hbar v_0/W$ reflects 1D confinement [32]. Confinement effects become apparent for low fields, approximately when the cyclotron diameter becomes greater than W. This causes deviation from linearity in the Landau plot as seen in Fig. 6. The LK analysis confirms the effect: for high magnetic fields the energy separation between the Landau levels increases with increasing field, while for low field the energy separation saturates and is determined by the quantum confinement. Note that this analysis does not require a determination of the locations of the MR peaks [12].

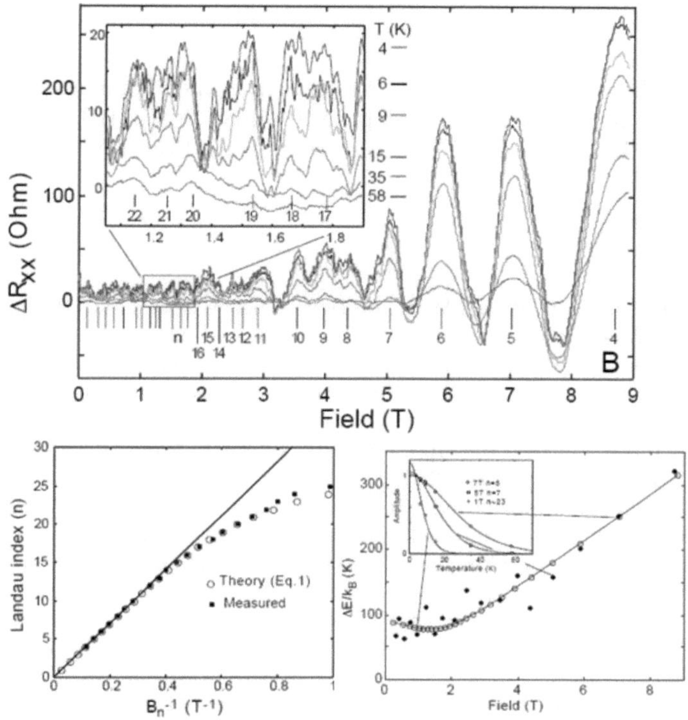

Fig. 6. Narrow Hall bar 0.5 μm × 6 μm. The zero field resistance is 1125 Ω. (**a**) MR oscillations for temperatures ranging from 4–58 K after subtraction of a smooth background. Conducatance fluctuations are superimposed on the SdH oscillations (**b**) Landau plot of the MR peaks. The deviation from linearity is due to quantum confinement. (**c**) The energy gap between the Fermi level and the lowest unoccupied Landau level is found from the Lifshitz–Kosevich analysis (inset) of the peaks

8 Conclusion

Magnetotransport and infrared spectroscopy clearly demonstrate the chiral nature of particles in epitaxial graphene on SiC. EG is clearly not graphite, which has a different Landau level spectrum and an entirely different electronic structure. This difference reflects that epitaxial graphene does not have the Bernal stacking that would lift the A-B sublattice degeneracy. Surprisingly, transport measurements show no evidence for quantum Hall plateaus. This may be related to the presence of multi layers. However the most intense SdH peaks in 2D samples are seen in the most defective samples, and the SdH amplitude decreases sharply with the EG ribbon width. It is possible that defects, specifically in the "bulk" of the sample (i.e., away from the edges) are required for large amplitude SdH peaks, and hence for the QHE [34].

Acknowledgements

D. Maud and C. Naud are warmly thanked for their help in the high magnetic lab. This work was supported by NSF-NIRT grant 0404084, NSF-MRI grant 0521041, a grant from Intel, and a USA-France travel grant from CNRS.

References

[1] S. Frank, P. Poncharal, Z. L. Wang, W. A. de Heer, Science **280**, 1744 (1998).
[2] S. J. Tans, R. M. Verschueren, C. Dekker, Nature **393**, 49 (1998).
[3] C. Berger, Z. M. Song, T. B. Li, X. B. Li, A. Y. Ogbazghi, R. Feng, Z. T. Dai, A. N. Marchenkov, E. H. Conrad, P. N. First, W. A. de Heer, *et al.* J. Phys. Chem. B **108**, 19912 (2004).
[4] T. Ando, T. Nakanishi, J. Phys. Soc. Jpn. **67**, 1704 (1998); T. Ando, T. Nakanishi, R. Saito, J. Phys. Soc. Jpn. **67**, 2857 (1998).
[5] V. P. Gusynin, S. G. Sharapov, Phys. Rev. B **71**, 125124 (2005); N. M. R. Peres, F. Guinea, A. H. Castro Neto, Phys. Rev. B **73**, 125411 (2006).
[6] For instance: R. Rosei, M. de Crescenzi, F. Sette, C. Quaresima, A. Savoia, P. Perfetti, Phys. Rev. B **28**, 1161 (1983).
[7] I. Forbeaux, J. M. Themlin, A. Charrier, F. Thibaudau, J. M. Debever, Appl. Surf. Sci. **162**, 406 (2000); A. J. van Bommel, J. E. Crombeen, A. van Tooren, Surf. Sci. **48**, 463 (1975).
[8] X. Lu, M. Yu, H. Huang, R. Ruoff, Nanotechnology **10**, 269 (1999).
[9] K. S. Novoselov, A. K. Geim, S. V. Morozov, D. Jiang, M. I. Katsnelson, I. V. Grigorieva, S. V. Dubonos, A. A. Firsov, Nature **438**, 197 (2005).
[10] Y. B. Zhang, Y. W. Tan, H. L. Stormer, P. Kim, Nature **438**, 201 (2005).
[11] K. S. Novoselov, Z. Jiang , Y. Zhang, S. V. Morozov, H. L. Stormer, U. Zeitler, J. C. Maan, G. S. Boebinger, P. Kim, A. K. Geim, Science **315**, 1379 (2007).
[12] C. Berger, Z. M. Song, X. B. Li, X. S. Wu, N. Brown, C. Naud, D. Mayo, T. B. Li, J. Hass, A. N. Marchenkov, E. H. Conrad, P. N. First, W. A. de Heer, Science **312**, 1191 (2006).
[13] F. Varchon, R. Feng, J. Hass, X. Li, B. N. Nguyen, C. Naud, P. Mallet, J. Y. Veuillen, C. Berger, E. H. Conrad, L. Magaud, cond-mat/0702311.
[14] Th. Seyller , K. V. Emtsev , K. Gao , F. Speck , L. Ley , A. Tadich , L. Broekman, J. D. Riley, R. C. G. Leckey, O. Rader, C. A. Varykhalov, A. M. Shikin: Surface Science **600**, 3906 (2006).
[15] Y.Zhang, Z.Jiang, J. P. Small, M. S. Purewal, Y. W. Tan, M.Fazlollahi, J. D. Chudow, J. A. Jaszczak, H. L. Stormer, P.Kim, Phys. Rev. Lett. **96**, 136806 (2006).
[16] F. Guinea, cond-mat/0611185.
[17] M. L. Sadowski, G. Martinez, M. Potemski, C. Berger, W. A. de Heer, Phys. Rev. Lett. **97** 266405 (2006); M. L. Sadowski, G. Martinez, M. Potemski, C. Berger, W. A. de Heer, Solid State Comm. (in press).
[18] J. Hass, R. Feng, J. Millán-Otoya, X. Li, M. Sprinkle, P. N. First, C. Berger, W. A. de Heer, E. H. Conrad, cond-mat/0702540, Phys. Rev. B (in press).
[19] P. Mallet, F. Varchon, C. Naud, L. Magaud, C. Berger, J.-Y. Veuillen, cond.mat/0702406.

[20] G. M. Rutter, N. P. Guisinger, J. N. Crain, E. A. A. Jarvis, M. D. Stiles, T. Li, P. N. First, J. A. Stroscio (to be published).

[21] E. Rollings, G.-H. Gweon, S. Zhou, B. Mun, J. McChesney, B. Hussain, A. Fedorov, P. First, W. de Heer, A. Lanzara, J. Phys. Chem. Solids **67**, 2172 (2006).

[22] A. Bostwick, T. Ohta, T. Seyller, K. Horn, E. Rotenberg, Nature Phys.**3**, 36 (2007); T. Ohta, A. Bostwick, T. Seyller, K. Horn, E. Rotenberg, Science **313**, 951 (2006).

[23] J. Hass, R. Feng, T. Li, X. Li, Z. Zong, W. A. de Heer, P. N. First, E. H. Conrad, C. A. Jeffrey, C. Berger, Appl. Phys. Lett. **89**, 143106 (2006).

[24] D. Ugarte, unpublished result.

[25] J. Hass, J. Millán-Otoya, M. Sprinkle, P. N. First, C. Berger, W. A. de Heer, E. H. Conrad (to be published).

[26] J. M. B. Lopes dos Santos, N. M. R. Peres, A. H. Castro Neto, cond-mat/ 0704.2128. L. Magaud *et al.* (to be published).

[27] Y. A. Bychkov, G. Martinez (to be published).

[28] I. M. Lifshitz, A. M. Kosevich, Sov. Phys. JETP **2**, 636 (1956).

[29] X. S. Wu, X. B. Li, Z. M. Song, C. Berger, W. A. de Heer, Phys. Rev. Lett. **98**, 136801 (2007).

[30] E. McCann, K. Kechedzhi, V. I. Fal'ko, H. Suzuura, T. Ando, B. L. Altshuler, Phys. Rev. Lett. **97**, 146805 (2006).

[31] C. Berger, Z. Song, X. Li, Xi. Wu, N. Brown, D. Maud, C. Naud, Walt A. de Heer, Phys. Sta. Sol. (a) 1–5 (2007) (in press).

[32] D.Mayou (unpublished); N. M. R. Peres, A. H. C. Neto, F. Guinea, Phys. Rev. B **73**, 241403 (2006).

[33] C. W. J. Beenakker, H. van Houten: *Quantum Transport in Semiconductor Nanostructures*, vol. 44 (Academic Press, New York, 1991), and reference therein (also in cond-mat/0412664).

[34] W. A. de Heer, C. Berger, X. Wu, P. N. First, E. H. Conrad, X. Li, T. Li, M. Sprinkle, J. Hass, M. L. Sadowski, M. Potemski, G. Martinez, Solid State Comm. (in press).

Photoemission Studies of Graphene on SiC: Growth, Interface, and Electronic Structure

A. Bostwick[1], K. V. Emtsev[2], K. Horn[3], E. Huwald[4], L. Ley[2], J. L. McChesney[2], T. Ohta[1,3], J. Riley[4], E. Rotenberg[1], F. Speck[2], and Th. Seyller[2]

[1] Advanced Light Source, Lawrence Berkely National Laboratory, USA
[2] Lehrstuhl für Technische Physik, Friedrich-Alexander-Universität Erlangen-Nürnberg, Germany
Thomas.Seyller@physik.uni-erlangen.de
[3] Abteilung Molekülphysik, Fritz-Haber-Institut, Germany
[4] Department of Physics, La Trobe University, Australia

Abstract. The possibility to grow well ordered graphitic films on SiC(0001) surfaces with thicknesses down to a single graphene layer is promising for future applications. Photoelectron spectroscopy (PES) is a versatile technique for investigating a variety of fundamentals and technologically relevant properties of this system. We survey results from recent PES studies with a focus on the growth of graphene and few layer graphene, the electronic and structural properties of the interface to the SiC substrate, and the electronic structure of the films.

1 Introduction

Graphene, a single sheet of sp^2-bonded carbon arranged in a honeycomb lattice, represents the elemental building block of carbon structures such as graphite, carbon nanotubes, and fullerenes. While the latter have been studied both experimentally and theoretically, the properties of graphene were merely a matter of theoretical interest. This situation changed suddenly when the discovery of graphene was reported by *Novoselov* and co-workers in 2004 [1], who – by using rather simple techniques of mechanical exfoliation of graphite – were able to prepare extremely thin graphite layers, frequently reffered to as few layer graphene (FLG), with thicknesses down to a single graphene sheet [2]. Transport measurements on single layer and bilayer graphene have revealed their unusual electronic properties which arise from the linear dispersion of the π- and π^*-bands in the vicinity of the K-point of the hexagonal Brillouin zone [3–6]. A recent review by *Geim* and *Novoselov* [7] highlights the properties which make graphene interesting for electronic applications: (i) a high carrier mobility; (ii) an insensitivity of the mobility on carrier concentration and temperature; (iii) low contact resistance. New high-speed electronic devices based on a ballistic transport in the sub-micrometer range at room temperature are envisioned to be developed in the near future [7].

R. Haug (Ed.): Advances in Solid State Physics,
Adv. in Solid State Phys. **47**, 159–170 (2008)

Besides mechanical exfoliation, a second approach to the fabrication of few layer graphene (FLG) is by solid state graphitization of silicon carbide (SiC) surfaces. It has long been known that annealing of SiC surfaces in vacuum at temperatures above 1100°C leads to a depletion of Si leaving behind carbon rich surfaces [8–14] and eventually a closed layer of graphite. Compared to exfoliation of HOPG this seems to be an elegant method for large scale production of graphene. FLG layers grown on SiC were investigated for their electronic transport properties by *Berger* et al. [15–17] who observed quantum confinement of electrons, phase coherence lengths of at least 1 μm at 4 K, and mobilities of $2.5\,\mathrm{m^2V^{-1}s^{-1}}$ at 4 K, pointing towards a possible application of this so-called epitaxial graphene in electronic devices.

Photoelectron spectroscopy (PES) is a surface sensitive technique which is exceptionally well qualified to probe various properties of thin films ranging from composition over electronic structure to their interface with the substrate. Furthermore, by carefully probing individual stages of its formation it is also possible to gain insight into the growth of the film. Therefore we have applied photoelectron spectroscopy in several of its varieties (XPS, SXPS, ARPES) to learn more about the fascinating new material graphene. This paper sketches the main results of our recent work [18–24].

2 Experimental

Unless otherwise stated we have used on-axis oriented, Nitrogen doped 6*H*-SiC(0001) purchased from SiCrystal AG with a doping concentration of $1.5 \pm 0.5 \times 10^{18}\,\mathrm{cm^{-2}}$. The growth of graphene and FLG was carried out by solid state graphitization, i.e., heating of the samples to temperatures above 1150°C as described in the individual studies. In agreement with the phase diagram of SiC(0001) [25] we utilized two different pathways to grow FLG layers on SiC. One way is to anneal the sample in a flux of silicon at 950°C which results in the Si-rich (3×3) structure thereby removing oxygen. Then Si is removed in further annealing steps. At 1050°C the Si-rich $(\sqrt{3} \times \sqrt{3})R30°$ structure is formed which converts to the carbon rich $(6\sqrt{3}\times 6\sqrt{3})R30°$ structure at 1150°C. The latter is a precursor to graphene growth as discussed below. Further annealing leads to growth of graphene and FLG layers with (1×1) periodicity of graphite, which are highly ordered on SiC(0001). Figure 1 shows low-energy electron diffraction (LEED) patterns observed for the individual stages of FLG growth.

The second pathway employs *ex-situ* H-etched surfaces which are covered with the silicate adlayer reconstruction [26]. Annealing of this reconstruction *in vacuo* induces desorption of oxygen and formation of the Si-rich $(\sqrt{3} \times \sqrt{3})R30°$ structure which then transforms into the carbon rich surface as described above [25]. The advantage of hydrogen etched samples is that they are extremely flat as shown by atomic force microscopy (AFM) with wide terraces and residual steps caused by the unintentional miscut (see Fig. 1).

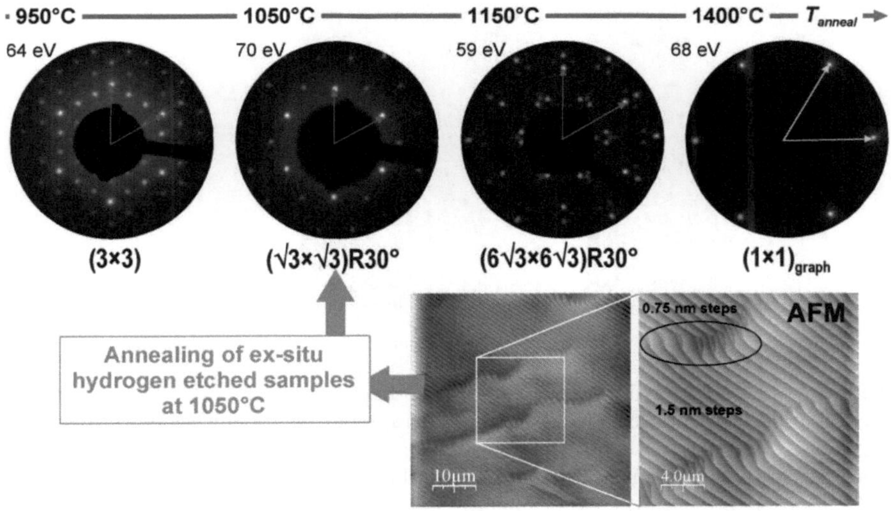

Fig. 1. Typical LEED patterns observed during various stages of FLG growth on SiC(0001). Annealing *ex-situ* H-etched SiC(0001) surfaces *in vacuo* also leads to the Si-rich $(\sqrt{3} \times \sqrt{3})R30^{\circ}$ structure. Hydrogen etching removes polishing damage and leads flat surfaces with micrometer sized terraces and a step height of single unit cell height, as shown by AFM

Photoelectron spectroscopy measurements were carried out either at the storage ring BESSY II, using a toroidal electron analyzer for angle-resolved valence band studies or a hemispherical analyzer (Phoibos 150) for core level studies [18, 19, 22, 23], or at the Electronic Structure Factory at beamline 7 of the Advanced Light Source [20, 21, 24]. For a detailed description the reader is referred to the original papers.

In this paper we concentrate on the properties of graphene grown on SiC(0001), i.e., the Si-face, because these layers are aligned with respect to the substrate. The quasi single crystalline behavior allows us to use momentum resolved photoelectron spectroscopy to study the electronic structure of graphene and FLG. The properties of graphene and FLG on the SiC(000$\bar{1}$) surface (C-face) will be dealt with in forthcoming publications.

3 Results

3.1 The $(6\sqrt{3} \times 6\sqrt{3})R30^{\circ}$ Reconstruction

The $(6\sqrt{3} \times 6\sqrt{3})R30^{\circ}$ reconstruction is a precursor to graphene growth on SiC(0001) and thus deserves special attention. Several models have been suggested for this reconstruction. For instance, it was suggested that it is a

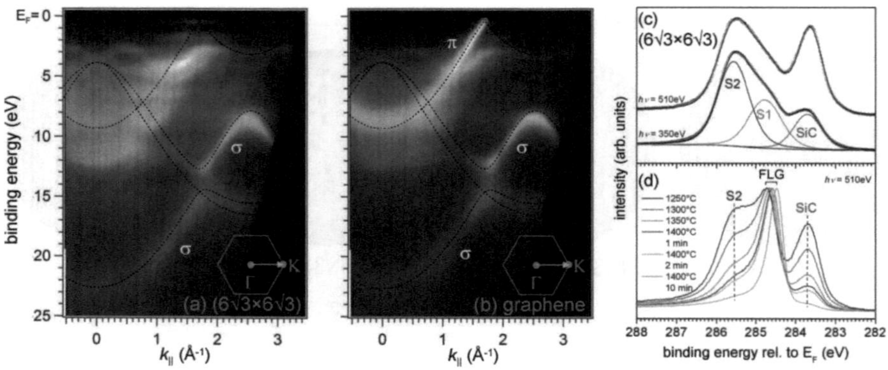

Fig. 2. Photoelectron intensity map vs. binding energy and parallel momentum (**a**) SiC(0001)-($6\sqrt{3} \times 6\sqrt{3})R30°$ and (**b**) graphene on top of SiC(0001)-($6\sqrt{3} \times 6\sqrt{3})R30°$, respectively [23]. The *dashed line* indicates theoretical results for free-standing graphene taken from the literature [31] which was stretched in energy by 13% in order to match the experimental band width. The insets show the direction of $k_{\|}$ within the hexagonal Brillouin zone of graphene. (**c**) C 1s core level spectra of SiC(0001)-($6\sqrt{3} \times 6\sqrt{3})R30°$ [23]. (**d**) Evolution of the C 1s core level with continuing FLG growth [23]

single graphene layer weakly bound by dispersion forces onto the unreconstructed SiC(0001)-(1×1) surface [8, 27] or onto the Si-rich $(\sqrt{3} \times \sqrt{3})R30°$ reconstruction [11, 28, 29]. A recent model [30] proposes a structure in which nano-islands of ca. 1 nm diameter form an ordered array due to self organization. On account of the large unit cell the atomic structure cannot be determined by conventional methods of surface crystallography like LEED. However, using PES it is possible to gain significant information about this reconstruction [23].

Figures 2a and b compare the valence band structure of SiC(0001)-($6\sqrt{3} \times 6\sqrt{3})R30°$ with that of graphene [23]. The characteristic σ-bands with an undistorted dispersion are clearly visible for both surfaces which appears to be incompatible with the proposed nano-islands [30]. On the other hand, this seems to indicate that the $(6\sqrt{3} \times 6\sqrt{3})R30°$ reconstruction contains sp^2 bonded carbon atoms with a C-C distance comparable to that of graphene. In contrast to that it is evident, however, that a graphene-like π-band is not developed in the $(6\sqrt{3} \times 6\sqrt{3})R30°$ reconstruction. Instead, structures are seen which follow the general expected shape of the graphene π-bands, but which also seem to have band gaps due to zone folding effects. Furthermore, there are no occupied states at the Fermi energy. These observations may be explained by a partial hybridization of p_z orbitals with dangling bonds of the SiC(0001) surface. A similar behavior was predicted by theory for H adsorption on free standing graphene [32]. The experimental evidence is in contradiction to the proposed weak van der Waals interaction [8, 11, 27–29].

The C 1s core level spectrum of the $(6\sqrt{3} \times 6\sqrt{3})R30°$ reconstruction (see Fig. 2c) contains a bulk component at 283.70±0.10 eV and two surface components $S1$ (284.75±0.10 eV) and $S2$ (285.55±0.10 eV) with an intensity ratio of $S1/S2 = 1/2$ independent of probing depth [23]. The C atoms leading to these signals are thus located in the same layer. With a C-C bond length similar to that in graphene, we estimate a layer thickness of 2.4±0.3 for the $(6\sqrt{3} \times 6\sqrt{3})R30°$ reconstruction [23], which suggests that the structure is made up from one monolayer of sp^2 bonded carbon atoms.

Considering these observations we propose a model [23] for the $(6\sqrt{3} \times 6\sqrt{3})R30°$ reconstruction which contains a single graphene-like layer with undistorted σ-bands. The p_z orbitals interact partially with the dangling bonds of the underlying SiC(0001) surface. Note that the area density of carbon in graphene is about three times that of a SiC bilayer, which explains the $S1/S2$ ratio: only one third of the p_z orbitals interacts with the substrate.

Another interesting observation is that the surface components S1/S2 remain present in the spectrum during continuing growth of the FLG film (see Fig. 2d) [23]. Furthermore, they are attenuated in the same fashion as the SiC bulk component. Similar observations are made for the $(6\sqrt{3} \times 6\sqrt{3})R30°$ derived structures in the ARPES spectra as well as for the corresponding LEED spots. From this we conclude [23] that the $(6\sqrt{3} \times 6\sqrt{3})R30°$ reconstruction remains at the interface between SiC(0001) and graphene/FLG where it passivates the SiC surface such that the subsequently growing layers possess properties which are – concerning photoelectron spectroscopy – practically identical to free-standing graphene and FLG.

3.2 Barriers Between SiC and FLG

The relative position of the bands of SiC and graphene/FLG is of importance if these films are to be used in electronic devices. The barriers were derived from core level photoelectron spectroscopy measurements performed for FLG films with a thickness of 5–6 monolayers on n-type 6H-SiC(0001), p-type 6H-SiC(0001), and n-type 6H-SiC(000$\bar{1}$) [19, 22].

Figure 3a shows a set of core level spectra. The signal from the FLG layer is clearly visible and is located at 284.42±0.07 eV. The SiC bulk shows up as a minor peak only due to attenuation of the photoelectrons. For SiC(0001) its position is independent of the doping type 283.70±0.07 eV. Using previously gathered information about the position of the C 1s core level of SiC with respect to the valence band maximum E_v [33] we can derive the position of E_F at the interface with respect to E_v as indicated by the top energy scale in 3a. This leads to the band diagrams shown in 3b [22, 33].

We note that the barrier for electrons on n-type 6H-SiC(0001) is rather small (0.3±0.1 eV). Since the valence band offset between different SiC polytypes is basically zero, we estimate that this barrier is increased by 0.3 eV to 0.6±0.1 eV when using n-type 4H-SiC(0001). The barrier is also strongly face specific [22] so that it is much larger on the C-face of n-type 6H-SiC.

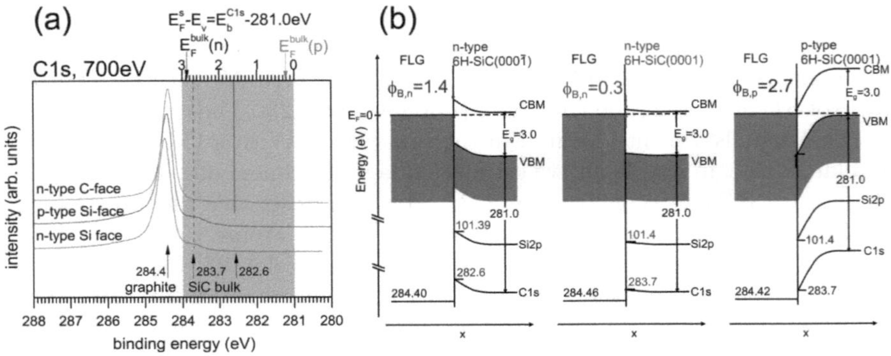

Fig. 3. (**a**) C 1s core levels of FLG grown on n-type $6H$-SiC(0001) (on axis, $1 \times 10^{18}\,\mathrm{cm}^{-2}$), p-type $6H$-SiC(0001) (3.5 degree off axis, epitaxial layer, $1 \times 10^{16}\,\mathrm{cm}^{-2}$), and n-type $6H$-SiC(000$\bar{1}$) (on axis, $1 \times 10^{18}\,\mathrm{cm}^{-2}$). The lower energy scale refers to E_F. The upper scale gives the position of the surface Fermi level of SiC with respect to E_v. For details see [19, 22, 33]. (**b**) Band diagrams determined by core level photoelectron spectroscopy

This could be an advantage of C-face oriented SiC when it comes to building devices, although we have to note that the FLG films on SiC(000$\bar{1}$) are not aligned with the substrate.

3.3 From Graphene to Graphite: Evolution of the Band Structure

In order to understand the interplay between electronic structure, stacking order, and interlayer coupling in graphene and FLG the development of the band structure of one to four monolayers of graphene was studied using ARPES [24]. Figure 4 demonstrates the evolution of the π-band in the vicinity of the Dirac point E_D with increasing layer thickness. The observed bands show a complex structure. The fact that the E_D point is below the E_F is attributed to a charge transfer from the SiC substrate to the FLG layers. The position of E_D with respect to E_F can be varied systematically by adsorption of alkali atoms such as potassium, which result in a charge transfer to the graphene/FLG layer. The bands of the monolayer are strongly normalized, i.e., they deviate considerably from the expected linear behavior, due to many-body interactions (see below).

In order to analyze the observed bands, they were modeled [24] using a generalized tight binding Hamiltonian [36–38] which accounts for the possibility of different stacking (Bernal vs. Rhomohedral), a layer dependent on-site Coulomb potential E_i and adjustable interlayer coupling constants γ_1 and band velocities v. Comparison of the calculated and measured bands led us to the conclusion that in the trilayer domains with Bernal-type and rhombohedral-type stackings coexist [24]. In contrast to that the quadlayer is dominated by Bernal-type stacking. This indicates the importance of second-

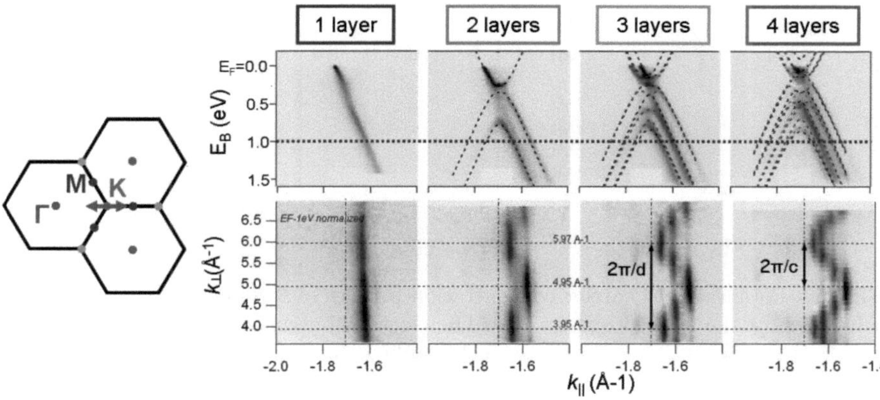

Fig. 4. Development of the bandstructure of FLG [24]. *Top*: Photoelectron intensity map vs. binding energy and parallel momentum showing the evolution of the electronic structure close to the Dirac point while growing FLG with a thickness of up to 4 monolayers. Note that the π-band of the monolayer is strongly renormalized due to many particle interactions [21]. The *dashed lines* are from a calculated tight binding band structure, with band parameters adjusted to reproduce measured bands. *Red* and *orange lines* are for Bernal-type (ABAB and ABAC) stackings, while *blue lines* are for rhombohedral-type stackings. *Bottom*: corresponding constant initial state spectra with the initial energy set to 1 eV binding energy below the Fermi level. The photon energy scale has been converted to perpendicular momentum assuming a free electron final state and an inner potential of 16.5 eV [24]. The sketch on the left hand site indicates the direction of $k_{||}$

Table 1. TB band parameters to reproduce measured bandstructure for $N = 1$–4 layers graphene [24] and $N = \infty$ (graphite, [34, 35]). The electron density n is measured in 10^{-3} electrons per 2D unit cell

N	v ($10^6 \frac{m}{s}$)	n	E_D	E_1	E_2 (eV)	E_3	E_4	γ_1
1	1.10	6.0	−0.44	−0.44				
2	1.05	8.1	−0.30	−0.35	−0.24			0.48
3	1.06	8.0	−0.21	−0.34	−0.16	−0.14		0.44
4	1.06	7.7	−0.15	−0.37	−0.10	−0.06	−0.05	0.44
∞	0.91							~ 0.35

nearest neighbor interaction for the stabilization of Bernal stacking in bulk graphite. The best fit parameters are compiled in Table 1. The total charge in the FLG layer was determined by measuring the size of the Fermi surface.

According to Table 1 the total amount of charge is fairly independent of the layer thickness wich agrees well with the fact that the C 1s core level binding energy of the SiC substrate remains constant for all studied FLG

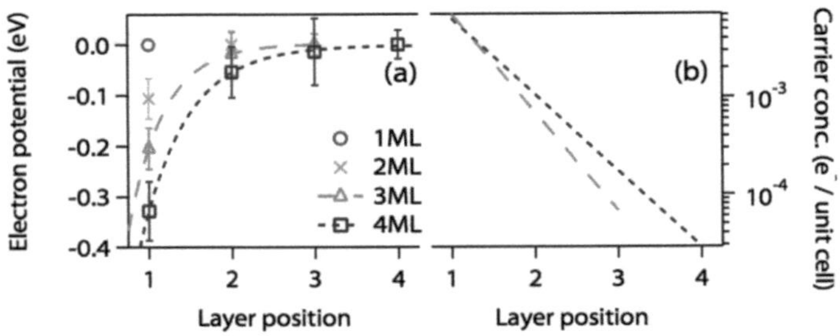

Fig. 5. Potential and carrier concentration profiles of the multilayer graphene as a function of the layer positions [24]. The electron potentials are shifted in such a way that the potential of the outermost graphene layer is at zero

thicknesses. The on-site Coulomb potential decreases from layer to layer in agreement with the observation that the C 1s core level assumes its ideal value after growth of about 5 monolayers of graphene (see above). The experimental E_i values determined by modeling the bands using the tight binding description mentioned above are shown in Fig. 5 together with fitted exponential decay functions. Using the latter we can apply Poisson's equation to determine the charge distribution, which is also shown in that figure. Apparently the charge density decays by about one order of magnitude from one layer to the next. The screening lengths are estimated to be 1.9 and 1.4 Å for triple and quadruple layers, respectively. This is much higher than in graphite (3.5 Å) due to higher charge density in the interface near layers. For the same reason the interlayer hopping integrals γ_1 are larger than that of bulk graphite.

The intensities of the bands of FLG show a characteristic dependence on the photon energy. This is evident from the constant initial state spectra displayed in Fig. 4 [24]. For the monolayer which is a pure 2D system and for which $k_\perp$ is meaningless, a smooth decay of the transition matrix element is observed, once the beamline efficiency profile is normalized out of the data. However, for 2, 3, and 4 monolayers we observe a periodic intensity oscillation with a period in $k_\perp$ that coincides fairly well with the reciprocal interlayer distance of graphite. Similar oscillations are reported for quantized thin films [39, 40] or oscillations of photoemission cross section of surface states [41, 42] with maxima near vertical transitions in the bulk. The absence of similar oscillations for monolayer graphene supports our previous conclusion that the underlying $(6\sqrt{3} \times 6\sqrt{3})R30°$ lacks fully developed π-bands and it corroborates our conclusion that the surface is well passivated by that reconstruction.

3.4 Switching the Gap of Bilayer Graphene

Bilayer graphene is a special case of FLG with intriguing properties such as massive Dirac Fermions and an unconventional quantum Hall effect [5]. In the

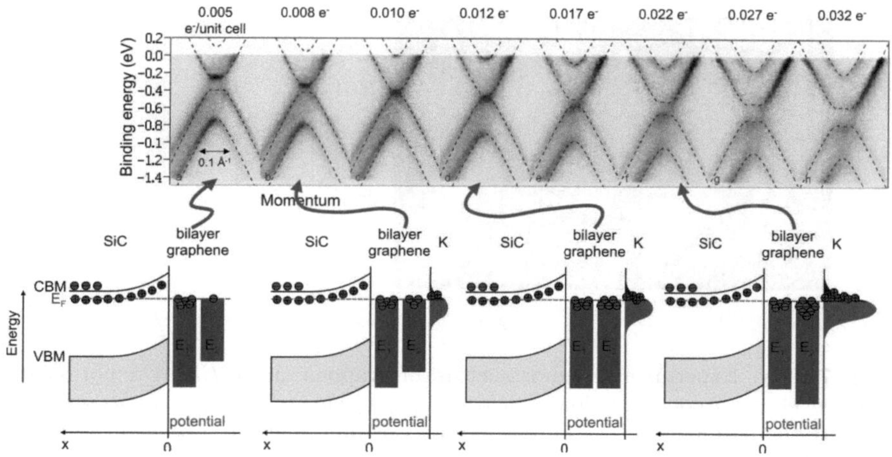

Fig. 6. Gap closing and opening in bilayer graphene induced by postassium adsorption [20]. The experimental bands and fitted tight binding bands [36] are shown for three different electron concentrations which are given at the top. The *left panel* shows the as-prepared bilayer. The sketches illustrate the effect of doping

vicinity of the K-point of the hexagonal Brillouin zone the band structure of bilayer graphene contains four bands [36]. The lower lying bands are degenerate at the K-point but the degeneracy is lifted when the symmetry between the two graphene layers is broken [36]. In order to test this experimentally the electronic structure of bilayer graphene was studied using ARPES [20].

As already discussed in Sect. 3.3 the first few graphene layers are charged by electrons from the substrate. Thereby the electron concentration decreases from layer to layer. Considering a situation with two layers this means that these layers carry different amounts of charge and thus feel different on-site Coulomb potentials E_i as sketched at the bottom of Fig. 6. Consequently, the electronic structure of bilayer graphene is expected to possess a gap at the K-point, which is observed by ARPES, indeed (see Fig. 6). As also sketched in that figure, adsorbed potassium donates electrons into the top layer which eventually balance the charge in the first layer. This leads to gap closing. Finally, an overweight of charge in the top layer reopens the gap. That this effect is at all visible is due to the fact that the screening length is short (see above). As a consequence, a similar effect can be observed for graphite where a surface gap is opened upon alkali metal adsorption [43].

Our results demonstrate that the gap in bilayer graphene can be manipulated in a controlled manner by varying the potential across the layers which could be the key for a switching functionality provided one manages to arrange the potential in such a way that the Fermi level is in the band gap.

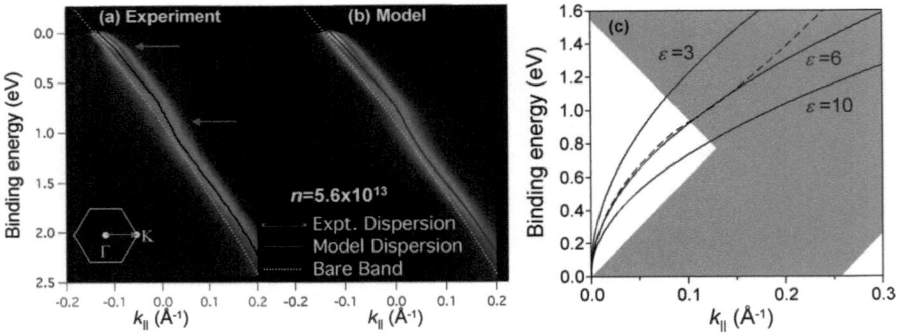

Fig. 7. (**a**) Experimental bandstructure of graphene near E_F at a potassium induced electron concentration of $n = 5.6 \times 10^{13}\,\mathrm{cm}^{-2}$. The *yellow dashed lines* indicate the bare bands. The kink at 200 meV is due to electron-phonon coupling. The second kink shifts with doping concentration and is thus caused by electronic excitations. (**b**) Simulated spectral function. (**c**) Energy diagram of the electronic excitations in graphene [21]. The shaded region shows the possible e-h pair excitations for graphene, computed for a conical band structure with $n = 5.6 \times 10^{13}\,\mathrm{cm}^{-2}$. The lines show plasmon dispersion relations calculated for dielectric constants of 3, 6, and 10 together with the relativisitic mass taken from transport measurements [2, 4]. The calculated plasmon dispersion of *Hwang* et al. ([52], *dashed line*) is shown for comparison

3.5 Many-Body Interactions in Graphene

As mentioned above, the bands of the monolayer are strongly renormalized, i.e., they deviate considerably from the expected linear behavior, due to many-body interactions such as electron-phonon, electron-electron, and electron-plasmon coupling [21]. This is evident for undoped graphene (see Fig. 4) as well as for graphene doped with potassium as shown in Fig. 7a [21]. Such many-body interactions are of interest for understanding the apparent superconductivity in carbon nanotubes [44, 45], alkali-doped C_{60} crystals [46], and graphite intercalation compounds [47–49] for which graphene stands as a model system. Thus, the quasi-particle dynamics in the band structure of as-grown and K-doped graphene was analyzed [21] in the k-independent approximation [50, 51].

It was found that the observed kink structure is derived from a complicated energy dependence of the scattering rate. In order to describe the experimentally determined behavior of the bands it is necessary to include three different decay channels: (i) electron-phonon coupling, which is independent of doping; (ii) electron-hole pair creation; (iii) electron plasmon coupling. The latter is necessary to account for that region of the band structure ($E_D \leq E \leq 2E_D$) where electron-hole pair creation is less favourable due to the small momentum change associated with that decay channel. The electronic excitations show a marked dependence on electron concentration, which is well reproduced within our model [21]. The energy diagram of these excitations is shown in Fig. 7c. The results imply that the fact that electrons in graphene have to be treated as Dirac Fermions does not alter the quasi-particle behavior of the charge carriers. Similar effects are expected for graphite and metallic carbon nanotubes.

4 Conclusions

The present paper provides a summary of our recent photoemission studies of epitaxial graphene and FLG on SiC(0001) surfaces. As expected, this method yields detailed information about multiple aspects of this new material. Although we have just scratched the surface, the results sketched above should prove helpful for further understanding and possible application of epitaxial graphene and FLG on SiC.

References

[1] K. S. Novoselov, A. K. Geim, S. V. Morozov, *et al.*: Science **306**, 666 (2004)
[2] K. S. Novoselov, D. Jiang, F. Schedin, *et al.*: Proc. Natl. Acad. Sci. **102**, 10451 (2005)
[3] K. S. Novoselov, A. K. Geim, S. V. Morozov, *et al.*: Nature **438**, 197 (2005)
[4] Y. Zhang, Y.-W. Tan, H. L. Stormer, P. Kim: Nature **438**, 201 (2005)
[5] K. S. Novoselov, E. McCann, S. V. Morozov, *et al.*: Nature Physics **6**, 177 (2006)
[6] K. S. Novoselov, Z. Jiang, Y. Zhang, *et al.*: Science **315**, 1379 (2007)
[7] A. K. Geim, K. Novolselov: Nature Materials **6**, 183 (2007)
[8] A. J. Van Bommel, J. E. Crombeen, A. Van Tooren: Surf. Sci. **48**, 463 (1975)
[9] L. I. Johansson, F. Owman, P. Mårtensson: Phys. Rev. B **53**, 13793 (1996)
[10] U. Starke, J. Schardt, M. Franke: Appl. Phys. A **65**, 587 (1997)
[11] I. Forbeaux, J.-M. Themlin, J.-M. Debever: Phys. Rev. B **58**, 16396 (1998) doi:10.1103/PhysRevB.58.16396
[12] L. I. Johansson, P. A. Glans, N. Hellgren: Surf. Sci. **405**, 288 (1998)
[13] U. Starke, M. Franke, J. Bernhardt, *et al.*: Mater. Sci. Forum **264–268**, 321 (1998)
[14] I. Forbeaux, J. Themlin, J. Debever: Surf. Sci. **442**, 9 (1999)
[15] C. Berger, Z. M. Song, T. B. Li, *et al.*: J. Phys. Chem. B **108**, 19912 (2004)
[16] C. Berger, Z. M. Song, X. B. Li, *et al.*: Science **312**, 1191 (2006)
[17] W. A. de Heer, C. Berge, X. Wu, *et al.*: cond-mat/0704.0285 (2007)
[18] T. Seyller, K. V. Emtsev, K. Gao, *et al.*: Surf. Sci. **600**, 3906 (2006)
[19] T. Seyller, K. Emtsev, F. Speck, *et al.*: Appl. Phys. Lett. **88**, 242103 (2006)
[20] T. Ohta, A. Bostwick, T. Seyller, *et al.*: Science **313**, 951 (2006)
[21] A. Bostwick, T. Ohta, T. Seyller, *et al.*: Nature Physics **3**, 36 (2007)
[22] T. Seyller, K. Emtsev, F. Speck, *et al.*: Mater. Sci. Forum, in print (2007)
[23] K. V. Emtsev, T. Seyller, F. Speck, *et al.*: Mater. Sci. Forum, in print (2007)
[24] T. Ohta, A. Bostwick, J. McChesney, *et al.*: Phys. Rev. Lett., in print (2007)

[25] U. Starke: Atomic structure of SiC surfaces. In: *Recent Major Advances in SiC*, ed. by W. Choyke, H. Matsunami (Springer Scientific 2003) p. 281
[26] J. Bernhardt, J. Schardt, U. Starke, K. Heinz: Appl. Phys. Lett. **74**, 1084 (1999)
[27] M. H. Tsai, C. S. Chang, J. Dow, I. S. T. Tsong: Phys. Rev. B **45**, 1327 (1992)
[28] J. E. Northrup, J. Neugebauer: Phys. Rev. B **52**, R17001 (1995)
[29] V. Van Elsbergen, T. Kampen, W. Mönch: Surf. Sci. **365**, 443 (1996)
[30] W. Chen, H. Xu, L. Liu, *et al.*: Surf. Sci. **596**, 176 (2005)
[31] S. Latil, L. Henrard: Phys. Rev. Lett. **97**, 036803 (2006)
[32] E. J. Duplock, M. Scheffler, P. J. D. Lindan: Phys. Rev. Lett. **92**, 225502 (2004)
[33] T. Seyller: Appl. Phys. A **85**, 371 (2006)
[34] S. Y. Zhou, G.-H. Gweon, J. Graf, *et al.*: Nature Physics **2**, 595 (2006)
[35] J.-C. Charlier, X. Gonze, J.-P. Michenaud: Phys. Rev. B **43**, 4579 (1991)
[36] E. McCann, V. I. Fal'ko: Phys. Rev. Lett. **96**, 086805 (2006)
[37] E. McCann: Phys. Rev. B **74**, 161403 (2006)
[38] F. Guinea, A. H. C. Neto, N. M. R. Peres: Phys. Rev. B **73**, 245426 (2006)
[39] E. D. Hansen, T. Miller, T. C. Chiang: J. Phys. Condens. Mat. **9**, L435 (1997)
[40] A. Mugarza, J. E. Ortega, A. Mascaraque, *et al.*: Phys. Rev. B **62**, 12672 (2000)
[41] S. G. Louie, P. Thiry, R. Pinchaux, *et al.*: Phys. Rev. Lett. **44**, 549 (1980)
[42] P. Hofmann, C. Søndergaard, S. Agergaard, *et al.*: Phys. Rev. B **66**, 245422 (2002)
[43] M. Pivetta, F. Patthey, I. Barke, *et al.*: Physical Review B **71**, 165430 (2005)
[44] Z. K. Tang, L. Zhang, N. Wang, *et al.*: Science **292**, 2462 (2001)
[45] M. Kociak, A. Y. Kasumov, S. Guéron, *et al.*: Phys. Rev. Lett. **86**, 2416 (2001)
[46] A. F. Hebard, M. J. Rosseinsky, R. C. Haddon, *et al.*: Nature **350**, 600 (1991)
[47] N. B. Hannay, T. H. Geballe, B. T. Matthias, *et al.*: Phys. Rev. Lett. **14**, 225 (1965)
[48] N. Emery, C. Herold, M. d'Astuto, *et al.*: Phys. Rev. Lett. **95**, 087003 (2005)
[49] T. E. Weller, M. Ellerby, S. S. Saxena, *et al.*: Nature Physics **1**, 39 (2005)
[50] A. Kaminski, H. M. Fretwell: New J. Phys. **7**, 98 (2005)
[51] A. A. Kordyuk, *et al.*: Phys. Rev. B **71**, 214513 (2005)
[52] E. H. Hwang, S. Das Sarma: cond-mat/0610561 (2006)

Raman Imaging and Electronic Properties of Graphene

F. Molitor, D. Graf, C. Stampfer, T. Ihn, and K. Ensslin

Laboratory for Solid State Physics, ETH Zurich,
8093 Zurich, Switzerland
ensslin@phys.ethz.ch

Abstract. Graphite is a well-studied material with known electronic and optical properties. Graphene, on the other hand, which is just one layer of carbon atoms arranged in a hexagonal lattice, has been studied theoretically for quite some time but has only recently become accessible for experiments. Here we demonstrate how single- and multi-layer graphene can be unambiguously identified using Raman scattering. Furthermore, we use a scanning Raman set-up to image few-layer graphene flakes of various heights. In transport experiments we measure weak localization and conductance fluctuations in a graphene flake of about 7 monolayer thickness. We obtain a phase-coherence length of about 2 µm at a temperature of 2 K. Furthermore we investigate the conductivity through single-layer graphene flakes and the tuning of electron and hole densities via a back gate.

1 Introduction

The interest in graphite has been revived in the last two decades with the advent of fullerenes [1] and carbon nanotubes [2]. In a pioneering series of experiments it has become possible to transfer single- and few-layer graphene to a substrate [3]. Transport measurements revealed a highly-tunable two-dimensional electron/hole gas which mimics relativistic Dirac Fermions embedded in a solid-state environment [4, 5]. The term "relativistic" describes the linear energy-wave vector relation which gives graphene its exceptional electronic properties. Going to few-layer graphene, however, disturbs this unique system in such a way that the usual parabolic energy dispersion is recovered. The large structural anisotropy makes few-layer graphene therefore a promising candidate to study the rich physics at the crossover from bulk to purely two-dimensional systems. Turning on the weak interlayer coupling while stacking a second layer onto a graphene sheet leads to a branching of the electronic bands and the phonon dispersion at the K point. Double-resonant Raman scattering [7] which depends on electronic and vibrational properties turns out to be an ingenious tool to probe the lifting of that specific degeneracy. Here we show scanning Raman images of graphene flakes and compare them with scanning force images. We evaluate the intensity, position and width of various Raman lines in order to quantify the numbers of monolayers in a given flake. Furthermore we determine the inelastic mean free path in a

R. Haug (Ed.): Advances in Solid State Physics,
Adv. in Solid State Phys. **47**, 171–176 (2008)

few-layer graphene wire and estimate the mobility in a single-layer graphene flake.

2 Raman Spectrum of Graphene

The energy gain and loss of scattered photons is related to the creation and annihilation of phonons at specific points in the phonon spectrum. For graphite the electron is excited from the valence band π to the conduction band π^* close to the K point in the electronic bandstructure. In graphene the unit cell is composed of two atoms leading to 6 phonon branches: Three acoustic branches starting at zero frequency and three optical branches at higher energies. The phonon spectrum of graphite has been calculated (see, e.g., [8]).

In Fig. 1a we start by presenting an image of a graphene flake taken with a scanning force microscope (SFM). The flakes were prepared following the method described in [3]. The number of monolayers is marked by a number in the respective flake area. Figure 2 shows Raman spectra taken with a laser spot within one of the respective areas corresponding to one and two layers of graphene. The Raman spectrum of graphite has four prominent peaks. For a recent review see [6]. The G line is a standard Raman signal arising from the E_{2g} in-plane vibration of the atoms. In first approximation its intensity increases monotonously with the amount of material. The D line (D for defect) usually has a pronounced intensity if a material has defects [10] or is strongly bend like in the case of carbon nanotubes. The absence of the D line in Fig. 2 is already a good indication for the structural quality of our graphene flakes. In the case of the D' line a second phonon excitation instead of an elastic backscattering is required [7]. The D' line is highly sensitive to the underlying energy dispersion. Consequently the D'-line is an ideal indicator to discriminate mono- from double layers in a given flake. The two monolayer areas show a single narrow peak, see Fig. 2, while the double layer area shows a broadened peak with substructure. The width of the D' peak or – at high resolution – its splitting into different sub-peaks is explained in the framework of the double-resonant Raman model [7]. A very similar analysis of the Raman lines of few-layer graphene was presented in [12].

3 Raman Imaging of Graphene

The most prominent difference in the spectra of single-layer, few-layer, and bulk graphite lies in the D' line: the integrated intensity of the D' line stays almost constant, even though it narrows to a single peak at lower wave number at the crossover to a single layer (Fig. 2). The width of the D' line is plotted in Fig. 1b for the same area of the SFM scan in Fig. 1a. The outline of the flake and in particular the difference between the single and double

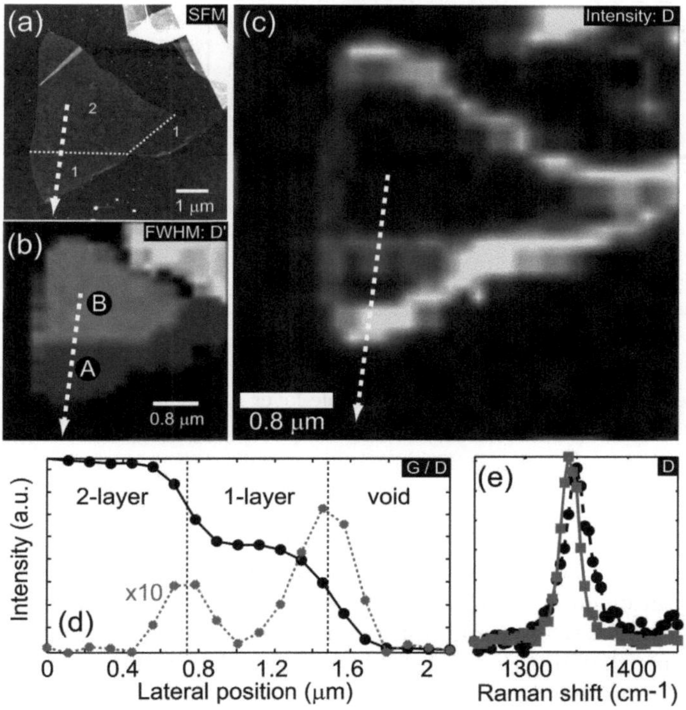

Fig. 1. (**a**) SFM micrograph of a graphitic flake consisting of one double- and two single-layer sections (*white dashed line* along the boundaries), highlighted in the Raman map (**b**) showing the FWHM of the D' line. (**c**) Raman mapping of the integrated intensity of the D line: A strong signal is detected along the edge of the flake and at the steps from double- to single-layer sections. (**d**) Raman cross section (*white dashed arrow* in (c)): Staircase behavior of the integrated intensity of the G peak (*solid line*) and pronounced peaks at the steps for the integrated intensity of the D line (*dashed line*). (**e**) Spatially averaged D peak for the crossover from double to single layer (*disk, dashed line*) and from single layer to the SiO_2 substrate (*square, solid line*). Taken from [9]

layer areas can clearly be observed. We conclude that the width of the D' line is an excellent indicator to identify single-layer graphene.

From cross-correlating the SFM micrograph in Fig. 1a with the Raman map of the integrated D line (1300–1383 cm^{-1}) intensity in Fig. 1c we infer directly that the edges of the flake and also the borderline between sections of different height contribute to the D band signal whereas the inner parts of the flakes do not. This is somewhat surprising since for thinner flakes the influence of a nearby substrate on the structural quality should be increasingly important. In the cross-section Fig. 1d we see clearly that the D line intensity is maximal at the section boundaries, which can be assigned to translational symmetry breaking or to defects. However, we want to emphasize that the

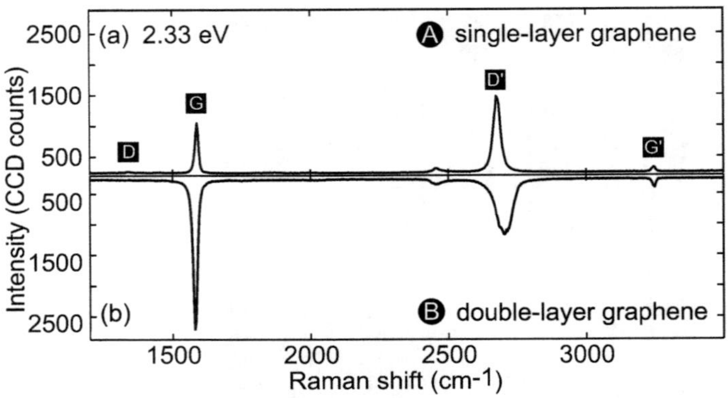

Fig. 2. Raman spectra of (**a**) single- and (**b**) double-layer graphene (collected at spots A and B, see Fig. 1b. Taken from [9]

D line is still one order of magnitude smaller than the G line. In Fig. 1e spatially averaged D mode spectra from the two steps shown in Fig. 1d are presented. The frequency fits well into the linear dispersion relation of peak shift and laser excitation energy found in earlier experiments [11]. In addition, we find that the peak is narrower and down-shifted at the edge of the single layer while it is somewhat broader and displays a shoulder at the crossover from the double to the single layer.

4 Transport Through Few-Layer Graphene

Figure 3 shows transport measurements through a graphitic flake of several µm length, 320 nm in width and about 7 monolayers in height. Ohmic contacts and four in-plane gates as indicated by the gold colored areas were fabricated by electron beam lithography (SFM scan, Fig. 3a). We measured the resistance down to temperatures of 1.7 K as a function of back gate voltage (Fig. 3b) and side gate voltage (Fig. 3c) [13]. All features in the resistance traces are reproducible. The metallic in-plane gates basically change the Fermi energy in a similar way as the homogeneous back gate, however, with a significantly reduced lever arm (Fig. 3d). We envision that in-plane gates could therefore serve well for the electrostatic tuning of nanostructures fabricated on graphene.

We also measured the resistance as a function of magnetic field and found well developed features corresponding to weak localization and conductance fluctuations [13]. The data could be quantitatively analyzed in the framework of diffusive one-dimensional metals. We obtained a phase-coherence length of about 2 µm at a temperature of about 2 K. The temperature dependence is consistent with carrier-carrier scattering being the dominant dephasing mechanism.

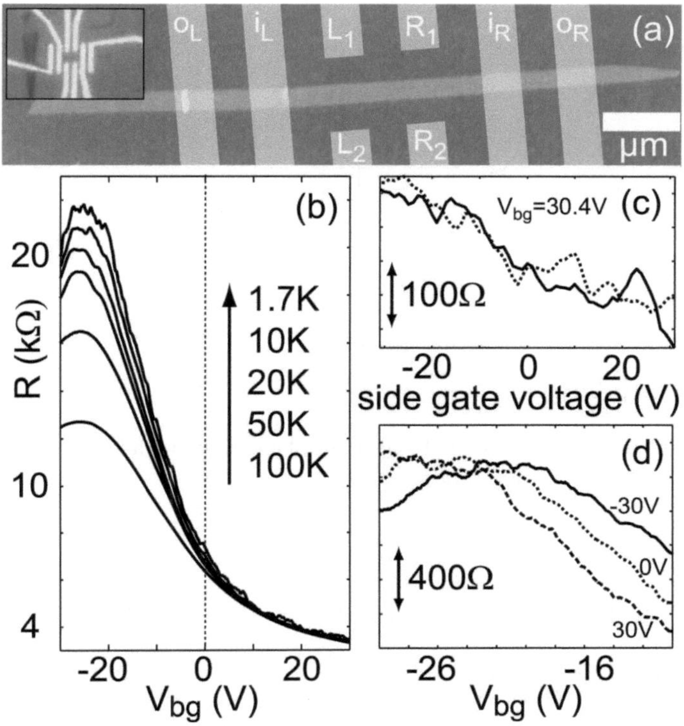

Fig. 3. (**a**) SFM micrograph of a graphite wire resting on a silicon oxide surface with a schematic of the four contacts(iL, iR, oL, oR) and four side gates (L1, L2, R1, R2). *Inset*: Optical microscope image of the structure. (**b**) Four-terminal resistance as a function of back gate for different temperatures. (**c**) Resistance change as a function of the side gates L1+L2 (*solid line*) and R1+R2 (*dotted line*) at 1.7 K. (**d**) Resistance change as a function of back gate for different side gate voltages (L1+L2+R1+R2) at 1.7 K. Taken from [13]

Figure 4 shows the resistance of a single-layer graphene flake. The maximum resistance, i.e., the charge neutrality point, is close to zero back gate voltage indicating that doping is relatively weak. Using a plate-capacitor model we can infer the electron and hole densities for positive and negative gate voltages. We find a maximum mobility of about $7000\,\mathrm{cm}^2/\mathrm{Vs}$.

Acknowledgements

We thank A. Jungen and C. Hierold for a fruitful collaboration during the measurement of the Raman spectra and L. Wirtz for theoretical support. Financial support from the Swiss National Science Foundation and NCCR Nanoscience is gratefully acknowledged.

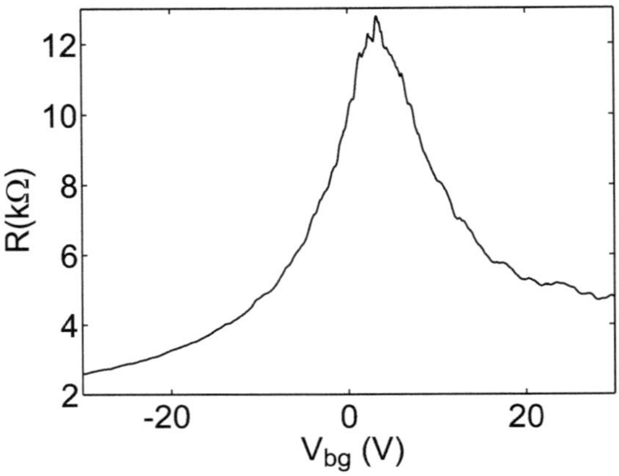

Fig. 4. Two-terminal resistance of a single-layer graphene flake as a function of back gate voltage

References

[1] H. W. Kroto, J. R. Heath, S. C. Obrien, R. F. Curl, R. E. Smalley: Nature **318**, 162 (1985)
[2] S. Iijima: Nature **354**, 56 (1991)
[3] K. S. Novoselov, A. K. Geim, S. V. Morozov, D. Jiang, Y. Zhang, S. V. Dubonos, I. V. Grigorieva, A. A. Firsov: Science **306**, 666 (2004)
[4] K. S. Novoselov, A. K. Geim, S. V. Morozov, D. Jiang, M. I. Katsnelson, I. V. Grigorieva, S. V. Dubonos, A. A. Firsov: Nature **438**,197 (2005)
[5] Y. Zhang, Y.-W. Tan, H. L. Stormer, P. Kim: Nature **438**, 201 (2005)
[6] S. Reich, C. Thomsen: Phil. Trans. R. Soc. Lond. A **362**, 2271 (2004)
[7] C. Thomsen and S. Reich: Phys. Rev. Lett. **85**, 5214 (2000)
[8] L. Wirtz and A. Rubio: Solid State Comm. **131**, 141 (2004)
[9] D. Graf, F. Molitor, K. Ensslin, C. Stampfer, A. Jungen, C. Hierold, and L. Wirtz: Nano Letters **7**, 238 (2007)
[10] F. Tuinstra and J. L. Koenig: J. Chem. Phys. **53**, 1126 (1970)
[11] R. P. Vidano, D. B. Fischbach, L. J. Willis, and T. M. Loehr: Solid State Commun. **39**, 341 (1981)
[12] A. C. Ferrari, J. C. Meyer, V. Scardaci, C. Casiraghi, M. Lazzeri, F. Mauri, S. Piscanec, Da Jiang, K. S. Novoselov, S. Roth, A. K. Geim: Phys. Rev. Lett. **97**, 187401 (2006)
[13] D. Graf, F. Molitor, T. Ihn, K. Ensslin: cond-mat/0702401

Part V

THz-Physics

Interaction of Semiconductor Laser Dynamics with THz Radiation

Carsten Brenner, Stefan Hoffmann, and Martin R. Hofmann

AG Optoelektronische Bauelemente und Werkstoffe, Ruhr-University Bochum, D-44780 Bochum, Germany
martin.hofmann@rub.de

Abstract. We discuss the generation and detection of THz radiation with semiconductor diode lasers. First, we analyze the generation of THz radiation by investigating a semiconductor laser in an external cavity arrangement that supports two color operation with tunable difference frequency. Second, the opposite process, i.e., THz detection with diode lasers is investigated. For that purpose, we inject THz radiation into the active region of a diode laser and analyze its dynamics under this injection. We observe a voltage variation over the p-n-junction depending on the injected THz power and compare the measured signal to the response of a standard Golay cell. Finally, we review our results with particular emphasis on completely diode-laser based THz imaging or spectroscopy systems.

Terahertz (THz) technology has been investigated extensively concerning its potential for numerous applications including spectroscopy, medical imaging, packaging control, security, and many others. Impressive results have been demonstrated in the last decade for many of those research fields in laboratory experiments. But a major breakthrough of THz technology into mass markets has not yet been realized. This is due to the fact that today's state of the art THz technology even if it is the most powerful concept for a specific application is too expensive and too complex. In this article, we review the current THz technology, in particular THz sources and THz detectors. Then we discuss our concepts for THz generation and THz detection with diode lasers and finally we give an outlook to compact diode-laser based THz systems.

1 Introduction: Current THz Technology

The major building blocks of a THz system are the source of THz radiation and the THz detector.

1.1 THz Sources

Terahertz sources might be subdivided into three major categories sorted by the generation process of the radiation.

R. Haug (Ed.): Advances in Solid State Physics,
Adv. in Solid State Phys. **47**, 179–190 (2008)

The first and most straightforward category contains the direct THz sources. In such systems the THz radiation is directly generated by an incoherent or coherent process. The most important incoherent direct THz source is heat radiation. Terahertz sources based on heat radiation are relatively easy accessible and may be used in applications but they provide weak intensity and broadband incoherent emission only.

Coherent direct THz emitters include the free electron laser [1], gas lasers [2], the THz quantum cascade laser [3] and the p-Germanium laser [4]. The free electron laser is an extremely powerful THz radiation source. It provides peak powers of 10^3–10^6 W [1] and even an average power of 20 W has been demonstrated [5] but it introduces enormous effort and is therefore only practical for selected scientific applications. Terahertz gas lasers make use of transitions between rotational levels in gases. Such lasers are usually pumped by CO_2-lasers and provide several mW of output power in the THz range [2]. But THz gas lasers are still complex and expensive and therefore not attractive for industrial mass applications. Semiconductor THz lasers as the THz quantum cascade laser [3] and the p-Germanium laser [4] are, in principle, compact and cost effective. But both concepts require cryogenic cooling which adds complexity and cost. In detail, THz quantum cascade lasers have been demonstrated in the frequency regime between 2 and 4 THz [3, 6, 7] with output powers of up to a few mW [3, 6, 7] and with maximum operation temperatures below 200 K [8]. The tunability of THz quantum cascade lasers is rather limited. The p-Germanium laser, in contrast, is tunable between 1 and 4 THz and provides an enormous peak output power of up to 10 W [9]. But it operates at liquid Helium temperature which requires even more cryogenic effort than cooling of the THz quantum cascade laser.

The second class of THz sources includes concepts with frequency multiplication. Such systems usually use fast electronic oscillators in combination with cascaded nonlinear frequency doubling or tripling units as, e.g., nonlinear Schottky diodes [10]. Such systems provide moderate THz powers in the range of 0.1 mW up to frequencies of about 1 THz with decreasing performance at higher frequencies. A major disadvantage of such concepts is the limited tunability of the order of less than 10 % of the center frequency.

The third class of THz sources is in fact responsible for the fast development of the THz technology in the last decade. It relies on THz generation from optical sources either via parametric generation [11], optical rectification [12], difference frequency generation [13], or photomixing [14]. Parametric THz generation requires a powerful laser source in the optical or near infrared (NIR) range, like a pulsed Nd:YAG laser. The output is sent onto a nonlinear crystal (for example $LiNbO_3$) that converts an optical or NIR photon into a photon with slightly smaller energy *and* a THz photon [11, 15]. Tuning of the THz frequency can be obtained by tilting the crystal. With an additional cavity for the THz photons, such a system can provide peak powers of about 100 mW [11, 15].

Optical rectification can be used in time domain THz systems to create broadband THz pulses. The fundamental idea is to send pulses of a femtosecond NIR laser onto a nonlinear crystal that – roughly speaking – creates the envelope of the optical pulse which corresponds to a broadband THz pulse [16].

Another often used alternative is electrooptic conversion in photoconductive switches: the pulses of a femtosecond NIR laser are sent onto a photoconducting switch that generates a femtosecond current pulse which acts as the source for emission of a broadband THz pulse. A particular advantage of time domain THz systems is the possibility of using electrooptic sampling for detection [12].

This detection concept provides an extremely good signal to noise ratio favorable over other detection concepts as will be discussed in the next section. Time domain THz systems are well established in laboratories and have provided the impressive results in many areas of research we have mentioned above.

However, a time domain THz system requires an expensive femtosecond laser and is complex and expensive. Accordingly, the mentioned impressive laboratory studies cannot be directly transfered into industrial applications.

Photomixing is a cost effective alternative for the generation of continuous wave THz radiation. Usually, two laser sources with well defined frequency separation are either sent onto a nonlinear crystal or, more commonly, onto a photoconducting antenna [14] which creates the difference frequency of the two optical frequencies. Tuning of one of the optical frequencies with respect to the other allows to tune the generated THz frequency in a wide range. This concept is particularly attractive in combination with cost effective diode lasers [17]. A THz system based on photomixing of two diode lasers is probably the most cost effective room temperature THz system today. Some complexity is introduced in photomixing systems based on two laser diodes because their frequencies have to be stabilized with respect to each other. It can be reduced when either monolithical [18] or external cavity two color lasers [19] are used. Monolithical devices offer the advantage of extreme compactness but are only tunable over a limited range while external cavity systems are more complex but tunable almost without restrictions [19].

1.2 THz Detection

For the detection of THz radiation again several different concepts can be used. The most common concepts are bolometers [20], golay cells [21], and heterodyne [22] or homodyne [14] detection concepts. Bolometers are broadband thermal detectors. In most cases, bolometers have to be cooled with liquid Helium to provide good signal to noise ratio for low THz signals. Golay cells are also thermal detectors but operate at room temperature [21]. Both, bolometers and Golay cells provide a broadband sensitivity and usually exhibit rather slow response times due to the thermal detection principle. Both concepts allow to detect THz intensities only.

Heterodyne and homodyne detection principles, in contrast, allow for the detection of THz field amplitudes and usually offer much better signal to noise ratio of the detection at the price of a higher complexity as compared to the thermal detectors.

In heterodyne detectors, the THz signal is usually mixed with the signal of a local oscillator operating at a similar frequency. Both are mixed in a nonlinear element as, for example, a Schottky diode and the difference frequency signal is then analyzed. Heterodyne detection is the method of choice in astronomy [23].

When the THz radiation is generated from optical radiation via one of the concepts mentioned in the last section, homodyne detection has been shown to be the most efficient detection concept. In a time domain spectroscopy or imaging system, the generated THz signal propagates via the sample under investigation and the reflected or transmitted THz signal is sent onto a nonlinear element where the THz signal is mixed with a pulse split off from the femtosecond laser used for the THz generation. When the THz signal is temporally delayed with respect to the laser pulse with an optomechanical delay stage, even the complete THz waveform can be measured [12]. The nonlinear mixing element can either be a nonlinear crystal or a photoconducting receiver antenna. Both concepts, electrooptic sampling with a nonlinear crystal and detection with an antenna, are compared in detail in [16].

A similar homodyne detection principle can be used for continuous wave THz systems. Again, the THz signal transmitted through the sample is mixed with a part of the optical or NIR radiation used for the THz generation in a photoconducting antenna. When both the THz signal and the optical or NIR beat signal hit the antenna, a photocurrent is generated that is directly related to the received THz field signal.

Both homodyne detection principles have a large advantage as compared to the thermal detectors: they are much less sensitive to broadband background signals and exhibit a much better signal to noise performance.

1.3 THz Systems for Industrial Mass Applications

All the discussed THz sources and detectors have their specific advantages and disadvantages. While for scientific applications properties like price and complexity are of minor importance, they are crucial for industrial applications. With this respect concepts involving complex laser systems or cryogenic cooling are much less attractive than compact room temperature systems. With the current state of technology, the best compromise in terms of complexity, price and performance might be a system based on photomixing of a two-color laser in a photoconducting antenna [24] with a homodyne detection scheme.

Such a system however, still exhibits a considerable complexity and will probably not allow to be miniaturized to a mobile handheld system as would be desireable for entering mass markets. Therefore, it is worthwile to search for alternative concepts that provide similar performance but with far reduced complexity and price. In this article, we investigate the potential of diode lasers for generation and detection of THz radiation by analyzing the interaction of THz radiation with the carrier system in diode lasers.

2 Direct Generation of Thz Radiation with Two Color Diode Lasers

The photomixers used in most Terahertz systems are based on semiconductors. This gave rise to the idea that the active region of a semiconductor laser might be operated itself as the nonlinear element performing the photomixing. Before analyzing this effect we like to review the carrier dynamics in semiconductor lasers as a basis for the later discussion of our experimental results.

2.1 Carrier Dynamics in the Semiconductor

The carrier dynamics in semiconductor lasers are extremely rich and take place on timescales of seconds down to several tens of femtoseconds. The dynamic effects have been extensively studied using numerous different experimental approaches. Down to the nanosceond time scale the most effective characterization techniques rely on measuring the electrical modulation response. Typical values of the modulation bandwidth are in the few GHz range, record values reach 30 GHz [25]. In this regime, the carrier dynamics are still dominantly governed by the interband recombination dynamics. However, at high frequencies, massive gain nonlinearities occur due to ultrafast intraband carrier dynamics including effects like carrier heating and spectral hole burning [26]. Such processes typically happen on a time scale of several picoseconds. The dynamics in the picosecond and subpicosecond range cannot be accessed experimentally by electrical characterization methods. The most prominent experimental techniques used to study the ultrafast intraband dynamics in semiconductor lasers were femtosecond pump-probe [27–30] and highly degenerate four wave mixing (FWM) [31] studies on semiconductor laser amplifiers. Non degenerate four wave mixing accesses the ultrafast nonlinear gain dynamics in the frequency domain in contrast to pump-probe studies which are performed in the time domain. In FWM studies, usually two light fields at slightly different frequencies are injected into the semiconductor optical amplifier. The different frequencies beat with each other in the semiconductor amplifier waveguide. If the beat frequency is not too large, the gain of the amplifier is modulated with this frequency due to nonlinearities and four wave mixing sidebands are generated. When the difference

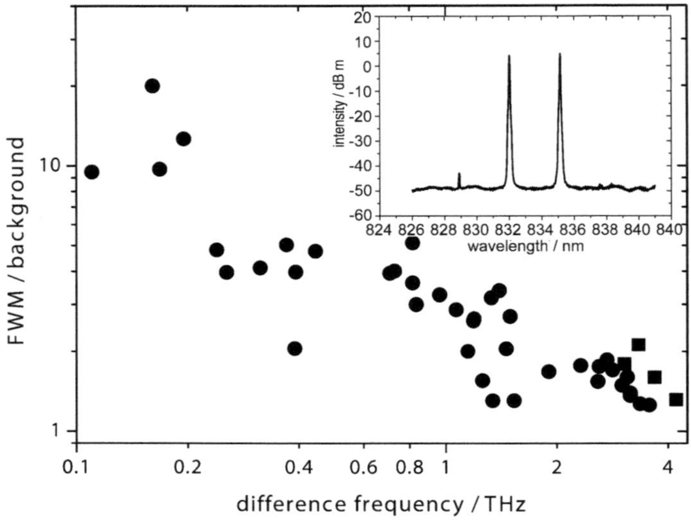

Fig. 1. Four wave mixing response of two color laser as a function of difference frequency. The *inset* shows a typical FWM spectrum

frequency of the two light fields is varied and the intensity of the FWM sidebands is measured as a function of the difference frequency, one can access the ultrafast frequency response of the carrier plasma. The results from FWM experiments, though more difficult to interpret, are qualitatively consistent with those of pump-probe experiments. Four wave mixing responses up to 4.3 THz have been reported [32].

In summary, the carrier plasma in the semiconductor laser amplifier exhibits a rich nonlinear response on picosecond and subpicosecond time scales corresponding to the THz frequency range.

2.2 THz Generation: Proof of Principle

Since the carrier plasma of semiconductor lasers exhibits a nonlinear response up to the THz frequency range one may expect that the ultrafast redistribution of carriers is associated with an acceleration of carriers acting as a source for (incoherent) electromagnetic radiation. Accordingly, a semiconductor amplifier operated in a FWM experiment should emit radiation at the difference frequency of the two light fields. Such a FWM setup would not provide a major advantage in terms of reduced complexity though. A two color laser, in contrast, is much less complex. We have analyzed FWM of the two colors of a two color external cavity laser and some results are shown in Fig. 1.

A typical emission spectrum with FWM sidebands is shown in the inset of Fig. 1. The measured FWM intensity of the high energy FWM sideband is plotted in the main graph as a function of difference frequency. The response

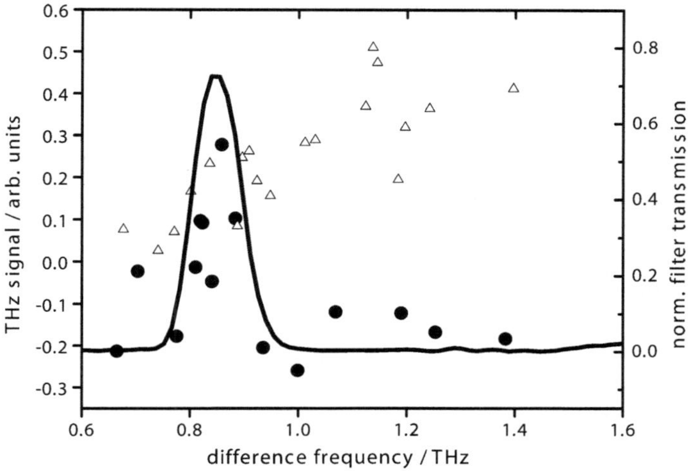

Fig. 2. Normalized THz signal emitted out of the two color laser. (**a**) without bandpass filter (*triangles*). (**b**) with bandpass filter (*circles*). The filter characteristic is also shown by the graph (*right axis*)

decreases rapidly below 1 THz but extends well up to more than 4 THz. This confirms that the carrier plasma in a two color laser also responds in the THz range. Such a two color laser should therefore also emit radiation at the difference frequency of the two colors, i.e., in the THz range. To verify this, we operated the two color laser with variable difference frequencies in the THz range and detected the emitted THz radiation. We used a commercial diode laser in a special external cavity arrangement that supports two color operation with tunable difference frequency [33]. A bolometer acted as a broadband detector for THz radiation and was placed behind the uncoated end facet of the diode laser without additional optical elements except for filters. An absorbing filter was used to suppress the strong near infrared laser emission. The triangles in Fig. 2 show the detected signal as a function of the difference frequency. Note that we plot the normalized difference signal between two color and single color operation [34]. The direct THz emission as well as the observed weak increase of the signal with increasing THz frequency were reproduced by numerical simulations based on a microscopic many body model [34, 35]. To further confirm the direct THz emission, we placed an additional filter between laser diode and bolometer. This was a THz bandpass filter with transmission characteristics shown in Fig. 2 as a straight line (right axis). The measured data (circles) are also shown in Fig. 2. The measured THz signal clearly follows the transmission characteristics of the bandpass filter and finally confirms the direct emission of THz radiation at the difference frequency of our two color laser.

2.3 THz Generation: Potential for Improvement

The THz power emitted directly out of our two color diode laser is in the regime of about 100 pW and thus by far too low for applications [36]. Nevertheless, one has to be aware that the laser diode in use was a commercial device that had by no means been optimized for the generation of THz radiation. We suspect that the propagation of a THz wave through our device is impossible due to the large THz absorption of the highly doped substrate of our diode laser. Accordingly, our THz signal is generated in the close vicinity of the output facet only.

We have analyzed the potential for an improvement of the emitted THz power by technical modifications of the device and the setup in detail [19]. According to this analysis an improvement scheme should necessarily include an improvement of the spectral filter capability of the two color cavity, the reduction of substrate absorption, the use of high power laser diodes and a controlled use of electronic resonances by band structure engineering of the active region. Further improvement can be obtained by adding antenna structures to the device and, in particular, by proper design of the waveguide: in addition to guiding the near infrared wave also the THz wave should be guided. Proper THz waveguide designs might be copied e.g., from those of the THz quantum cascade lasers [3].

The overall improvement potential associated with the listed technical modifications is enormous: we expect that the generation of THz output powers of the order of 10 μW and above are accessible [19]. These values are comparable with the best values obtained by state of the art photomixing systems. Such optimized sources could therefore be extremely attractive for applications. This holds in particular, when they are combined with effective detection concepts.

3 Detection of THz Radiation with Diode Lasers

Our discussions in the previous sections have shown that the carrier system of diode lasrs interacts with THz radiation. Accordingly, one might expect that diode lasers do not only allow for the generation of THz radiation but also are sensitive to the injection of THz radiation. If this sensitivity leads to a variation of the laser diode properties, one could make use of it to design a THz detection scheme. In the following section we present such a scheme.

3.1 Electronical Detection

When the free carrier system of a semiconductor laser absorbs injected THz radiation the carrier distribution functions will necessarily be modified. Any changes of the carrier distribution functions can be sensitively detected by

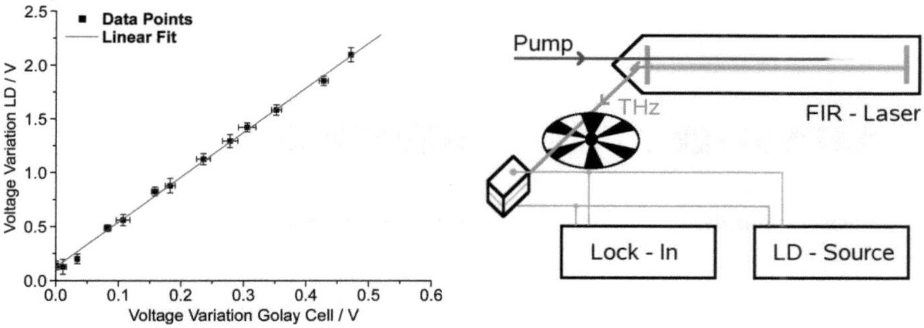

Fig. 3. THz detection: (**a**) Measured voltage variation over the p-n-junction as a function of THz power as measured with a reference Golay cell (*left*). (**b**) The setup for electronic detection of THz radiation with diode lasers (*right*)

voltage variations over the p-n-junction of the laser diode. In quasi equilibrium (without THz injection) the voltage over the p-n-junction is the applied voltage which is equal to the quasi Fermi level splitting divided by the elementary charge e. Variations of the quasi Fermi level splitting due to THz injection, for example, should therefore lead to voltage variations over the p-n-junction which can be electronically detected. To demonstrate that this concept for electronic THz detection with a diode laser works in principle, we injected a THz beam from a THz gas laser pumped by a CO_2-laser into the active region of a diode laser [37]. This is shown schematically in Fig. 3b.

The THz beam is focused onto the active region of a commercial laser diode. The laser diode is operated with a constant current source and the voltage variations over the p-n-junction upon THz injection are measured in lock-in technique.

The measured voltage variation as a function of injected THz power is shown in Fig. 3a. We find a linear dependence that clearly confirms the room temperature THz detection capability of the laser diode [37]. More detailed experimental and theoretical studies [38] reveal that the physical mechanisms responsible for this THz detection are complex. They will be analyzed in detail elsewhere. In the following we restrict the discussion to the application potential of this effect.

3.2 Perspectives for Improvement of the Detection Scheme

The injected THz power used for the data in Fig. 3a ranges from several mW up to a few tens of mW. To make this scheme attractive for applications, the efficiency of the detection should be improved at least by three orders of magnitude. Then, THz signals generated with photomixing or by direct emission out of an optimized two color laser could be detected and a really compact THz system could be realized. Though the physical mechanisms responsible for the detection have not yet been completely analyzed and understood a

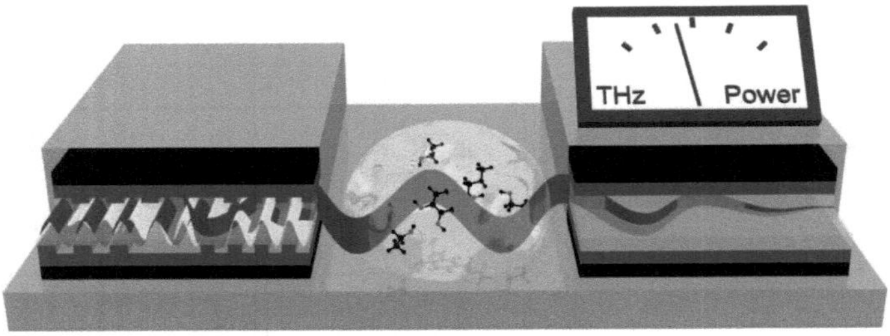

Fig. 4. Compact room temperature diode laser based THz spectroscopy system

few aspects for a possible improvement are quite straightforward. Again, the THz absorption in the substrate is a severe problem because it does not allow propagation of the injected THz field through the device. Most of the injected power is absorbed in the substrate and does not contribute to the detected signal. Replacement of the highly doped substrate by a semi insulating substrate is again the method of choice to remove this problem and promises the same improvement factor of the order of hundred as for the THz generation. In addition, the introduction of an additional THz waveguide into the device and, in particular, the implementation of an antenna for optimum coupling of the injected THz radiation into the active region promise further considerable improvement. In summary, the required improvement factor of three orders of magnitude for the detection sensitivity is realistic and might even be increased electronically by using specifically stabilized current sources for the diode laser.

4 Conclusions and Outlook

We have shown that two color lasers emit THz radiation at the difference frequency of the two colors and that diode lasers can detect THz radiation. Both effects were weak though and have to be optimized in order to become attractive for applications. Our analysis has provided an enormous potential for improvement of both effects provided that considerable technological effort is put into the appropriate device design. The combination of an according improvement of the emitted THz power with the improvement of the detection efficiency would rise the possibility to construct a compact diode laser based THz system for spectroscopic applications. Such a system is shown schematically in Fig. 4.

The THz source consists of a monolithic two color laser. It could be realized with a multiple section device with two Bragg sections supporting the two different wavelengths. An antenna structure is implemented to provide optimal outcoupling of the generated radiation. The emitted THz radiation is sent through a sample. The transmitted THz radiation is then coupled into a second laser diode with a detection antenna and the voltage variation over its active region is detected as a measure of the incoming THz radiation. The price of such a room temperature THz system would only be a few thousands of Dollars, i.e., a small fraction of the price for a state of the art time-domain THz system. Provided that the technological improvements for the THz sources and the THz detectors are successfully realized, such a concept promises to bring THz technology into our everyday life.

Acknowledgements

This work would not have been possible without the support of numerous collaborators. The laser diode devices were provided by A. Klehr, G. Erbert, and G. Tränkle from the Ferdinand Braun Institute in Berlin, Germany. The THz generation experiments were performed in cooperation with S. Saito, K. Sakai (NICT, Japan) and E. Bründermann from the University of Bochum. The THz detection studies were done at the Technical University of Braunschweig in cooperation with M. Salhi and M. Koch. In addition, we thank J. T. Steiner, M. Kira and S. W. Koch from the University of Marburg for theoretical support and stimulating discussions. Financial support by the Japan Society for the Promotion of Science (JSPS), the German Science Foundation (DFG) and by the Stiftung Industrieforschung is gratefully acknowledged.

References

[1] M. Sherwin, Nature **420**, 131 (2002)
[2] P. Siegel, IEEE Trans. Microwave Theory Tech. **50**(3), 910 (2002)
[3] R. Köhler, A. Tredicucci, C. Mauro, F. Beltram, H. E. Beere, E. H. Linfield, A. G. Davies, D. A. Ritchie, Appl. Phys. Lett. **84**, 1266 (04)
[4] E. Bründermann, D. R. Chamberlin, E. E. Haller, Appl. Phys. Lett. **76**(21), 2991 (2000)
[5] G. Carr, M. Martin, W. McKinney, K. Jordan, G. Neil, G. Williams, Nature **420**, 153 (2002)
[6] B. Williams, S. Kumar, Q. Hu, J. Reno, Electron. Lett. **40**, 431 (2004)
[7] B. S. Williams, H. Callebaut, S. Kumar, Q. Hu, Appl. Phys. Lett. **82**, 1015 (2003)
[8] B. Williams, S. Kumar, Q. Hu, J. Reno, Opt. Express **13**, 3331 (2005)
[9] E. Bründermann, in: *Long-wavelength Infrared Semiconductor Lasers* (Wiley and Sons, New York, 2004)
[10] T. Crowe, W. Bishop, D. Porterfield, J. Hesler, Proceedings of the joint 29th Int. Conf. on Infrared and Millimeter Waves and 12th Int. Conf. on Terahertz Electronics p. 85 (2004)
[11] K. Kawase, J. Shikata, H. Ito, J. Phys. D, Appl. Phys. **35**, R1 (2002)
[12] K. Sakai (ed.), *Terahertz Optoelectronics*, *Topics in Applied Physics*, vol. 97 (Springer, 2005)
[13] Y. Sasaki, A. Yuri, K. Kawase, H. Ito, Appl. Phys. Lett. **81**, 3323 (2002)
[14] S. Verghese, K. McIntosh, S. Calawa, W. Dinatale, E. Duerr, K. Molvar, Appl. Phys. Lett. **73**(26), 3824 (1998)

[15] B. Saleh, *Fundamentals of Photonics* (John Wiley and Sons Inc., New York, 1991)
[16] Y. Cai, I. Brener, J. Lopata, J. Wynn, L. Pfeiffer, J. Stark, Q. Wu, X. Zhang, J. Federici, Appl. Phys. Lett. **73**(4), 444 (1998)
[17] M. Tani, P. Gu, M. Hyodo, K. Sakai, T. Hidaka, Optical and Quantum Electronics **32**, 503 (2000)
[18] S. Roh, T. Yeoh, R. Swint, A. Huber, J. Woo, J. Coleman, IEEE Phot. Tech. Lett. **12**, 1307 (2000)
[19] S. Hoffmann, M. R. Hofmann, Laser and Photon. Rev. **1**, 44 (2007)
[20] P. Richards, J. Appl. Phys. **76**(1), 1 (1994)
[21] P. Golay, Rev. Sci. Instr. p. 357 (1946)
[22] H. Hartfuss, T. Geist, M. Hirsch, Plasma Phys. Control. Fusion **39**, 1693–1769 (1997)
[23] O. Hachenberg, B. Vowinkel, *Technische Grundlagen der Radioastronomie* (Bibliographisches Institut, 1982)
[24] T. Kleine-Ostmann, P. Knobloch, M. Koch, S. Hoffmann, M. Breede, M. Hofmann, G. Hein, K. Pierz, M. Sperling, K. Donhuijsen, Electronics Letters **37**(24), 1461 (2001)
[25] Y. Matsui, H. Murai, S. Arahira, S. Kutsuzawa, Y. Ogawa, IEEE Photon. Technol. Lett. **9**, 25 (1997)
[26] R. Nagarajan, M. Ishikawa, T. Fukushima, R. S. Geels, J. E. Bowers, IEEE J. Quantum Electron. **28**, 1992 (1992)
[27] M. Stix, M. Kesler, E. Ippen, Appl. Phys. Lett. **48**, 1722 (1986)
[28] K. L. Hall, G. Lenz, E. P. Ippen, U. Koren, G. Raybon, Appl. Phys. Lett. **61**, 2512 (1992)
[29] J. Mark, J. Mørk, Appl. Phys. Lett. **61**, 2281 (1992)
[30] C. K. Sun, H. K. Choi, C. A. Wang, J. G. Fujimoto, Appl. Phys. Lett. **62**, 747 (1992)
[31] A. Mecozzi, J. Mørk, IEEE J. Selected Topics of Quantum Electronics **3**, 1190 (1997)
[32] A. D'ottavi, E. Iannone, A. Mecozzi, S. Scotti, P. Spano, R. Dall-Ara, G. Guekos, J. Eckner, Appl. Phys. Lett. **65**, 2633-2635 (1994)
[33] M. Breede, S. Hoffmann, J. Zimmermann, J. Struckmeier, M. Hofmann, T. Kleine-Ostmann, P. Knobloch, M. Koch, J. Meyn, M. Matus, S. Koch, J. Moloney, Opt. Comm. **207**, 261 (2002)
[34] S. Hoffmann, M. Hofmann, M. Kira, S. Koch, Semiconductor Science and Technology **20**, 205 (2005)
[35] M. Matus, M. Kolesik, J. Moloney, M.Hofmann, S. Koch, JOSA B **21**, 1758 (2004)
[36] S. Hoffmann, M. Hofmann, E. Bründermann, M. Havenith, M. Matus, J. V. Moloney, A. S. Moskalenko, M. Kira, S. W. Koch, S. Saito, K. Sakai, Appl. Phys. Lett. **84**(18), 3585 (2004)
[37] C. Brenner, S. Hoffmann, M. R. Hofmann, M. Salhi, M. Koch, in *CLEO/QELS* (2006), p. CTuL6
[38] J. T. Steiner et al. in this issue.

Ultrafast THz Spectroscopy of Excitons in Multi-Component Carrier Gases

R. A. Kaindl[1], M. A. Carnahan[1], D. Hägele[1,2], and D. S. Chemla[1]

[1] Department of Physics, University of California at Berkeley and Materials Sciences Division, E. O. Lawrence Berkeley National Laboratory, 1 Cyclotron Road, Berkeley, CA 94720, USA
rakaindl@lbl.gov

[2] present address: Ruhr-Universität Bochum, Universitätsstraße 150, 44780 Bochum, Germany

Abstract. We discuss time-resolved experiments that study photo-excited *e-h* pairs in GaAs quantum wells via their THz response. Resonant generation of excitons leads to characteristic THz spectra dominated by 1*s*-2*p* intra-excitonic transitions, which provide a direct density gauge. Fundamental differences between the intra-excitonic and free-carrier conductivity enable quantitative analysis of multi-component *e-h* gases. We examine exciton ionization dynamics, where the exciton fraction approaches a quasi-equilibrium value in agreement with the Saha equation. A Saha model that takes into account thermal excitation of both heavy- and light-hole pairs is derived. Such experiments identify non-equilibrium phases of *e-h* gases and their underlying quasi-particle densities and dynamics.

1 Introduction

Collective properties of complex many-body systems are among the most prevalent topics of condensed matter physics today. Coulomb interactions can lead to low-energy quantum states, with exciton gases in semiconductors and Cooper-pair condensates in superconductors as prominent examples. In such materials, important microscopic interactions occur on ultrafast timescales and arise from the same forces responsible for the ground state many-body interactions. This has motivated numerous time-resolved experiments using femtosecond visible or near-IR pulses to photoexcite non-equilibrium states and directly resolve relaxation and dephasing dynamics in solids [1, 2].

Such optical light pulses are, however, not resonant to the numerous fundamental low-energy excitations around the Fermi energy. These excitations are commensurate with the thermal energy scale and thus most revelant for low-temperature condensed matter properties. Electromagnetic waves with corresponding photon energies are in the terahertz (THz) spectral range, since 1 THz corresponds to $\hbar\omega \approx 4.1\,\text{meV}$. Recent time-resolved experiments have started to explore ultrafast changes in the THz electromagnetic response of complex materials [3–10]. This is achieved via optical-pump THz-probe

R. Haug (Ed.): Advances in Solid State Physics,
Adv. in Solid State Phys. **47**, 191–202 (2008)

schemes that allow for direct field-resolved detection of the electromagnetic waves.

In this paper, we review aspects of recent THz studies of exciton dynamics in GaAs quantum wells [6, 11–13]. Absorption and luminescence around the semiconductor bandgap is often employed for exciton spectroscopy, though only a fraction of exciton states around center-of-mass momentum $K \approx 0$ couple strongly to near-IR photons [1]. Moreover, the currently active search for excitonic Bose–Einstein condensates focuses on "dark excitons" with a vanishing interband dipole moment [14, 15]. Therefore, determining total exciton densities from luminescence is difficult. In contrast, "intra-excitonic" transitions between exciton levels at THz frequencies represent a fundamentally different exciton probe. Intra-excitonic transitions are not influenced by interband dipole moments, and can occur for all excitons irrespective of their center-of-mass momenta. This has strongly motivated the use of THz probes as new tools for exciton spectroscopy, as supported by microscopic models [16]. While only few experiments were reported early on [17–19], recent advances in ultrafast THz and mid-IR techniques provide powerful new ways for intra-excitonic spectroscopy of exciton populations and kinetics [6, 15].

2 Optical-Pump THz-Probe Spectroscopy

The experiments described here are carried out using a customized optical-pump THz-probe experimental setup operating at 250-kHz repetition rate. At the outset, a Ti:sapphire regenerative amplifier (Coherent RegA) delivers 150-fs pulses at 800 nm wavelength with 7 μJ pulse energy. About 80 % of the output is used as a pump beam after attenuation and tuning of center wavelength and bandwidth. Generation of sub-ps THz probe pulses covering the 2–12 meV spectral range is achieved with the remaining amplifier output via optical rectification in a 500-μm thick ZnTe crystal. The THz beam is focused through the sample and refocused on a second, 500-μm thick ZnTe crystal for electro-optic detection. The coherent THz electric field $E(t)$ is then sampled directly in the time domain by a co-propagating fs pulse that measures the Pockels-effect polarization retardance [20]. The multi-kHz amplified scheme allows for an ideal balance between pump power (needed to fill out the large THz probe foci) and sensitivity, where field detection with a signal-to-noise ratio exceeding 10^4 is routinely achieved.

Figure 1a illustrates the timing of signals in the experiment. The sample's electronic system exhibits a dynamics evolving with pump-probe delay Δt after photo-excitation. At a given timepoint Δt, the equilibrium THz dielectric function $\epsilon(\omega)$ of the sample has changed by an amount $\Delta\epsilon(\omega)$. The photoinduced change of the THz field $\Delta E(t)$ is sampled by changing the generation time delay t with respect to the electro-optic sampling pulse. The delay between the pump pulse and the electro-optic sampling pulse remains fixed

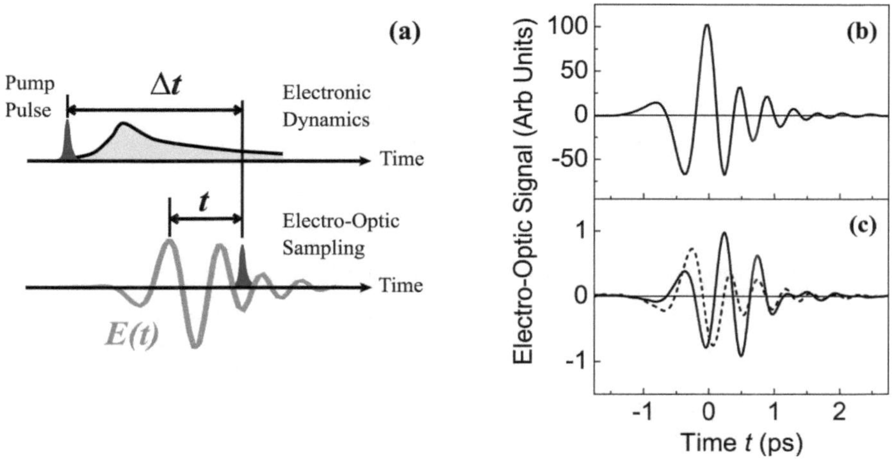

Fig. 1. (**a**) Optical-pump THz-probe scheme with definitions of pump-probe delay Δt and electro-optic sampling time t. (b),(c) Electro-optically detected THz signals in transmission through GaAs multiple quantum wells at $T = 60\,\mathrm{K}$: (**b**) Full transmitted THz electric field without photo-excitation. (**c**) Photo-induced change of the THz field at $\Delta t = 1\,\mathrm{ps}$ (*solid line*) and 40 ps (*dashed line*)

and, hence, $\Delta\epsilon(\omega)$ is sampled at this fixed pump-probe delay Δt [21]. Typical THz signals from the experiment are shown in Fig. 1b,c. Both the direct field $E(t)$ without photo-excitation (Fig. 1b) and its photo-induced change $\Delta E(t)$ (Fig. 1c) are recorded. A Fourier transform of the time-domain traces then yields the complex-valued transmission ratio

$$t^*(\omega) \equiv \frac{E(\omega)}{E(\omega) + \Delta E(\omega)} \tag{1}$$

in the frequency domain. The photo-induced change is obtained by expressing t^* in the transfer matrix approach for a given sample geometry and numerically solving for $\Delta\epsilon(\omega)$ [22]. In the following, the response will be written in terms of the optical conductivity

$$\sigma(\omega) = \sigma_1(\omega) + i\omega\epsilon_0[1 - \epsilon_1(\omega)]\,, \tag{2}$$

which represents the current response $J = \sigma E$ of the many-body system to the incident THz field. The real part of the conductivity, $\sigma_1(\omega)$, measures the absorbed power density. The real part of the dielectric function, $\epsilon_1(\omega)$, in turn represents the inductive out-of-phase response. The concurrent availability of both parts of the response is crucial to separate different parts of a multi-component carrier response.

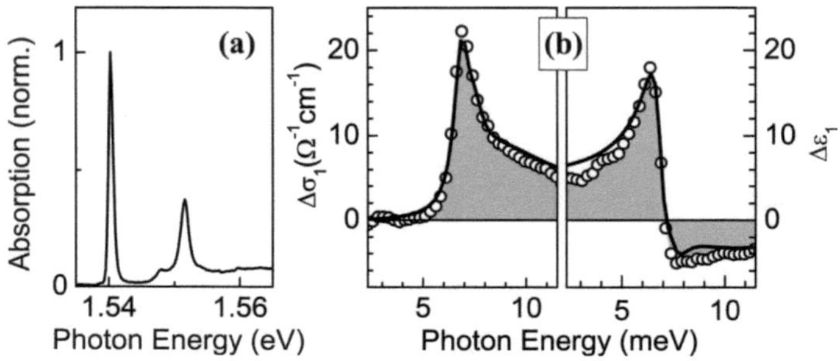

Fig. 2. Intra-excitonic spectroscopy of GaAs quantum wells at $T = 6\,\text{K}$. (**a**) Near-IR absorption spectrum. To avoid undesirable excitation of the GaAs substrate, it was etched off after attaching the QW side to a MgO substrate. (**b**) THz response after resonant $1s$ HH excitation. *Open dots*: induced conductivity and dielectric function change at $\Delta t = 10\,\text{ps}$. *Lines*: intra-excitonic response calculated for exciton density $n_X = 2.7 \cdot 10^{10}\ \text{cm}^{-2}$, with $\lambda = 0.678$, $\varepsilon_s = 13.2$, and $\mu = 0.054\,m_0$ (see text)

3 Intra-Excitonic Transitions

In this section, we discuss THz spectra of resonantly excited quasi-2D exciton gases in GaAs multiple quantum wells. The investigated sample contains ten undoped 14-nm thick wells separated by 10-nm thick AlGaAs barriers. The low-temperature near-IR absorption spectrum is shown in Fig. 2a. It is dominated by the $1s$ heavy-hole (HH) exciton peak at $1.54\,\text{eV}$, followed with increasing photon energy by the $2s$-HH exciton peak, the onset of the interband continuum and the $1s$ light-hole (LH) exciton. Figure 2b shows the THz response after resonant excitation at the $1s$ HH line, for a lattice temperature $T = 6\,\text{K}$. The observed conductivity change $\Delta\sigma_1$ exhibits a strong peak around $7\,\text{meV}$ followed by a higher-energy shoulder. A strongly dispersive dielectric function change $\Delta\epsilon_1$ appears at the same energy, consistent with the far-IR oscillator that has been created. After dephasing of interband polarizations within a few ps, the THz response is explained by intra-excitonic transitions of a $1s$-HH population. In particular the peak around $7\,\text{meV}$ arises from transitions between the $1s$ and $2p$ exciton levels.

For quantitative analysis, we model the intra-excitonic THz dielectric response $\epsilon(\omega)$ in absolute units. Summing all dipole transitions between the $1s$ level to higher bound and continuum final states yields the induced dielectric response

$$n_X \Delta\epsilon_X(\omega) = n_X \frac{2e^2}{\hbar\varepsilon_0 d_{QW}} \sum_f \frac{\omega_{1s,f}}{\omega_{1s,f}^2 - \omega^2 - i\omega\Gamma_f} \left|\langle\psi_f| x |\psi_{1s}\rangle\right|^2 . \tag{3}$$

Here, n_X is the 2D exciton sheet density which is multiplied by the response of a single exciton, $\Delta\epsilon_X(\omega)$. The quantum well width is given by d_{QW}. For the $1s$ ground state and final state wavefunctions (ψ_{1s} and ψ_f) we follow the approach of *Ekenberg* [23] which accounts for the fractional, quasi-2D dimensionality of the confined carriers by scaling the Coulomb interaction with a parameter λ. The exciton reduced mass μ and the static dielectric constant ε_s also enter the eigenenergies and wavefunctions. The choice of λ simply follows the condition that the known binding energy (or $1s$-$2p$ spacing) must be reproduced. In the above equation $\omega_{1s,f} \equiv \omega_f - \omega_{1s}$ are the transition energies and Γ_f are the broadening parameters for transitions between $1s$ and the final states f. All bound-bound transitions are broadened with a parameter Γ_b to reproduce the observed line width. Transitions to continuum states are broadened with $\Gamma_c = 2\,\text{meV}$ to account for enhanced continuum dephasing, in agreement with typical Drude widths observed for non-resonantly excited free carriers. As evident from Fig. 2b, the calculation (solid lines) agrees very closely with the experimental results. In particular, the calculation confirms that the peak stems from the $1s$-$2p$ transition while the shoulder arises from transitions into higher levels and the continuum.

We emphasize that a central advantage of measuring THz-frequency transport lies in the ability to extract absolute carrier densities. Our model confirms (as expected in a parabolic band picture) that the partial oscillator strength sum rule

$$\int_0^{E_C} \sigma_1(\omega) d\omega = \frac{n}{\mu} \cdot \frac{\pi e^2}{2 d_{QW}} \tag{4}$$

holds true, where n is the 2D pair density and E_C is a cut-off energy just below the interband transitions. Hence, the pair density n can be determined directly from the area underneath $\sigma_1(\omega)$ for a given reduced mass which is well known in many semiconductor materials.

4 Exciton Ionization and Thermal Quasi-Equilibria

The THz response and dynamics strongly depend on the lattice temperature. Figure 1c shows the induced THz field change $\Delta E(t)$ at $T = 60\,\text{K}$ for two different time delays after resonant excitation at the $1s$-HH exciton line[1]. As is clear from Fig. 1c, significant reshaping and a phase shift occurs as the delay is increased from $\Delta t = 1\,\text{ps}$ (solid line) to 40 ps (dashed line). Corresponding THz spectra are shown in Fig. 3a. The conductivity spectra are well described by an intra-excitonic lineshape immediately after excitation ($\Delta t = 1\,\text{ps}$), but they broaden notably with increasing delay as conductivity builds up in the low-frequency region. The rate and depth of transformation increases strongly with increasing lattice temperature (not shown). At the same time,

[1] note that the near-IR exciton peak wavelength shifts with temperature; we monitor its position in transmission through the sample with a fiber spectrometer in order to optimize the pump pulse center wavelength.

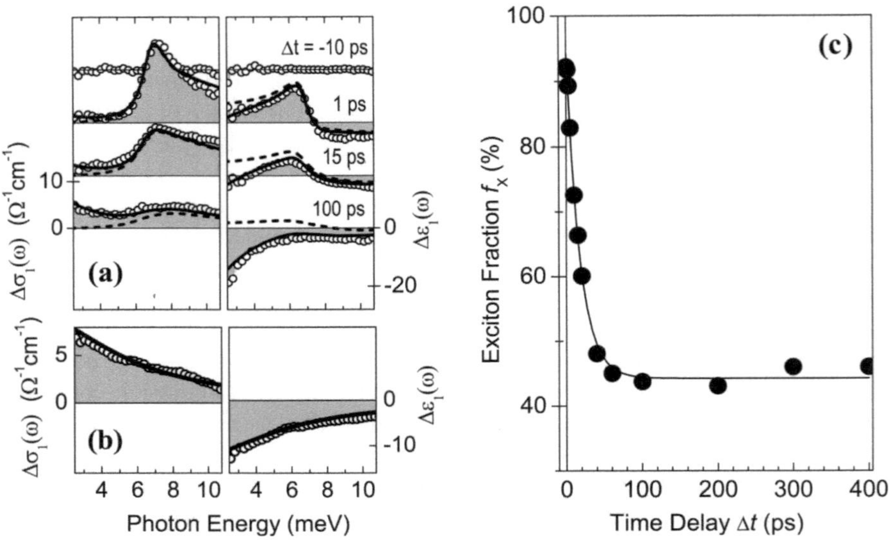

Fig. 3. Exciton ionization. (**a**) Measured THz response (*open dots*) at 60 K for different time delays Δt after resonant 1*s*-HH excitation. *Solid lines*: two-component model (see text), *dashed line*: exciton model alone. The initial density is $n = 3 \times 10^{10}\ \mathrm{cm}^{-2}$. In each panel, curves are shifted vertically yet scaled equally. (**b**) THz response (open dots) after continuum excitation at 300 K, and Drude fit (*line*). (**c**) Exciton fraction at 60 K determined from two-component fits for different Δt

$\Delta\epsilon_1$ flattens out and approaches a purely negative, $-1/\omega^2$ functional shape. These features at long delay times are the hallmark of a Drude response. For comparison, spectra obtained after continuum excitation at 300 K (to assure minimal exciton survival) are well described by a pure Drude response

$$n_{eh}\Delta\epsilon_{eh}(\omega) = n_{eh}\left(\frac{e^2}{\mu\varepsilon_0 d_{QW}} \cdot \frac{-1}{\omega^2 + i\omega\Gamma_D}\right) \tag{5}$$

as shown in Fig. 3b. Here, n_{eh} and Γ_D are the free carrier sheet density and Drude broadening, respectively. Given that the spectra in Fig. 3a approach this Drude-like shape with increasing delay time, the dynamics is explained by exciton ionization via LO phonon absorption. However, even at the longest delays the response at 60 K is not purely Drude-like but still contains a broad excitonic peak.

The Drude and excitonic models on their own are obviously insufficient to reproduce the response, as illustrated by the intra-excitonic fit in Fig. 3a (dashed lines). To quantitatively describe the ionization process, we need a model which can describe the multi-component response. In the following, we approximate the dielectric response of the fairly dilute mixture of excitons and unbound pairs by a two-component model

$$\Delta\epsilon(\omega) = n_X\Delta\epsilon_X(\omega) + n_{eh}\Delta\epsilon_{eh}(\omega)\ . \tag{6}$$

Fits with this model to the THz spectra in Fig. 3a are shown as solid lines, and exhibit excellent agreement with the experiment. The parameters are strongly constrained due to fundamental spectral differences in the response of excitons and free carriers, and by the need to explain $\Delta\sigma_1$ and $\Delta\epsilon_1$ simultaneously over a broad energy range.

Equipped with the knowledge of n_{eh} and n_X from the above model fits, we can calculate the exciton fraction $f_X \equiv n_X/(n_{eh} + n_X)$ of the multi-component *e*-*h* gas. The dynamics in Fig. 3c reveal that f_X decays with increasing pump-probe delay and reaches a quasi-equilibrium value at long times. Repeating this for datasets at different lattice temperatures produces the curve in Fig. 4 (diamonds) which indicates the steady-state exciton fraction as a function of lattice temperature.

We can model the quasi-equilibrium between excitons and free carriers with the well-known Saha equation. It describes a thermal statistical equilibrium of a system consisting of bound and unbound pairs including their center-of-mass dispersion. For a 2D *e*-*h* gas, it reads

$$\frac{n_{eh}^2}{n_X} = \frac{k_B T}{2\pi\hbar^2}\,\mu\, e^{-E_b/k_B T}, \tag{7}$$

where E_b is the exciton binding energy. Equation (7) can be solved for n_X and hence f_X given the total pair density $N = n_{eh} + n_X$. As shown in Fig. 4 (solid line), a pair density of $N = 2 \times 10^{10}$ cm^{-2} yields excellent agreement with the experiment. At the same time, this density agrees well with the *absolute* value from the two-component fits at these delay times (not shown).

The confluence of theory and experiment corroborates our approach of using the two-component model to analyze the response of a dilute mixture of excitons and free carriers, and the absolute densities derived. Even beyond any specific model, it seems worth to re-emphasize the stark differences in the reponse of excitons and free carriers which allow for analysis of their share in the *e*-*h* gas. By definition, an exciton is a bound state. Due to the correlated motion of electron and hole in the charge-neutral pair, the exciton response is insulating and thus exhibits a vanishing low-frequency conductivity. Any conductivity far below the 1*s*-2*p* transition therefore must arise from other excitations, e.g., from free carriers. Secondly, the dielectric function change $\Delta\epsilon_1$ is always positive below the center frequency of an oscillator (the 1*s*-2*p* transition in the case of excitons), while it is always negative for a Drude response. Therefore, the sign $\Delta\epsilon_1$ at these frequencies directly indicates the presence of a predominant species.

5 Multicomponent Saha Equation

An interesting question concerns the extent to which light-hole excitons (and unbound electron – light hole pairs) will modify the above simplified Saha

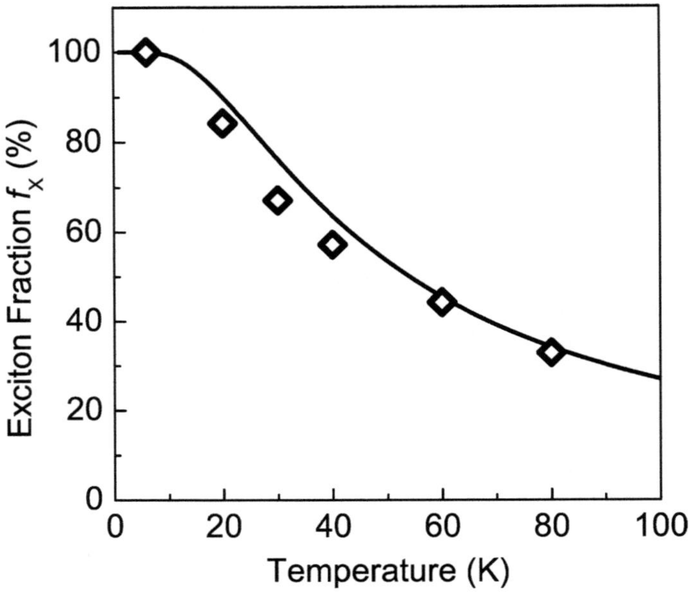

Fig. 4. Temperature dependence of the ionized state, compared with the Saha model. *Diamonds*: quasi-equilibrium exciton fraction obtained from two-component fits to the experiment. *Line*: Saha equation result for total pair density $N = 2 \times 10^{10}\,\mathrm{cm}^{-2}$, $\mu = 0.054\, m_0$, and $E_b = 7.7\,\mathrm{meV}$

equation model. To study this point, we calculate an expanded model which includes both HH and LH species of excitons and unbound pairs. The chemical potentials for these different e-h pair species are given by

$$\mu_{eh}^{\mathrm{H}} = k_B T \, \ln\left(\frac{n_e n_h^{\mathrm{H}}}{m_e m_h^{\mathrm{H}}} \frac{\pi^2 \hbar^4}{(k_B T)^2} \right) , \tag{8}$$

$$\mu_{eh}^{\mathrm{L}} = k_B T \, \ln\left(\frac{n_e n_h^{\mathrm{L}}}{m_e m_h^{\mathrm{L}}} \frac{\pi^2 \hbar^4}{(k_B T)^2} \right) + \Delta_{\mathrm{H-L}} , \tag{9}$$

$$\mu_X^{\mathrm{H}} = k_B T \, \ln\left(\frac{n_X^{\mathrm{H}}}{m_e + m_h^{\mathrm{H}}} \cdot \frac{2\pi\hbar^2}{4 k_B T} \right) - E_b^{\mathrm{H}} \quad \text{and} \tag{10}$$

$$\mu_X^{\mathrm{L}} = k_B T \, \ln\left(\frac{n_X^{\mathrm{L}}}{m_e + m_h^{\mathrm{L}}} \cdot \frac{2\pi\hbar^2}{4 k_B T} \right) - E_b^{\mathrm{L}} + \Delta_{\mathrm{H-L}} . \tag{11}$$

These potentials are derived by integrating the total pair density for each species within Boltzmann statistics from the 2D density of states $(m/2\pi\hbar^2$

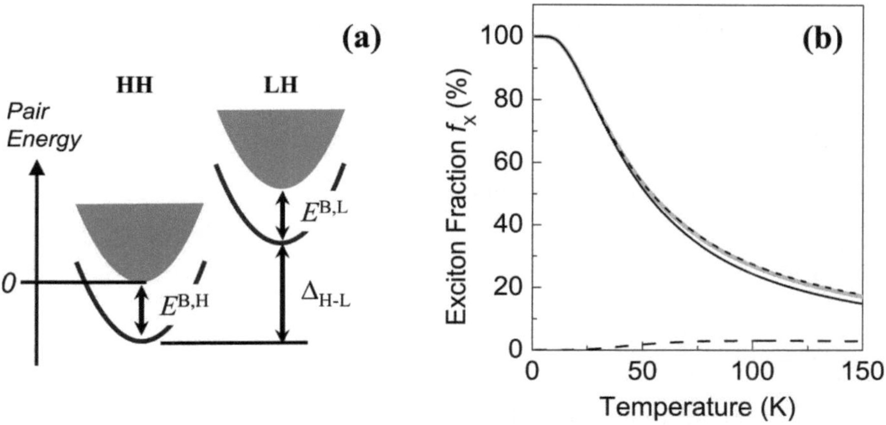

Fig. 5. Saha equation result for a multi-component mixture of HH and LH excitons, and HH and LH unbound pairs (*free carriers*). (**a**) Energy diagram showing the relations of the various species. (**b**) Temperature dependence of exciton fractions in the multi-component mixture, for total pair density $N = 2 \times 10^{10}\,\text{cm}^{-2}$. *Solid line*: HH exciton fraction, *dashed line*: LH exciton fraction, *short-dashed line*: total exciton fraction, *thick gray line*: Saha model with HH excitons only (for comparison)

per spin) and by taking into account a degeneracy of $g = 4$ for excitons. The result for unbound pairs follows from the sum of electron and hole chemical potentials, where $g = 2$ for each free carrier type. In the above, H, L designates heavy- and light-hole species, respectively, while eh and X indicates unbound pairs and excitons. Furthermore, $E_b^{\mathrm{H,L}}$ are the binding energies and Δ_{H-L} is the HH-LH energy splitting as illustrated in Fig. 5a. Since we are interested in the equilibrium solution, the chemical potentials of all e-h pair species must be the same:

$$\mu_{eh}^{\mathrm{H}} = \mu_{eh}^{\mathrm{L}} = \mu_X^{\mathrm{H}} = \mu_X^{\mathrm{L}} \tag{12}$$

because otherwise these different types of e-h pairs would interconvert. By equating (8) and (10), we obtain

$$\frac{n_e n_h^{\mathrm{H}}}{n_X^{\mathrm{H}}} = \frac{k_B T}{2\pi\hbar^2} \cdot \frac{m_e m_h^{\mathrm{H}}}{m_e + m_h^{\mathrm{H}}}\, e^{-E_b^{\mathrm{H}}/k_B T}\,, \tag{13}$$

which reproduces the previous Saha equation (7) if we assume only HH species, so that the total pair density is restricted to $N = n_{eh} + n_X^{\mathrm{H}}$ and $n_{eh} = n_e = n_h^{\mathrm{H}}$. However, to study the effect of light hole species, we include these in the following model. The total pair density then reads:

$$N = n_{eh} + n_X^{\mathrm{H}} + n_X^{\mathrm{L}}\,. \tag{14}$$

Here, the unbound pair density is $n_{eh} = n_e = n_h^{\mathrm{H}} + n_h^{\mathrm{L}}$, i.e., it is equal to both the total hole and the electron density due to the symmetry imposed by the interband photo-excitation process. By equating (9) and (11), we further obtain

$$\frac{n_e n_h^{\mathrm{L}}}{n_X^{\mathrm{L}}} = \frac{k_B T}{2\pi\hbar^2} \cdot \frac{m_e m_h^{\mathrm{L}}}{m_e + m_h^{\mathrm{L}}}\, e^{-E_b^{\mathrm{L}}/k_B T} \,. \tag{15}$$

Finally, equating (8) and (9) results in

$$\frac{n_{eh}}{n_h^{\mathrm{H}}} = 1 + \frac{m_h^{\mathrm{L}}}{m_h^{\mathrm{H}}}\, e^{-\Delta_{\mathrm{H-L}}} \quad \text{and} \tag{16}$$

$$\frac{n_{eh}}{n_h^{\mathrm{L}}} = 1 + \frac{m_h^{\mathrm{H}}}{m_h^{\mathrm{L}}}\, e^{\Delta_{\mathrm{H-L}}} \,. \tag{17}$$

In the next steps, we multiply (13) with (16) to obtain

$$\frac{n_X^{\mathrm{H}}}{n_{eh}^2} = \frac{2\pi\hbar^2}{k_B T}\, e^{E_b^{\mathrm{H}}/k_B T} \left(\frac{m_e m_h^{\mathrm{H}}}{m_e + m_h^{\mathrm{H}}} + \frac{m_e m_h^{\mathrm{L}}}{m_e + m_h^{\mathrm{H}}}\, e^{-\Delta_{\mathrm{H-L}}} \right)^{-1} \equiv C_H(T) \,, \tag{18}$$

and similarly multiplying (15) and (17) results in

$$\frac{n_X^{\mathrm{L}}}{n_{eh}^2} = \frac{2\pi\hbar^2}{k_B T}\, e^{E_b^{\mathrm{L}}/k_B T} \left(\frac{m_e m_h^{\mathrm{L}}}{m_e + m_h^{\mathrm{L}}} + \frac{m_e m_h^{\mathrm{H}}}{m_e + m_h^{\mathrm{L}}}\, e^{\Delta_{\mathrm{H-L}}} \right)^{-1} \equiv C_L(T) \,. \tag{19}$$

To solve these equations for the various pair densities, we identify $N = n_{eh} + n_X^{\mathrm{H}} + n_X^{\mathrm{L}} = n_{eh} + (C_H + C_L) n_{eh}^2$ to obtain the solution for the unbound pair density

$$n_{eh} = -\frac{1}{2(C_H + C_L)} + \sqrt{\left(\frac{1}{2(C_H + C_L)}\right)^2 + \frac{N}{C_H + C_L}} \tag{20}$$

and from this the exciton densities

$$n_X^{\mathrm{H}} = C_H n_{eh}^2 \quad \text{and} \tag{21}$$

$$n_X^{\mathrm{L}} = C_L n_{eh}^2 \,. \tag{22}$$

The results of the calculations are illustrated in Fig. 5b. Here, we have assumed the same binding energy of 7.7 meV for both species, and carrier masses $m_e = 0.0665\ m_0$, $m_h^{\mathrm{H}} = 0.28\ m_0$, and $m_h^{\mathrm{L}} = 0.09\ m_0$. Also, $\Delta_{H-L} = 11$ meV is chosen to reproduce the observed splitting in the near-IR absorption spectrum. As evident from Fig. 5b, the statistics does not favor excitation of the LH excitons for our parameters. Instead, at temperatures where LH excitons would be significantly populated, they can also ionize into the *e*-*h* continuum. Therefore, our previous analysis in terms of only HH species and unbound pairs is justified.

6 Conclusions

In summary, we have discussed time-resolved experiments that study photoexcited e-h gases in GaAs quantum wells via their low-energy THz response. Resonant excitation of 1s-HH excitons leads to characteristic "intra-excitonic" THz spectra dominated by the 1s-2p interlevel transition. Fundamental differences of the intra-excitonic response and the free-carrier Drude response enable separation of different components of a multi-component e-h gas. We reported the application of a two-component model to the determination of the exciton fraction during ionization at elevated lattice temperatures. The exciton fraction approaches a quasi-equilibrium value in close agreement with a statistical thermal equilibrium described by the Saha equation. This result also corroborates a major advantage of intra-excitonic spectroscopy, namely the ability to determine absolute pair densities. A discussion and derivation of the Saha model was presented, including multi-component mixtures of light and heavy-hole pairs. Ultrafast THz spectroscopy constitutes a powerful probe of exciton physics, and its ability to couple to excitons even with vanishing interband dipole moment makes it a particularly promising tool in the search for new collective quantum phases.

Acknowledgements

We thank J. Reno for providing quantum well samples. This work was supported by the Director, Office of Science, Office of Basic Energy Sciences, of the U.S. Department of Energy under Contract No. DE-AC02-05CH11231. R. Kaindl and D. Hägele acknowledge support from the Deutsche Forschungsgemeinschaft and Alexander von Humboldt Foundation, respectively.

References

[1] J. Shah: *Ultrafast Spectroscopy of Semiconductors and Semiconductor Nanostructures* (Springer Verlag 1999)
[2] T. Kobayashi, *et al.* (Eds.): *Ultrafast Phenomena XIV - Proceedings of the 14th International Conference* (Springer Verlag 2005)
[3] R. A. Kaindl, R. D. Averitt, in: *Terahertz Spectroscopy: Principles and Applications* (CRC Press 2007 - to appear)
[4] M. C. Beard, G. M. Turner, C. A. Schmuttenmaer: Phys. Rev. B **62**, 15764 (2000)
[5] R. D. Averitt, G. Rodriguez, A. I. Lobad, J. L. W. Siders, S. A. Trugman, A. J. Taylor: Phys. Rev. B **63**, 140502 (2001)
[6] R. A. Kaindl, M. A. Carnahan, D. Hägele, R. Lövenich, D. S. Chemla: Nature **423**, 734 (2003)
[7] C. W. Luo, K. Reimann, M. Woerner, T. Elsaesser, R. Hey, K. H. Ploog: Phys. Rev. Lett. **92**, 047402 (2004)

[8] R. A. Kaindl, M. A. Carnahan, D. S. Chemla, S. Oh, J. N. Eckstein: Phys. Rev. B **72**, 060510(R) (2005)
[9] T. Kampfrath, L. Perfetti, F. Schapper, C. Frischkorn, M. Wolf: Phys. Rev. Lett. **95**, 187403 (2005)
[10] R. Bratschitsch, A. Leitenstorfer: Nature Materials **5**, 855 (2006)
[11] R. Huber, R. A. Kaindl, B. A. Schmid, D. S. Chemla: Phys. Rev. B **72**, 161314(R) (2005)
[12] R. A. Kaindl, D. Hägele, M. A. Carnahan, D. S. Chemla: J. Nanoelectron. Optoelectron. **2**, 83 (2007)
[13] R. A. Kaindl, D. Hägele, M. A. Carnahan, D. S. Chemla: to be published
[14] L. V. Butov, L. S. Levitov, A. V. Mintsev, B. D. Simons, A. C. Gossard, D. S. Chemla: Phys. Rev. Lett. **92**, 117404 (2004)
[15] M. Kubouchi, K. Yoshioka, R. Shimano, A. Mysyrowicz, M. Kuwata-Gonokami: Phys. Rev. Lett. **94**, 016403 (2005)
[16] M. Kira, W. Hoyer, T. Stroucken, S. W. Koch.: Phys. Rev. Lett **87**, 176401 (2001)
[17] T. Timusk: Phys. Rev. B **13**, 3511 (1976)
[18] R. H. M. Groeneveld, D. Grischkowsky: J. Opt. Soc. Am. B **11**, 2502 (1994)
[19] J. Černe, J. Kono, M. S. Sherwin, M. Sundaram, A. C. Gossard, G. E. W. Bauer: Phys. Rev. Lett. **77**, 1131 (1996)
[20] Q. Wu, X.-C. Zhang: Appl. Phys. Lett. **68**, 1604 (1996)
[21] J. T. Kindt, C. A. Schmuttenmaer: J. Chem. Phys. **110**, 8589 (1999)
[22] M. Born, E. Wolf: *Principles of optics* (University Press, Cambridge 1999)
[23] U. Ekenberg, M. Altarelli: Phys. Rev. B **35**, 7585 (1987)

Terahertz Near-Field Microscopy

Roland Kersting[1], Federico F. Buersgens[1], Guillermo Acuna[1], and Gyu Cheon Cho[2]

[1] Photonics and Optoelectronics Group, Department of Physics, University of Munich, Amalienstr. 54, 80799 Munich, Germany
[2] IMRA America Inc., 1044 Woodridge Ave., Ann Arbor, MI 48105, USA
roland.kersting@physik.uni-muenchen.de

Abstract. We report on apertureless terahertz (THz) microscopy and its application for semiconductor characterization. Extreme subwavelength resolutions down to 150 nm are achieved with few-cycle THz pulses having a bandwidth of 3 THz. The imaging mechanism is characterize by time-resolved THz techniques. We find that apertureless THz microscopy can be well described by the electronic resonance of the scanning-tip interacting with the sample's surface. The capacitance between tip and surface is a key parameter, which provides insight into the local high frequency permittivity of the semiconductor structure. Applying electromodulation techniques allows for imaging electronic charge distributions in microstructured semiconductors. The sensitivity of THz microscopy suffices to detect as few as about 5000 electrons.

1 Introduction

The trend in semiconductor technology towards increased bandwidth continues and many devices already reach frequencies beyond one hundred GHz. The design of future devices requires the detailed knowledge of the high frequency permittivity of semiconductor materials and dielectrics on a nano-scale dimension. Of particular interest are contributions due to the dynamical properties of charge carriers in semiconductors. The ultrafast dynamics became accessible with the advent of time-resolved spectroscopy [1], which provided experimental tools such as time-resolved pump-probe spectroscopy [2, 3], time-resolved luminescence spectroscopy [4, 5], or four-wave mixing techniques [6–8]. Many of these techniques address the dynamics of interband excitations, where electrons and holes are generated simultaneously in the semiconductor by an exciting laser pulse. In this case it can be challenging to deduce the dielectric response due to one type of charge carriers only. One reason is the strong Coulomb interaction between electrons and holes, which may induce for instance ultrafast dephasing processes.

1.1 Terahertz Spectroscopy and Microscopy

Compared to femtosecond optical techniques, the excitation energy in terahertz (THz) spectroscopy is much lower. Photon energies of a few meV only

R. Haug (Ed.): Advances in Solid State Physics,
Adv. in Solid State Phys. **47**, 203–222 (2008)

allow for intraband excitations and provide a very direct probe of the dielectric properties of charge carriers. Time-resolved THz experiments yield the spectral amplitude as well as the spectral phase from which the complex dielectric function or the optical conductivity can be obtained for a broad frequency range. The high frequency Drude conductivity of the charge carriers was determined in doped silicon [9, 10] and doped GaAs [11]. Similarly THz spectroscopy provides access to the electronic properties of quantum systems. Terahertz spectroscopy was used to study Landau systems in GaAs [12] and the dephasing of two-dimensional quantum systems [13–15]. Semiconducting organic materials are a new field where THz characterization is applied successfully as for instance on organic molecular crystals [16] or conducting polymers [17, 18]. Terahertz spectroscopy also provides insight into dynamical properties of charge carriers. Examples are the intervalley transfer of electrons in GaAs [19, 20] or the electronic response following photoexcitation in GaAs [21–23], in low-temperature grown GaAs [24], or in silicon on sapphire [25]. Fundamental questions on the transient electron hole interaction after photoexcitation were investigated by *Huber* [26].

The pioneering works on THz imaging used far-field techniques and diffraction limited resolutions of the order of 1 mm were demonstrated [27–29]. Much finer resolutions were achieved with apertures [30, 31]. The integration of apertures into THz sensors yielded lateral resolutions as small as 5 μm or $\lambda/300$ [32, 33]. However, aperture techniques suffer from the fact that the transmission of electromagnetic waves scales with the sixth power of the aperture's size [34], which leads to rapidly decaying signal intensities and makes this approach unsuitable when submicron resolutions are desired.

Extreme subwavelength resolutions have been demonstrated with apertureless scanning near-field optical microscopes (ANSOMs) operated in the visible part of the electromagnetic spectrum [35, 36]. Resolutions down to 1 nm have been achieved [37] using a sharp metallic tip in order to confine the field energy of the incident radiation. The field enhancement is highest at the apex of the scanning probe. The extreme imaging resolutions result from the enhanced interaction between radiation field and surface underneath the tip. Usually the field enhancement is understood in the quasistatic limit because in most cases the apex is much smaller than the wavelength of the radiation used [38]. Assuming the apex to be a point dipole with an image dipole in the sample underneath allows for calculating the scattering efficiency of the tip. However, quasistatic calculations as well as time-dependent calculations have shown that also the shape of the apex and its shaft may contribute significantly to the image contrast [39, 40]. Both results suggest that the antenna properties of the tip play an important role in apertureless imaging.

The field enhancement by metallic tips has been used successfully at THz frequencies [41, 42]. Spatial resolutions of about 10 μm were achieved by placing the sampling tip onto electro-optic crystals where the enhancement can be read out by electro-optic sampling techniques. Microscopic crystals of CsI were spectroscopically identified [43] and it was found that antenna effects play an important role in the imaging process [44]. Recently, we presented an apertureless near-field scanning optical microscope for the THz range (THz-ANSOM) with a spatial resolution of 150 nm [45]. We showed that the imaging process is well described by macroscopic parameters of the sampling metal tip such as its inductance, resistance, and capacitance with the sample underneath [46]. This finding is substantial for interpreting spectroscopic imaging data in order to conclude on the dielectric permittivity of the sample.

In the following section we will present the experimental setup for time-resolved THz microscopy. The basic properties of the THz-ANSOM will be shown in Sect. 3, which includes a discussion of the imaging mechanism in apertureless THz microscopy. In Sect. 4 we will show that THz microscopy can be used to characterize charge carrier distributions in semiconductor structures and devices. We will conclude with a summary and with final remarks in Sect. 5.

2 Experimental

The experimental setup for time-resolved THz spectroscopy and microscopy is illustrated in Fig. 1. A Ti:Sapphire laser delivers pulses of about 60 fs duration at a wavelength of 780 nm and at a pulse repetition rate of 80 MHz. About 700 mW of laser power are used for the generation of the THz radiation and about 50 mW are used for the laser beam that samples the THz pulses. Femtosecond laser excitation of the n-doped InAs crystal initiates coherent oscillations of the extrinsic electrons [47–49]. The resulting oscillating current leads to the emission of THz radiation, which is fed into the head of the THz-ANSOM.

The radiation that propagates through the THz-ANSOM is time-resolved by electro-optic sampling [50, 51] in a 1 mm thick ZnTe crystal. The THz radiation rotates the polarization of the probing laser pulse due to the Pockels effect. Changes of the polarization are detected by a pair of balanced photodiodes. A typical THz pulse is shown in Fig. 1. Dephasing of the plasmon in the InAs leads to the decay of the oscillating field within a few ps. The corresponding amplitude spectrum is centered at about 2 THz and the range usable for spectroscopy spans over about 2.5 octaves. The maximum field strength of the THz pulse can be obtained from the rotation angle [43, 52]. Typical is a rotation by 10^{-4} radians, which yields an electric field strength of the THz pulse of $E_{max} \approx 1.5\,\mathrm{V/cm}$. Signal to noise ratios of about $3 \cdot 10^4\ \mathrm{Hz}^{1/2}$ are achieved.

Figure 2 shows a schematic of the head of the THz-ANSOM. Dispersion free optics focus the THz pulses onto the sample at an angle of incidence of 60 degree. The diffraction limited spot has a diameter of $D \approx 4\lambda$. Spatial resolution is achieved with a sharp tungsten probe having a length $l \gg \lambda$

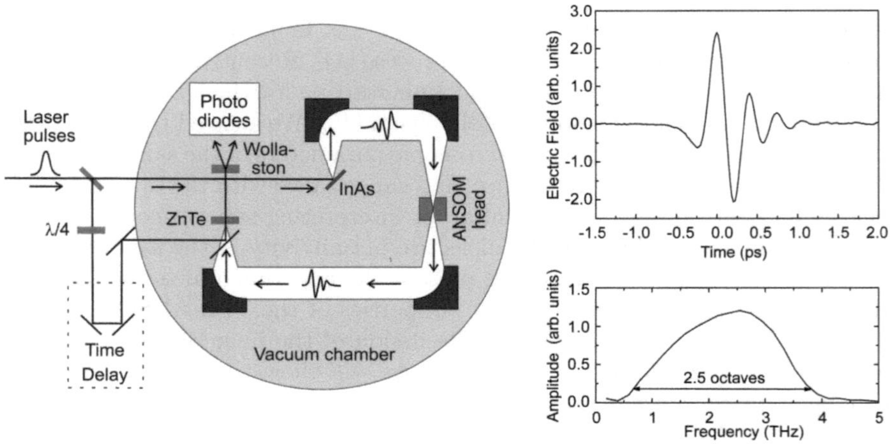

Fig. 1. *Left*: Schematic of the experimental setup for time-resolved THz spectroscopy. The ANSOM head is located at the focus of the THz radiation. *Right*: Time-resolved THz pulse and corresponding amplitude spectrum

and a cone angle of about 16 degree. The role of the tip is that of an antenna. It collects the incident power with a cross section $A \approx 0.1\lambda^2$ [53] and concentrates it to the area underneath the tip. The tungsten probes have apex curvatures between 100 nm and 2 μm, which leads to an enormous field enhancement when the tip is in close proximity to the sample's surface. The field enhancement also depends on the local dielectric permittivity. We will later show that permittivity of the sample's surface modifies the optical properties of the tip-surface system, which is the imaging mechanism in apertureless THz microscopy. Insight into the high frequency dielectric permittivity $\epsilon(\omega)$ is gained by time-resolving the THz radiation after passing through the microscope head. Many ANSOMs operated in the visible part of the spectrum detect radiation that is scattered by the tip. This is not the case in our THz-ANSOM where we exclusively detect and time-resolve the radiation that propagates through the head in a specular way as indicated in Fig. 2.

During imaging the sample is moved with respect to the incident THz beam and the probing metal tip. Scanning the surface of the sample is performed in two different modes:

i) When scanning at constant height the tip's position in z-direction is kept constant. In this mode the topology of the sample leads to a varying distance between tip and surface. Thus, changes of the sample's topology as well as changes of the sample's permittivity contribute to the image contrast.

ii) Scanning at constant distance between tungsten probe and sample surface. In this mode the image contrast mirrors exclusively the dielectric permittivity of the sample.

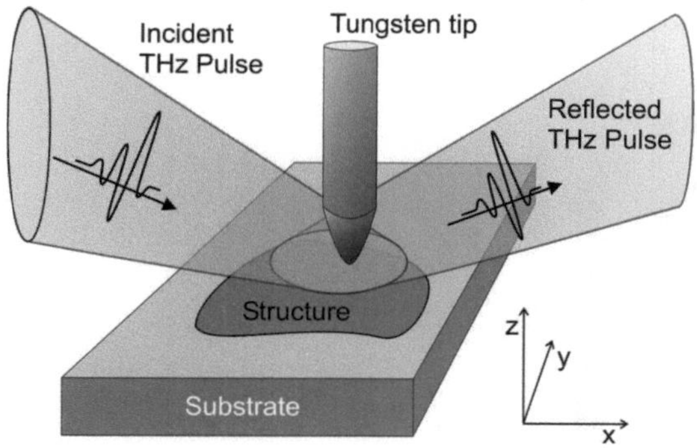

Fig. 2. Schematic of the head of the apertureless near-field scanning optical microscope. Only the THz radiation that propagates in specular direction through the head is collected and time-resolved

One method for controlling the tip-surface distance and for maintaining a constant distance is the application of a shear-force controlled feedback loop [54–56]. This technique samples the friction between tip and surface and maintains a constant value. Compared to other techniques that rely on tunneling currents between tip and surface, shear-force control is advantageous, because it can be applied on samples with low conductivity such as semiconductors or dielectrics. For our THz microscope the sampling tungsten tips are mounted on metallic tuning forks [57]. Driving the tuning forks with external piezos leads to an oscillatory motion of the tip across the sample's surface. Typical oscillation amplitudes are of the order of 1 nm, which is much smaller than the spatial resolution obtained in THz microscopy. The acoustic resonance frequency increases with the friction between tip and surface. An electronic feedback control monitors the amplitude of the oscillation and corrects the distance by moving the fork and the tip in z-direction. Typical distances between tip and surface are about 20 nm.

In many ANSOM techniques the scanning probe is dithered in order to achieve a modulation signal [37, 58, 59]. In our experiments we refrain from dithering the tip. The advantage is that time-resolved measurements of the image signal are expected to provide a very direct and unambiguous access to the complex permittivity of the surface. The trade-off is that the image contrast is small compared to the signal of the THz radiation that propagates through the THz-ANSOM without interacting with the tip-surface system. Furthermore, our technique allows for the precise characterization of the imaging process. For semiconductor characterization, electromodulation techniques are useful. Here, the charge carrier density under the tip is modulated by an alternating bias between tip and sample. The resulting differential THz signal carries information only from the mobile charges and is free from contributions due to the semiconductor itself. It provides insight into the high frequency permittivity $\epsilon(\omega)$ of the charge carriers.

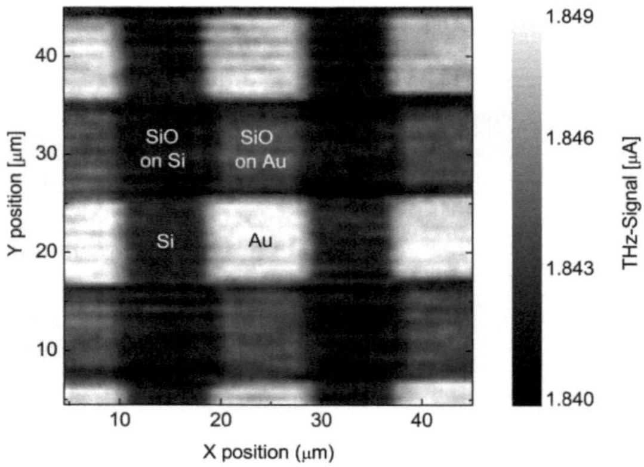

Fig. 3. Terahertz image of a two-dimensional semiconductor structure fabricated on semi-insulating silicon. The *gold lines* have vertical orientation, are 10 μm wide and have a thickness of 100 nm. The SiO *lines* have horizontal orientation, are 10 μm wide and have a thickness of 200 nm. The image was recorded at a surface to tip distance of about 20 nm utilizing shear-force control

3 Terahertz Microscopy

In order to investigate the spatial resolution of apertureless THz microscopy semiconductor structures were fabricated and imaged. Figure 3 shows the THz image of a silicon microstructure with vertical gold lines and perpendicular lines of SiO. The resulting pattern consists of 10 μm by 10 μm squares of Si, Au, SiO on Si, and SiO on Au. The thickness of the gold layers exceeds the skin depth at THz frequencies [60]. Thus, field lines are terminated by the gold layer and the silicon underneath the gold does not contribute to the image contrast. The THz image of Fig. 3 reproduces the pattern. The individual square structures show different signal intensities, which indicates that the dielectric permittivities of the materials can be distinguished.

The THz image in Fig. 3 indicates an extreme subwavelength resolution of about 1 μm, which corresponds to $\lambda/150$. We investigated the dependence of the resolution on the geometrical properties of the tip-surface system by one-dimensional scans across the edge of a metallic grating line as illustrated in Fig. 4. The grating line was located by running a small probe current through the needle and the abrupt decrease of the current indicates the edge of the metal. The lower part of Fig. 4 shows that the THz signal reproduces the edge of the grating line by a signal decrease of about 0.4 %. The data show a spatial resolution (10 % to 90 %) of about 150 nm. This value is close to the apex radius of the tungsten probe used ($R \approx 100$ nm). Further experiments with larger tips confirmed that the spatial resolution is limited by the apex radius [61]. It should be noted that the spatial resolution increases by about one order of magnitude when the tip-surface distance exceeds the tip radius [62]. This dependence shows the need for maintaining a nanometer

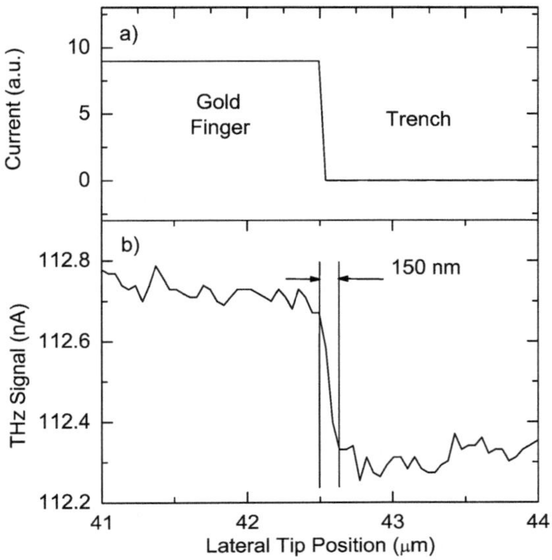

Fig. 4. Lateral resolution achieved with the THz-ANSOM. The upper curve in (**a**) illustrates the edge of the grating line. The data were recorded by a small probe current through the sampling tip. The lower curve in (**b**) shows the corresponding THz signal, which reproduces the edge of the grating line with a resolution of about 150 nm

distance between tip and surface in order to achieve a fine lateral resolution. Surprisingly, the image contrast does not decrease with decreasing tip size. This indicates that even finer spatial resolutions may be possible. The resolution limit may be set by the skin depth of the sampling tip, which is for many metals about 30 nm at THz frequencies [63].

3.1 Imaging Mechanism

Terahertz images such as Fig. 3 clearly resolve different materials, but this information is also accessible by other techniques such as optical microscopy. The scientific opportunity offered by THz techniques is the access to the dielectric constant $\epsilon(\omega)$ or optical conductivity $\sigma(\omega)$ at THz frequencies [9]. In order to extract the local permittivity from data obtained by THz microscopy, a detailed understanding of the imaging mechanism is necessary.

Apertureless microscopy has been demonstrated in many regions of the electromagnetic spectrum as for instance in the visible and in the near-infrared [36, 64]. The commonly used model to describe apertureless microscopy is that of a Mie-type scattering process [37,65–67]. In this framework the incident radiation generates an electric dipole within the sampling tip and a corresponding image dipole appears in the substrate. One important property of this model is that Mie scattering increases when the tip approaches the

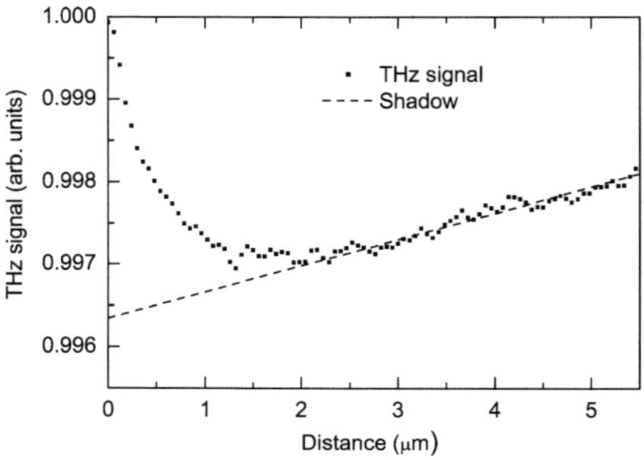

Fig. 5. Dependence of the THz signal on the distance between tip and gold surface. The *dashed line* shows the dependence as expected from the tip's shadow and diffraction

surface of the sample. In consequence, a reduced radiation intensity should be observed when detecting the specular components.

Our experimental findings at THz frequencies deviate from the common understanding of imaging by a Mie-type scattering mechanism [46]. The data of Figs. 3 and 4 show that an enhanced THz signal is observed when the tip is in proximity to a conducting surface, as for instance to a gold layer. Such an increase is incompatible with the picture of a Mie-type scatterer for which an even faster decay of the specular signal would be expected when the tip approaches the surface. Figure 5 illustrates this unexpected behavior in more detail. At large tip-sample distances the transmission of the specular system decreases with decreasing distance. This can be well explained by diffraction at the sampling tip and the resulting shadow cast. Close to the surface the behavior reverts. At distances smaller than about 2 μm the THz signal increases again.

Further insight into the imaging process can be obtained from the time-domain THz signal. Figure 6 shows the incident THz pulse as well as a differential signal ΔS, which was obtained from two measurements performed at different distances between the sampling tip and a gold surface. This differential signal shows the interaction between tip and sample, when the tip is in proximity to the surface. Although ΔS follows the exciting THz pulse at short time delays, its oscillation appears to be much slower. The corresponding amplitude spectra show that the response of the imaging system covers only the low frequency part of the exciting THz pulse.

Both observations, the fact that the THz signal increases when the tip approaches the sample as well as its spectral distribution, can be understood when considering the electronic properties of the tip-surface system. This should include not only the local interaction between the apex of the tip and the sample but also the electronic high frequency properties of the tip's shaft. Therefore we apply antenna theory and describe the properties of the imaging tip-surface system by its macroscopic parameters, which are given by the inductance L, input resistance R, and the capacitance C between the

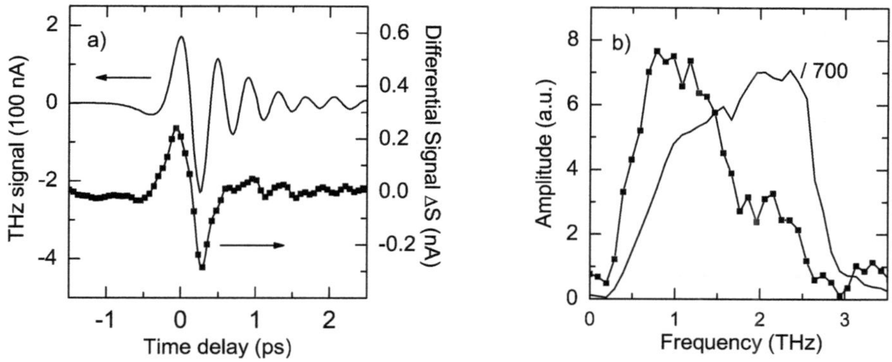

Fig. 6. (**a**) Terahertz pulses recorded in time-domain. The upper curve (*solid line*) shows the exciting THz pulse. The lower curve (*symbols*) is the difference of two measurements performed at a distance between tip and gold surface of 0.1 μm and a distance of 1.6 μm. (**b**) Corresponding amplitude spectra

apex and the surface. When radiation impinges onto the tip-surface system it is dissipated and scattered into all directions. The efficiency is given by the relation between dissipated power and incident power:

$$\eta = \frac{P_{diss}}{P_{inc}} = \frac{RZ_0}{4\pi\left[R^2 + \left(\omega L - \frac{1}{\omega C}\right)^2\right]}, \tag{1}$$

where Z_0 is the vacuum impedance. The inductance of the sampling tip is $L \approx 0.3$ nH, when considering an apex angle of the tip of 16° and a length of 0.5 mm [68, 69]. We measured the capacitance between tip and surface. Typical values for a tip radius of 1 μm are 600 aF and 250 aF for distances of 0.1 μm and 1.6 μm, respectively [46].

Changing the distance between tip and surface affects the tip-surface capacitance. In result, the resonance frequency of the tip-surface system shifts. For the undamped case this frequency is given by $\nu = \frac{1}{2\pi}/\sqrt{LC}$, which yields about 0.6 THz for the parameters presented above. The spectral dependence of the dissipation is illustrated in Fig. 7 for two tip-surface distances. Both curves are of similar shape, but the resonance for a distance of 0.1 μm is significantly lower than the response calculated for a distance of 1.6 μm. The resonance shift changes the overlap with the spectrum of the incident THz pulse. In consequence, less radiation is dissipated and more radiation passes the tip in specular direction when the tip is close to the surface. This property explains why the THz signal increases when the tip approaches a conducting surface as illustrated in see Fig. 5. The high frequency slope of the data shown in Fig. 7 can be well reproduced using the values for L and C as described above. For the input resistance we obtain values between 1000 Ω and 2000 Ω, which exceed the values expected for a monopole antenna by about one order of magnitude. We explain this difference by the fact that the tip-

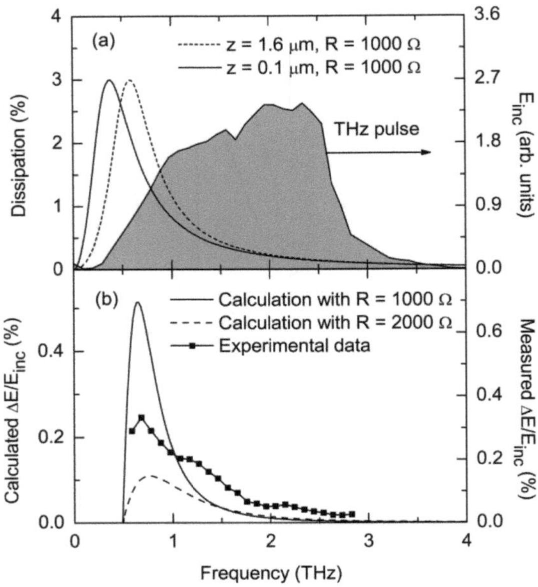

Fig. 7. Spectral dependence of the dissipation of the tip-surface system. The *upper graph* shows the calculated dissipation for two different distances between tip and surface and the spectrum of the incident THz radiation. The *lower graph* shows experimental data of a differential measurement (*symbols*) and calculated differential spectra (*lines*)

surface system is a highly directional antenna. The radiation pattern strongly depends on the latitude angle between incident beam and tip [53]. Angular THz patterns were measured by *Walther* [70]. For our angle of incidence a weakened coupling of the radiation to the antenna is expected. We deduce that the dissipation is reduced by about a factor of 10, which explains the increased input resistance observed above.

The dissipation mechanism allows for THz imaging of the sample's dielectric permittivity: Local changes of the permittivity across the sample's surface lead to changes of the tip-surface capacitance, which in turn shifts the resonance frequency and the overall dissipation of the incident THz radiation. It is worth to mention that the commonly used Mie-model and the antenna model lead to opposing results. The Mie-model predicts a decreasing signal transmission with increasing permittivity. The model of the resonant antenna yields an increase of the image signal when the permittivity increases. We verified this behavior on structures such as shown in Fig. 3. Many dielectric materials exhibit a flat dispersion at THz frequencies. This offers the opportunity to calibrate the signal dependence on the permittivity. One useful calibration specimen is for instance a structure of stacked MgF_2 layers.

4 Terahertz Imaging of Charge Carrier Distributions

The Drude-type response of charge carriers to a driving high frequency field contributes to the dielectric permittivity $\epsilon(\omega)$ as discussed in [9–11]. This electronic contribution can be mapped by time-resolved THz microscopy [71]. Time-resolved THz microscopy makes use of the fact that charge carriers within the semiconductor change the capacitance of the tip-surface system by their $\epsilon(\omega)$ and the resulting shift of the antenna's resonance frequency becomes visible in the THz signal.

Of particular importance are charge carrier accumulations in electronic devices. One example is the field effect transistor, where charge carrier densities can be controlled by applying a bias to the gate region [72]. The fundamental effect can be well described by a Schottky contact on the surface of a doped semiconductor, where a depletion zone arises at the semiconductor surface, which is free from mobile charge carriers. The depletion width

$$W = \sqrt{\frac{2\epsilon\epsilon_0}{eN_D}\left(V_{bi} - V_{ext}\right)} \tag{2}$$

depends on the doping density N_D, the built-in potential $U_{bi} = eV_{bi}$ between surface and electron gas, and the applied bias V_{ext}.

Most metal interfaces on GaAs have a built-in potential U_{bi} of about 0.5 eV [72]. In the case of GaAs a n-doping of $N_D = 10^{16}\,\mathrm{cm}^{-3}$ results in a depletion region of about 300 nm when no external bias is applied. In a similar way a tungsten tip forms a Schottky contact, when it is located close to the surface of a semiconductor. Thus, the electron density underneath the metallic probe can be controlled by applying an external bias V_{ext} between the tip and the electron gas in the bulk of the semiconductor. In the following we will first discuss model calculations on the electron density underneath the tip. We will then present experimental data and compare them to the results of the calculation.

4.1 Model Calculations

The goal of the model calculations is to obtain information on the charge carrier distribution underneath the tungsten tip. This is achieved by calculating the electron density in a volume element, which is located at the surface of the semiconductor as illustrated in Fig. 8. Several structures with different parameters are investigated but for clarity we will focus on the following example: The structure consists of a cylindrical disk having a diameter of 9 µm and a thickness of 550 nm. The volume element is homogeneously n-doped at $N_D = 10^{16}\,\mathrm{cm}^{-3}$. A metal contact with a radius of 500 nm is located in the center of the air-semiconductor interface.

The charge carrier distribution is obtained by finite differential time-domain calculations of the drift diffusion equation within a cylindrical lattice

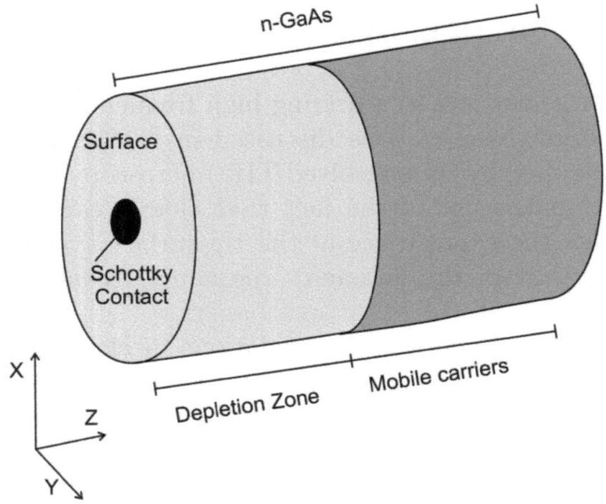

Fig. 8. Cylindrical volume element of n-doped GaAs with a Schottky contact on the surface

having 30^3 nodes. In principle, a similar result could be achieved by static calculations [73], but the time-domain approach offers the opportunity to investigate time-dependent phenomena, too.

The model treats the charge carrier transport and the resulting fields self-consistently. Charge transport between the nodes of the calculated semiconductor volume is considered by [47, 74]:

$$\frac{\partial n_e(z,t)}{\partial t} = \frac{\partial}{\partial z}\Big(D_e \frac{\partial n_e(z,t)}{\partial z}\Big) + \frac{\partial}{\partial z}\Big(\mu_e(z)E(z,t)n_e(z,t)\Big), \qquad (3)$$

where $n_e(z,t)$, $\mu_e(z)$ and D_e are, the electron density, mobility, and diffusion coefficient. For our purposes we use a constant diffusion coefficient. The drift velocity depends on the local field and should exhibit the well known saturation behavior [72]. We therefore consider the mobility to depend on the local field strength. Every charge transport results in a Coulomb field, which can be calculated from the densities of electrons n_e and the doping concentration N_D using Poisson's equation:

$$\frac{\partial E(z,t)}{\partial z} = \frac{e}{\epsilon\epsilon_o}\Big(N_D(z,t) - n_e(z,t)\Big). \qquad (4)$$

The calculations are performed in two steps. The goal of the first step is to achieve the equilibrium situation for an unbiased structure. The calculation shows how the built-in potential drives charges into the bulk of the structure. A closed loop circuitry between an ohmic substrate contact and the surface is considered. Charge carriers that leave the ohmic contact are added at the surface in order to maintain charge neutrality. An equilibrium is reached after simulating a time interval of about 5 ps. The calculation yields a depletion width of about 300 nm, as expected from (2). In a second step the external bias V_{ext} is applied and charge transport is limited either to an exchange

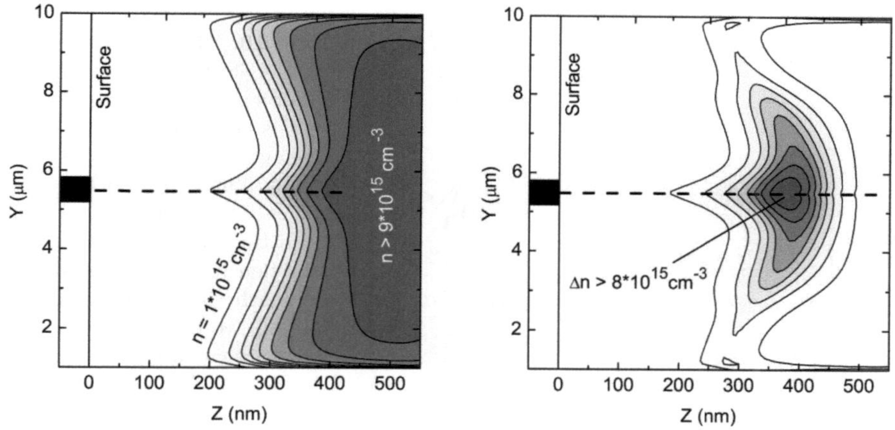

Fig. 9. Simulation of the electron density in a GaAs structure bulk doped at $N_d = 10^{16}\,\mathrm{cm}^{-3}$. The graphs show a cross section in the yz-plane through the cylinder. The symmetry axis is indicated by the *dashed line*. A metallic Schottky contact with a diameter of 500 nm is located at the surface (*black box*). At a depth of 550 nm the structure has an ohmic contact to the electron gas. (**a**) Charge carrier distribution when a bias of $V_{ext} = 0.3\,\mathrm{V}$ is applied to the Schottky contact. (**b**) Differential carrier distribution when the bias is switched from $+0.3\,\mathrm{V}$ to $-0.3\,\mathrm{V}$

between ohmic substrate and metal contact only, or along the air-surface interface.

The model calculation illustrated in Fig. 9a shows the electron distribution in a GaAs structure n-doped at $N_D = 10^{16}\,\mathrm{cm}^{-3}$. An external bias $V_{ext} = 0.3\,\mathrm{V}$ is applied to the metal contact. As expected, the semiconductor is depleted at the top surface and the depletion width is about 300 nm. The deviations at the outer surfaces of the cylinder result from the fact that the calculated volume is finite in radial direction, but these deviations have no significance for the discussion here. A kink of the charge carrier density appears underneath the Schottky contact due to the positive bias. When a negative bias is applied, the kink reverts to a further depletion. Figure 9b shows the differential carrier distribution when applying an alternating bias of $\Delta V = \pm 0.3\,\mathrm{V}$. Under these conditions the modulated volume is about $0.1\,\mu\mathrm{m} \times (1.5\,\mu\mathrm{m})^2$.

4.2 Terahertz Microscopy of Charge Carrier Distributions

The model calculation presented above indicates that the charge carrier distribution can be electronically controlled underneath the sampling tip. This fact allows for utilizing electromodulation techniques as illustrated in Fig. 10. Their advantage is that the recorded signal results exclusively from the modulated electron density underneath the tip [75]. Other THz signals as for instance due to the dielectric response of the semiconductor lattice do not

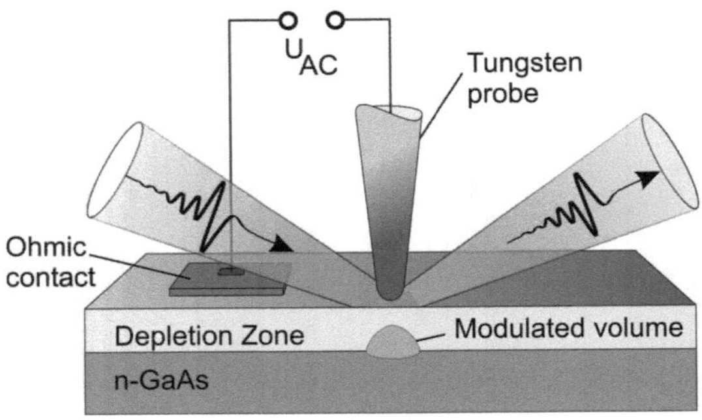

Fig. 10. Schematic of the THz-ANSOM in electromodulation mode. The ohmic contact is connected to the electron gas in the bulk of the GaAs. The tungsten tip over the GaAs surface forms a Schottky contact. The electron density is modulated underneath the tip by applying an AC voltage between tip and ohmic contact

contribute to the electro-modulation signal. Thus, electromodulation THz microscopy is expected to provide an image of the local distribution of mobile carriers.

Terahertz imaging of charge carrier distributions is investigated on n-doped GaAs structures. They consist of a 2 μm thick n-doped layer with $N_D = 2 \cdot 10^{16}\,\mathrm{cm}^{-3}$ on a semi-insulating substrate. Trenches of 10 μm width and 2 μm depth were etched into the surface layer removing the n-doped GaAs in the trench region. The electrons show a mobility of about $5000\,\mathrm{cm}^2/\mathrm{Vs}$, which corresponds to a momentum relaxation time of 190 fs. An ohmic contact to the electron gas is fabricated by alloying gold-germanium eutectic into the surface. The density of the electron gas is modulated by applying a voltage $V_{1,2}$ between the ohmic contact and the tip. In order to avoid a current in forward direction of the Schottky contact we set $V_1 = 0$ and thus $\Delta V = V_2$. AC voltages range between 0 V and 20 V at frequencies of 20 kHz. According to (2) the depletion zone underneath the tip is 240 nm and 1200 nm, respectively.

The spatial resolution of THz microscopy in electromodulation mode is deduced by scanning across the edge of a trench (see Fig. 11). During the imaging process the tip was held at constant height. In consequence, only the electron density within the fingers is modulated. When the tip is located above the trench, no modulation signal is expected. The THz data clearly reproduce the finger structure and resolve the edge of the finger. As expected, the modulation signal is zero, when the tip is over the trench. From the data we deduce a lateral resolution (10 % to 90 %) of 1.8 μm. This lateral resolution results from two contributions: On one hand the tip itself has a limited spatial resolution, which is about 1 μm in this experiment. On the other hand the modulated electron gas can be expected to extend further in lateral direction than the Schottky contact of the needle. The model calculations shown in Fig. 9b indicate this. Thus, improved resolutions require not only the use of submicron needles but also a small distance between modulated electron gas and surface.

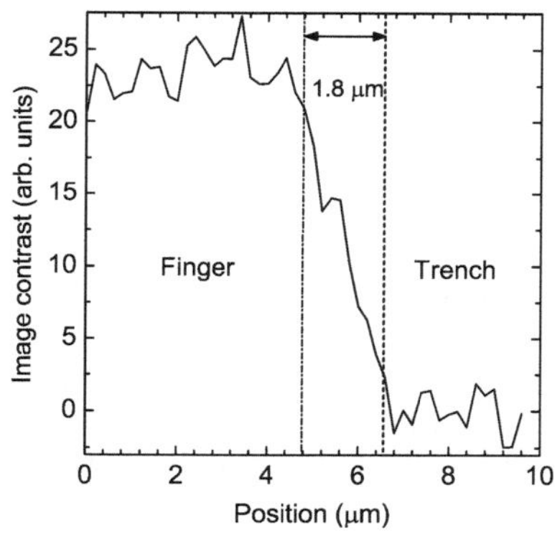

Fig. 11. Spatial resolution achieved by electromodulation. The data show the decrease of the THz signal when scanning across the edge of a finger structure etched into GaAs n-doped at $2 \cdot 10^{16}\,\mathrm{cm}^{-3}$

Let us consider the electric permittivity. Electrons within the GaAs structure contribute to the high frequency dielectric constant due to their plasma response to the driving THz field. This dielectric response is screened by the background dielectric constant $\epsilon_b \approx 13$ [76]. Assuming negligible damping gives the plasma angular frequency:

$$\omega_p = \sqrt{\frac{n_e e^2}{\epsilon_b \epsilon_0 m_e^*}}\,, \tag{5}$$

where m_e^* is the effective electron mass. Using the model of a Lorentz oscillator yields for the frequency dependence of the complex permittivity:

$$\epsilon(\omega) = \epsilon_b \left(1 + \frac{\omega_p^2}{\omega_p^2 - \omega^2 - i\omega\Gamma}\right)\,, \tag{6}$$

where Γ is the momentum relaxation rate. For the n-doped GaAs structure described above the plasma frequency $\nu_p = \omega_p/2\pi$ is about 1.5 THz. Below this frequency the real part of the permittivity increases due to the plasmon response and reveals a nearly constant value of about $\epsilon = 14$.

Time-resolved data obtained by THz electro-modulation microscopy are shown in Fig. 12. In this experiment the modulation voltage was switched between 0 V and 16 V. According to (2) the corresponding depletion widths under the tip are 180 nm and 1100 nm, respectively. The electrons are excited by a few-cycle THz pulse as illustrated in the upper part of the figure. The electro modulation data show the oscillatory response of the electron gas underneath the tip. Differences of the THz pulse shapes are evident. The electronic response appears to be dominated by low frequencies compared to the driving THz pulse, which is confirmed by the spectral amplitude distributions shown in Fig. 13. It should be mentioned that the signal due to the response of the electron gas is minute. The reason is that in first approximation the signal is given by the product of the imaging contrast (0.5 %) and

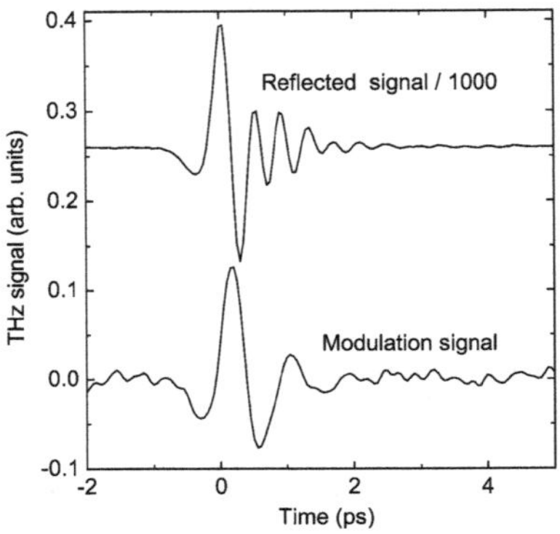

Fig. 12. Transient of the driving THz pulse after reflection on the sample's surface (scaled by a factor of 1000) and time-resolved modulation signal. The data were obtained on a bulk GaAs structure n-doped at $N_D = 2 \cdot 10^{16}\,\mathrm{cm}^{-3}$

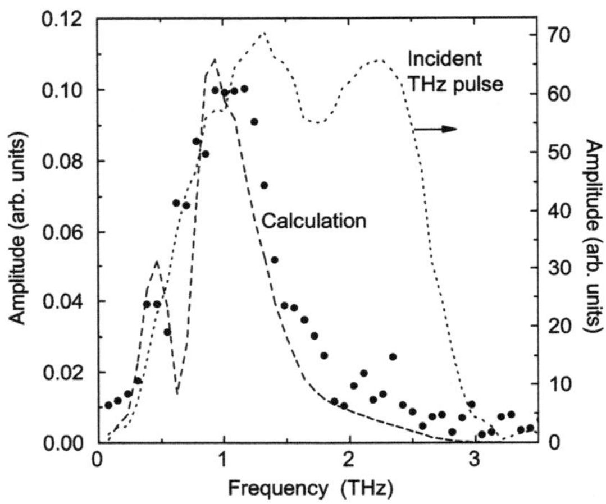

Fig. 13. Amplitude spectrum of the incident THz pulse (*dotted line*), modulation signal of the THz-ANSOM (*dots*), and model calculation of the modulation signal (*dashed line*)

the contribution of the electronic THz response, which is of the order of 10 % (see, e.g., [24]).

The overall signal amplitude due to the modulated electron gas depends in a nearly linear manner on the modulation amplitude ΔV as shown in Fig. 14. In order to understand this behavior we assume that the modulation data result from changes of the capacitance between sampling tip and electron gas underneath. The capacitance depends on the width of the depletion zone (see (2)) and yields for a parallel plate capacitor

$$C(\Delta U) = S\sqrt{\frac{e\epsilon(\omega)\epsilon_0 N_D}{2\,(U_{bi} + \Delta V)}}\,, \tag{7}$$

where S stands for the capacitor's surface. The model calculations shown in Fig. 9 support the use of the approximation of a plate capacitor, provided that the diameter of the sampling probe is much larger than the depletion width. Taking into account (1), (6), and (7) allows for reproducing the spectral data

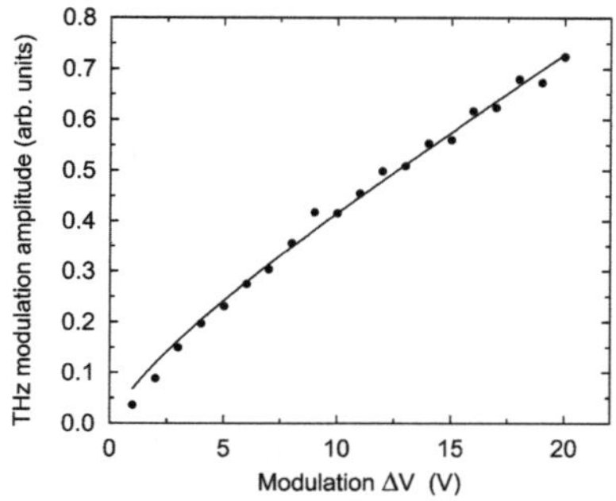

Fig. 14. Amplitude of the THz modulation signal in dependence on the amplitude of the AC-Bias applied to the tungsten tip. The *solid line* shows the result of the model calculation

(Fig. 13) as well as the dependence on the modulation voltage (Fig. 14) with the same set of parameters. These are the characteristic properties of the sampling tip, R and L as given above and a surface area of the capacitor of $S = 1.1 \cdot 10^{-12}\,\mathrm{m}^2$. The latter value is in good agreement with the spatial resolution as shown in Fig. 11.

One of the most important parameters of the technique is its sensitivity to detect charge carriers in semiconductors. The sensitivity can be expressed in terms of the minimum number of electrons that can be detected. Using the doping density and the volume, which is depleted under the tip provides this number. For a modulation voltage of $\Delta V{=}2\,\mathrm{V}$, a volume of $3 \cdot 10^{-13}\,\mathrm{cm}^3$ is depleted, which corresponds to the displacement of about 5000 electrons [71].

5 Concluding Remarks

In summary we have developed a microscope for the THz range, which merges the capabilities of THz time-domain spectroscopy with submicrometer spatial resolution. Imaging is achieved by a near-field interaction of the scanning metal probe with the sample surface. The interaction can be well described by an antenna model. Experimental data as well as the model deviate from the common understanding of a Mie-type scattering process for the imaging mechanism. We apply the THz-ANSOM for mapping the charge carrier concentration in n-doped GaAs structures. We show that charge carriers are sensed by their Drude–Lorentz response to the driving THz electric. The sensitivity of the THz-ANSOM is sufficient to detect as few as 5000 electrons.

Today, many electronic devices have dimensions as small as about 100 nm, close to the boundary of quantum mechanical size effects [77]. Future concepts in semiconductor device technology will incorporate quantum structures such as quantum points [79] and single electron transistors [78]. Conductance measurements in these nano devices are a fundamental challenge, because common probes, such as electrical leads, would influence the quantum system itself. Apertureless microscopy in general may serve here as a contactless tool in order to determine the intrinsic dielectric properties of these quantum systems. Furthermore we would like to emphasize that the application is not limited to the characterization of inorganic semiconductors as demonstrated in this paper. In particular the vast number of molecular electronic systems [80–82] may require techniques that probe the electronic properties in a contactless way.

Acknowledgements

We thank S. Manus for technical support with the development of the shear-force controlled feedback electronics. Numerical support by S. Heucke is gratefully acknowledged. The research was funded by National Science Foundation (ECS02-45461) and by the Deutsche Forschungsgemeinschaft (KE 516/1-1).

References

[1] J. Shah, *Ultrafast spectroscopy of semiconductors and semiconductor heterostructures* (Springer, 1996).
[2] J. L. Oudar *et al.*, Phys. Rev. Lett. **53**, 384 (1984).
[3] P. C. Becker, H. L. Fragnito, C. H. B. Cruz, J. Shah, and R. L. Fork, Appl. Phys. Lett. **53**, 2089 (1988).
[4] T. Elsaesser, J. Shah, L. Rota, and P. Lugli, Phys. Rev. Lett. **66**, 1757 (1991).
[5] D. W. Snoke, W. W. Rühle, Y.-C. Lu, and E. Bauser, Phys. Rev. Lett. **68**, 990 (1992).
[6] E. O. Göbel *et al.*, Phys. Rev. Lett. **64**, 1801 (1990).
[7] K. Leo *et al.*, Phys. Rev. Lett. **66**, 201 (1991).
[8] J. Feldmann *et al.*, Phys. Rev. **B 46**, 7252 (1992).
[9] M. van Exter and D. Grischkowsky, Appl. Phys. Lett. **56**, 1694 (1990).
[10] M. van Exter and D. Grischkowsky, Phys. Rev. B **41**, 12140 (1990).
[11] N. Katzenellenbogen and D. Grischkowsky, Appl. Phys. Lett. **61**, 840 (1992).
[12] D. Some and A. Nurmikko, Phys. Rev. B **50**, 5783 (1994).
[13] J. Heyman, R. Kersting, and K. Unterrainer, Appl. Phys. Lett. **72**, 644 (1998).
[14] R. Kersting, R. Bratschitsch, G. Strasser, K. Unterrainer, and J. Heyman, Opt. Lett. **25**, 272 (2000).
[15] R. Ascazubi, O. Akin, T. Zaman, R. Kersting, and G. Strasser, Appl. Phys. Lett. **81**, 4344 (2002).
[16] F. A. Hegmann, R. R. Tykwinski, K. P. H. Lui, J. E. Bullock, and J. E. Anthony, Phys. Rev. Lett. **89**, 227403 (2002).
[17] T.-I. Jeon and D. Grischkowsky, Appl. Phys. Lett. **79**, 4142 (2001).
[18] J. Lloyd-Hughes *et al.*, Appl. Phys. Lett. **89**, 112101 (2006).
[19] M. Nuss, D. Auston, and F. Capasso, Phys. Rev. Lett. **58**, 2355 (1987).
[20] P. N. Saeta, J. F. Federici, I. Greene, and D. R. Dykaar, Appl. Phys. Lett. **60**, 1477 (1992).
[21] B. I. Greene, J. F. Federici, D. R. Dykaar, A. F. J. Levi, and L. Pfeiffer, Opt. Lett. **16**, 48 (1991).
[22] R. H. M. Groeneveld and D. Grischkowsky, J. Opt. Soc. Am. B **11**, 2502 (1994).

[23] M. C. Beard, M. Turner, and C. A. Schmuttenmaer, Phys. Rev. B **62**, 15764 (2000).
[24] S. S. Prabhu, S. E. Ralph, M. R. Melloch, and E. S. Harmon, Appl. Phys. Lett. **70**, 2419 (1997).
[25] K. P. H. Lui and F. A. Hegmann, Appl. Phys. Lett. **78**, 3478 (2001).
[26] R. Huber *et al.*, Nature **414**, 286 (2001).
[27] T. S. Hartwick, T. Hodges, D. H. Barker, and F. B. Foote, Appl. Opt. **15**, 1919 (1976).
[28] B. B. Hu and M. C. Nuss, Opt. Lett. **20**, 1716 (1995).
[29] D. M. Mittleman, H. Jacobsen, and M. C. Nuss, IEEE J. Select. Top. Quant. Electr. **2**, 679 (1996).
[30] S. Hunsche, M. Koch, I. Brener, and M. C. Nuss, Opt. Comm. **150**, 22 (1998).
[31] Q. Chen, Z. Jiang, G. X. Xu, and X.-C. Zhang, Opt. Lett. **25**, 1122 (2000).
[32] O. Mitrofanov *et al.*, Appl. Phys. Lett. **78**, 252 (2001).
[33] O. Mitrofanov *et al.*, Appl. Phys. Lett. **79**, 907 (2001).
[34] H. A. Bethe, Phys. Rev. **66**, 163 (1944).
[35] J. Wessel, J. Opt. Soc. Am. B **2**, 1538 (1985).
[36] M. Specht, J. D. Pedaring, W. M. Heckl, and T. W. Haensch, Phys. Rev. Lett. **68**, 476 (1992).
[37] F. Zenhausern, T. Martin, and H. K. Wickramasinghe, Science **269**, 1083 (1995).
[38] W. Denk and D. W. Pohl, J. Vac. Sci. Technol. B **9**, 510 (1990).
[39] J. L. Bohn, D. J. Nesbitt, and A. Gallagher, J. Opt. Soc. Am. A **18**, 2998 (2001).
[40] O. J. F. Martin and C. Girard, Appl. Phys. Lett. **70**, 705 (1997).
[41] N. C. J. van der Valk and P. C. M. Planken, Appl. Phys. Lett. **81**, 1558 (2002).
[42] K. Wang, A. Barkan, and D. M. Mittleman, Sub-wavelength resolution using apertureless near-field microscopy, in: *Proceedings Conference of Lasers and Electro-Optics*, (2003).
[43] P. C. M. Planken, C. E. W. van Rijmenan, and R. N. Schouten, Sem. Sci. Techn. **20**, 121 (2005).
[44] K. Wang, D. M. Mittleman, N. C. J. van der Valk, and P. C. M. Planken, Appl. Phys. Lett. **85**, 2715 (2004).
[45] H.-T. Chen, R. Kersting, and G. C. Cho, Appl. Phys. Lett **83**, 3009 (2003).
[46] H.-T. Chen, S. Kraatz, G. C. Cho, and R. Kersting, Phys. Rev. Lett. **93**, 267401 (2004).
[47] R. Kersting, K. Unterrainer, G. Strasser, H. Kauffmann, and E. Gornik, Phys. Rev. Lett. **79**, 3038 (1997).
[48] R. Kersting, J. Heyman, G. Strasser, and K. Unterrainer, Phys. Rev. B **58**, 4553 (1998).
[49] M. P. Hasselbeck *et al.*, Phys. Rev. B **65**, 233203 (2002).
[50] Q. Wu and X.-C. Zhang, Appl. Phys. Lett. **67**, 3523 (1995).
[51] Q. Wu, M. Litz, and X.-C. Zhang, Appl. Phys. Lett. **68**, 2924 (1996).
[52] P. C. M. Planken, H.-K. Nienhuys, H. J. Bakker, and T. Wenckebach, J. Opt. Soc. Am. B **18**, 313 (2001).
[53] W. L. Stutzman and G. A. Thiele, *Antenna Theory And Design*, 2 ed. (John Wiley and Sons, 1998).
[54] K. Karrai and R. D. Grober, Appl. Phys. Lett. **66**, 1842 (1995).
[55] R. D. Grober *et al.*, Rev. Sci. Instr. **71**, 2776 (2002).

[56] F. J. Giessibl, Rev. Mod. Phys. **75**, 949 (2003).
[57] F. F. Buersgens, C. H. Lang, G. Acuna, S. Manus, and R. Kersting, Rev. Sci. Instr., submitted (2007).
[58] R. Hillenbrandt and F. Keilmann, Phys. Rev. Lett. **85**, 3029 (2000).
[59] R. Hillenbrandt and F. Keilmann, Appl. Phys. B **73**, 239 (2001).
[60] M. A. Ordal, R. J. Bell, L. L. Long, and M. R. Querry, Appl. Opt. **27**, 744 (1987).
[61] G. C. Cho, H.-T. Chen, S. Kraatz, N. Karpowicz, and R. Kersting, Sem. Sci. Techn. **20**, 286 (2005).
[62] R. Kersting, H.-T. Chen, N. Karpowicz, and G. C. Cho, J. Opt. A **7**, 184 (2005).
[63] M. A. Ordal, R. J. Bell, R. W. Alexander, L. A. Newquist, and M. R. Querry, Appl. Opt. **27**, 1203 (1988).
[64] E. Betzig, J. K. Trautmann, T. D. Harris, J. S. Weiner, and R. L. Kostalek, Science **251**, 1468 (1991).
[65] B. Knoll and F. Keilmann, Nature **399**, 134 (1999).
[66] B. Knoll and F. Keilmann, Appl. Phys. Lett. **77**, 3980 (2000).
[67] C. Bohren and D. Huffman, *Absorption and scattering by small particles* (John Wiley & Sons, 1983).
[68] K. Ogura and C. P. Steinmetz, Physical Review **25**, 184 (1907).
[69] E. Rosa, Bulletin of the Bureau of Standards **4**, 301 (1908).
[70] M. Walther, G. S. Chambers, Z. Liu, M. Freeman, and F. A. Hegmann, J. Opt. Soc. Am. B **22**, 2357 (2005).
[71] F. Buersgens, R. Kersting, and H.-T. Chen, Appl. Phys. Lett. **88**, 112115 (2006).
[72] S. Sze, *Semiconductor Devices* (John Wiley & Sons, 1985).
[73] G. L. Snider, I.-H. Tan, and E. L. Hu, J. Appl. Phys. **68**, 2849 (1990).
[74] T. Dekorsy, T. Pfeifer, W. Kütt, and H. Kurz, Phys. Rev. B **47** (1993).
[75] W. Hansen and H. Drexler, Spectroscopy of field-effect induced quantum wires and quantum dots, in: *Festkörperprobleme, Advances in Solid State Physics,* **35**, edited by R. Helbig (Vieweg, 1996).
[76] E. Palik, *Handbook of Optical Constants of Solids* (Academic Press, 1985).
[77] M. H. Devoret and R. J. Schoelkopf, Nature **406**, 1039 (2000).
[78] T. A. Fulton and G. J. Dolan, Phys. Rev. Lett. **59**, 109 (1987).
[79] B. J. van Wees *et al.*, Phys. Rev. Lett. **60**, 848 (1988).
[80] C. Joachim, J. K. Gimzewski, and A. Aviram, Nature **408**, 541 (2000).
[81] X. D. Cui *et al.*, Science **294**, 571 (2001).
[82] A. Nitzan and M. A. Ratner, Science **300**, 1384 (2003).

Interaction of THz Radiation with Semiconductors: Microscopic Theory and Experiments

J. T. Steiner[1], M. Kira[1], S. W. Koch[1], T. Grunwald[1], D. Köhler[1], S. Chatterjee[1], G. Khitrova[2], and H. M. Gibbs[2]

[1] Department of Physics and Material Sciences Center, Philipps-University, Renthof 5, 35032 Marburg, Germany
steiner@physik.uni-marburg.de
[2] College of Optical Sciences, University of Arizona, Tucson, AZ 85721, USA

Abstract. A microscopic theory is presented for the Coulomb-interacting electron-hole system in an optically excited semiconductor that is probed by weak THz radiation. The relevant equations for the THz response are summarized and numerically evaluated for different optical excitation conditions. Experimental results are presented for a high-quality InGaAs/GaAs quantum-well system under resonant excitation. Formation and decay of excitonic populations are investigated.

1 Introduction

The availability of radiation sources in the THz regime of the electromagnetic spectrum [1–4] make it possible to probe semiconductor many-body states generated by optical interband excitation. Examples of recent studies [5–11] include investigations of exciton and plasma populations as well as the dynamic build-up of a plasmon pole. A detailed microscopic analysis of these experiments requires a many-body theory of the Fermionic system of electron and hole excitations, including the Coulomb and phonon interactions as well as the coupling to optical and THz radiation. In the present manuscript, we outline such a theory and present numerical results for different optical excitation conditions. Additionally, we show experimental results for a high-quality InGaAs/GaAs quantum-well (QW) system.

We start in Sect. 2 with a discussion of the interaction Hamiltonian. We especially emphasize the intraband contributions relevant for the THz response and concentrate on the semiclassical regime where quantum-optical effects are not considered. In Sect. 3, we use an equation-of-motion approach to investigate the quantities needed to compute the THz absorption and refraction spectra. In Sect. 4, we evaluate the theory for different optical excitation conditions. In Sect. 5, we present examples of experimental results showing the formation and decay of incoherent exciton populations.

R. Haug (Ed.): Advances in Solid State Physics,
Adv. in Solid State Phys. **47**, 223–235 (2008)

2 Microscopic Theory of Terahertz Response

Our microscopic theory includes the Coulomb interacting Fermionic electron-hole system, its coupling to phonons and to fields in the optical and THz regimes of the electromagnetic spectrum. For the description of THz processes in semiconductors, it is particularly important that the relevant intra-band quantities are consistently included. We start our investigation with the minimal-substitution Hamiltonian in second quantization which describes the coupling of the carrier system to a classical electromagnetic field with vector potential $\boldsymbol{A}(\boldsymbol{r})$ [12–14]

$$
\begin{aligned}
\hat{H}_{\mathrm{LM}} &= \int \hat{\Psi}^{\dagger}(\boldsymbol{r}) \left\{ \frac{1}{2m_0} \left[\boldsymbol{p} - Q\boldsymbol{A}(\boldsymbol{r})\right]^2 + U(\boldsymbol{r}) \right\} \hat{\Psi}(\boldsymbol{r}) d\boldsymbol{r} \equiv \hat{H}_0 + \hat{H}_A \\
\hat{H}_0 &= \int \hat{\Psi}^{\dagger}(\boldsymbol{r}) \left[\frac{\boldsymbol{p}^2}{2m_0} + U(\boldsymbol{r}) \right] \hat{\Psi}(\boldsymbol{r}) d\boldsymbol{r} \\
\hat{H}_A &= \int \hat{\Psi}^{\dagger}(\boldsymbol{r}) \left[\frac{-Q}{m_0} \boldsymbol{A}(\boldsymbol{r}) \cdot \boldsymbol{p} + \frac{e^2}{2m_0} \boldsymbol{A}^2(\boldsymbol{r}) \right] \hat{\Psi}(\boldsymbol{r}) d\boldsymbol{r}.
\end{aligned} \tag{1}
$$

Here, $Q = -|e|$ is the electron charge, m_0 the free electron mass and $U(\boldsymbol{r})$ the periodic lattice potential. We separated $\hat{H}_{\mathrm{LM}}$ into an A-independent part, $\hat{H}_0$, and an A-dependent part, $\hat{H}_A$, within the Coulomb gauge. We focus on a planar QW system and assume that the light field is polarized in direction $\boldsymbol{e}_P$ and propagating along the z-axis, i.e., perpendicular to the QW. Then, the classical vector potential can be described by a scalar field, $A(z)\boldsymbol{e}_P \equiv \boldsymbol{A}(z)$.

The optically active electrons are assumed to be confined to one conduction band (c) and one valence band (v). Their properties follow from the field operator, $\hat{\Psi}(\boldsymbol{r}) = \sum_{\boldsymbol{k}_{\|},\lambda} \hat{a}_{\lambda,\boldsymbol{k}_{\|}} \phi_{\lambda,\boldsymbol{k}_{\|}}(\boldsymbol{r}_{\|}, z)$, where the Fermionic operator $\hat{a}_{\lambda,\boldsymbol{k}_{\|}}$ destroys an electron with momentum $\boldsymbol{k}_{\|}$ along the QW plane and λ is the combined band and spin index. We separate $\boldsymbol{r} = (\boldsymbol{r}_{\|}, z)$ into the two-dimensional component $\boldsymbol{r}_{\|}$ in the QW plane and the one-dimensional component z perpendicular to it. The wave functions, $\phi_{\lambda,\boldsymbol{k}_{\|}}(\boldsymbol{r}_{\|}, z)$, can be expressed in envelope approximation [15] as $\phi_{\lambda,\boldsymbol{k}_{\|}}(\boldsymbol{r}_{\|}, z) = \xi_\lambda(z) e^{i\boldsymbol{k}_{\|}\cdot\boldsymbol{r}_{\|}} w_{\lambda,\boldsymbol{k}_{\|}}(\boldsymbol{r})/\sqrt{\mathcal{S}}$ where $\xi_\lambda(z)$ is the confinement function, $w_{\lambda,\boldsymbol{k}_{\|}}(\boldsymbol{r})$ the lattice-periodic Bloch function and $\mathcal{S}$ the quantization area.

Inserting the electron wave functions into the interaction Hamiltonian, $\hat{H}_{\mathrm{LM}}$, we can evaluate the second quantization integrals treating the Bloch functions at the level of $\boldsymbol{k}\cdot\boldsymbol{p}$ theory. This way, we can write the light-matter interaction as [16]

$$
\begin{aligned}
\hat{H}_{\mathrm{LM}} = &\sum\nolimits_{\lambda,\boldsymbol{k}_{\|}} \left[\epsilon^{\lambda}_{\boldsymbol{k}_{\|}} - j_\lambda(\boldsymbol{k}_{\|}) A_{\mathrm{QW}} + \tfrac{e^2}{2m_0} A^2_{\mathrm{QW}} \right] \hat{a}^{\dagger}_{\lambda,\boldsymbol{k}_{\|}} \hat{a}_{\lambda,\boldsymbol{k}_{\|}} \\
&- \sum_{\substack{\lambda,\boldsymbol{k}_{\|} \\ (\lambda'\neq\lambda)}} d_{\lambda,\lambda'} A_{\mathrm{QW}} \hat{a}^{\dagger}_{\lambda,\boldsymbol{k}_{\|}} \hat{a}_{\lambda',\boldsymbol{k}_{\|}} \,.
\end{aligned} \tag{2}
$$

Here, we introduced the kinetic energy, $\epsilon^{\lambda}_{\boldsymbol{k}_{\|}} = \hbar^2 \boldsymbol{k}^2_{\|}/(2m_\lambda)$, the current-matrix element $j_\lambda(\boldsymbol{k}_{\|}) = -|e|\hbar\boldsymbol{k}_{\|}\cdot\boldsymbol{e}_P/m_\lambda$ and the dipole-matrix element $d_{\lambda,\lambda'} =$

$-|e|\boldsymbol{p}_{\lambda,\lambda'} \cdot \boldsymbol{e}_P/m_0$. Further, A_{QW} is the field at the QW position, z_{QW}. The interband matrix element

$$\boldsymbol{p}_{\lambda,\lambda'} \equiv \frac{1}{v_0} \int_{v_0} d^3r \, w^{\star}_{\lambda,\boldsymbol{k}_\parallel}(\boldsymbol{r}) \, \boldsymbol{p} \, w_{\lambda',\boldsymbol{k}_\parallel}(\boldsymbol{r}) \tag{3}$$

is defined via an integral over the unit-cell volume v_0 and non-vanishing only for different band indices. The terms in the first line of (2) describe electronic transition within a single band and consequently have *intraband* character. Note that the intraband term linear in A contains the effective mass for each band, m_λ, whereas the A^2-dependent term contains the bare electron mass, m_0. For a full description of semiconductor excitations, also the Coulomb and phonon interaction part has to be included. The standard form can be found, e.g., in [15].

The dynamics of the vector potential follows from the wave equation [14],

$$\left[\frac{\partial^2}{\partial z^2} - \frac{n^2}{c^2}\frac{\partial^2}{\partial t^2}\right] A = -\mu_0 \delta(z - z_{\mathrm{QW}}) \left(J' + \frac{\partial P}{\partial t}\right), \tag{4}$$

with the macroscopic intraband current and interband polarization

$$J' \equiv \frac{1}{\mathcal{S}} \sum_{\boldsymbol{k}_\parallel,\lambda} \left[j_\lambda(\boldsymbol{k}_\parallel) - \frac{e^2}{m_0} A_{\mathrm{QW}} \right] \langle \hat{a}^\dagger_{\lambda,\boldsymbol{k}_\parallel} \hat{a}_{\lambda,\boldsymbol{k}_\parallel} \rangle , \tag{5}$$

$$\frac{\partial P}{\partial t} \equiv \frac{1}{\mathcal{S}} \sum_{\substack{\boldsymbol{k}_\parallel,\lambda \\ (\lambda \neq \lambda')}} d_{\lambda,\lambda'} \langle \hat{a}^\dagger_{\lambda,\boldsymbol{k}_\parallel} \hat{a}_{\lambda',\boldsymbol{k}_\parallel} \rangle . \tag{6}$$

Further, c is the speed of light and n the background refractive index of the medium. The total source, $J' + \frac{\partial P}{\partial t}$, determines the response of the semiconductor material. It can be shown [16] that in the THz regime the A-dependent part of (5) and the interband polarization (6) do not lead to absorption. Thus, true absorption must be due to the remaining intraband current which we denote by J_{THz}. With the electron and hole distributions, $f^{\mathrm{e}}_{\boldsymbol{k}_\parallel} = \langle \hat{a}^\dagger_{\mathrm{c},\boldsymbol{k}_\parallel} \hat{a}_{\mathrm{c},\boldsymbol{k}_\parallel} \rangle$ and $f^{\mathrm{h}}_{\boldsymbol{k}_\parallel} = \langle \hat{a}_{\mathrm{v},\boldsymbol{k}_\parallel} \hat{a}^\dagger_{\mathrm{v},\boldsymbol{k}_\parallel} \rangle$, the THz current reads

$$J_{\mathrm{THz}} = \frac{1}{\mathcal{S}} \sum_{\boldsymbol{k}_\parallel} \left[j_{\mathrm{c}}(\boldsymbol{k}_\parallel) f^{\mathrm{e}}_{\boldsymbol{k}_\parallel} - j_{\mathrm{v}}(\boldsymbol{k}_\parallel) f^{\mathrm{h}}_{\boldsymbol{k}_\parallel} \right] . \tag{7}$$

Since $j_\lambda(\boldsymbol{k}_\parallel)$ is an odd function of $\boldsymbol{k}_\parallel$ and the carrier distributions are even functions of $\boldsymbol{k}_\parallel$ for a homogeneous system, the THz current vanishes for homogeneous systems. This means that a THz current cannot emerge spontaneously but can only be induced by an external THz field.

The basic measurable quantities for the linear response follow from the linear susceptibility

$$\chi(\omega) \equiv \frac{J_{\mathrm{THz}}(\omega)}{\epsilon_0 \omega^2 A(\omega)} . \tag{8}$$

This expression is obtained from the general relations $P_{\mathrm{THz}}(\omega) = iJ_{\mathrm{THz}}(\omega)/\omega$ between the THz polarization and the intraband current (7) and $E(\omega) = i\omega A(\omega)$ between the electrical field and the vector potential.

3 Linear Response in the Incoherent Regime

In order to analyze the THz response of Coulomb interacting electron-hole excitations, we have to determine the THz current J_{THz}. We now focus on the incoherent regime where all optical fields and interband coherences vanish (i.e., $\boldsymbol{P} = \mathbf{0}$). Under these conditions, J_{THz} only follows from incoherent quantities and the corresponding carrier dynamics has to be solved. Since $\hat{a}^{\dagger}_{\lambda,\boldsymbol{k}_\parallel}\hat{a}_{\lambda,\boldsymbol{k}_\parallel}$ commutes with the THz interaction Hamiltonian (2), the carrier distributions are changed only via the Coulomb interaction. The Heisenberg equations of motion for the electron density yield (similarly for the holes)

$$i\hbar\frac{\partial}{\partial t}f^{\mathrm{e}}_{\boldsymbol{k}_\parallel} = 2i\mathrm{Im}\left[+\sum_{\boldsymbol{k}'_\parallel,\boldsymbol{q}_\parallel} V_{\boldsymbol{q}_\parallel} c^{\boldsymbol{q}_\parallel,\boldsymbol{k}'_\parallel,\boldsymbol{k}_\parallel}_{\mathrm{c,c,c,c}} - \sum_{\boldsymbol{k}'_\parallel,\boldsymbol{q}_\parallel} V_{\boldsymbol{k}_\parallel-\boldsymbol{q}_\parallel-\boldsymbol{k}'_\parallel} c^{\boldsymbol{q}_\parallel,\boldsymbol{k}'_\parallel,\boldsymbol{k}_\parallel}_{\mathrm{X}}\right] \tag{9}$$

where the two-particle correlations are defined via

$$\begin{aligned} c^{\boldsymbol{q}_\parallel,\boldsymbol{k}'_\parallel,\boldsymbol{k}_\parallel}_{\lambda,\nu,\lambda,\nu} &\equiv \langle \hat{a}^{\dagger}_{\lambda,\boldsymbol{k}_\parallel}\hat{a}^{\dagger}_{\nu,\boldsymbol{k}'_\parallel}\hat{a}_{\lambda,\boldsymbol{k}'_\parallel+\boldsymbol{q}_\parallel}\hat{a}_{\nu,\boldsymbol{k}_\parallel-\boldsymbol{q}_\parallel}\rangle \\ &\quad - \langle \hat{a}^{\dagger}_{\lambda,\boldsymbol{k}_\parallel}\hat{a}^{\dagger}_{\nu,\boldsymbol{k}'_\parallel}\hat{a}_{\lambda,\boldsymbol{k}'_\parallel+\boldsymbol{q}_\parallel}\hat{a}_{\nu,\boldsymbol{k}_\parallel-\boldsymbol{q}_\parallel}\rangle_{\mathrm{S}}\,. \end{aligned} \tag{10}$$

In these two-particle correlations, the factorized single-particle contributions $\langle\ldots\rangle_{\mathrm{S}}$ have been subtracted; in practice, $\langle\ldots\rangle_{\mathrm{S}}$ is obtained from the Hartree–Fock factorization. Excitonic correlations between different bands are given by $c_{\mathrm{X}} \equiv c_{\mathrm{c,v,c,v}}$. Since the THz field does not directly enter (9), we also have to solve the dynamics of the two-particle correlations.

For a systematic treatment of the optically induced THz effects, we apply the cluster expansion for solids up to the doublet level where all one- and two-particle quantities are fully included [12, 17, 18]. Since the resulting equations are very lengthy, see, e.g., [12], we do not present them in their full form here, but rather discuss analytic aspects of the general form by expanding the electron-hole correlation, c_{X}, in an exciton basis which is defined via the generalized Wannier equation [19],

$$\left(\tilde{\epsilon}^{\mathrm{c}}_{\boldsymbol{k}_\parallel} - \tilde{\epsilon}^{\mathrm{v}}_{\boldsymbol{k}_\parallel}\right)\phi^{\mathrm{R}}_{\lambda}(\boldsymbol{k}_\parallel) - \left(1 - f^{\mathrm{e}}_{\boldsymbol{k}_\parallel} - f^{\mathrm{h}}_{\boldsymbol{k}_\parallel}\right)\sum_{\boldsymbol{l}_\parallel} V_{\boldsymbol{l}_\parallel-\boldsymbol{k}_\parallel}\phi^{\mathrm{R}}_{\lambda}(\boldsymbol{l}_\parallel) = E_\lambda\phi^{\mathrm{R}}_{\lambda}(\boldsymbol{k}_\parallel) \tag{11}$$

where we identified the energy of an exciton state, E_λ. Since this is not a hermitean eigenvalue problem, we obtain both right-handed $\phi^{\mathrm{R}}_{\lambda}(\boldsymbol{k}_\parallel)$ and left-handed $\phi_\lambda^{\mathrm{L}}(\boldsymbol{k}_{\|}) = \phi_\lambda^{\mathrm{R}}(\boldsymbol{k}_{\|})/(1 - f^{\mathrm{e}}_{\boldsymbol{k}_{\|}} - f^{\mathrm{h}}_{\boldsymbol{k}_{\|}})$ solutions with the normalization $\sum_{\boldsymbol{k}_{\|}} \phi_\lambda^{\mathrm{L}}(\boldsymbol{k}_{\|})\phi_\nu^{\mathrm{R}}(\boldsymbol{k}_{\|}) = \delta_{\lambda,\nu}$. These functions provide the transformation

$$c_{\mathrm{X}}^{\boldsymbol{q}_{\|},\boldsymbol{k}'_{\|},\boldsymbol{k}_{\|}} = \sum_{\lambda,\nu} \phi_\lambda^{\mathrm{R}}(\boldsymbol{k}_{\|} - \boldsymbol{q}_{\|}^{\mathrm{e}})\phi_\nu^{\mathrm{R}}(\boldsymbol{k}'_{\|} + \boldsymbol{q}_{\|}^{\mathrm{h}})\Delta N_{\lambda,\nu}(\boldsymbol{q}_{\|}) \tag{12}$$

where $\boldsymbol{q}_{\|}^{\mathrm{e(h)}} = \boldsymbol{q}_{\|}\, m_{\mathrm{e(h)}}/(m_{\mathrm{e}} + m_{\mathrm{h}})$ has been defined. The incoherent exciton populations are obtained from the diagonal elements, $\Delta N_{\lambda,\lambda}(\boldsymbol{q}_{\|})$. With help of this transformation, we can express the exciton correlation dynamics as [12, 20]

$$\begin{aligned} i\hbar\frac{\partial}{\partial t}\Delta N_{\lambda,\nu}(\boldsymbol{q}_{\|}) &= (E_\nu - E_\lambda)\,\Delta N_{\lambda,\nu}(\boldsymbol{q}_{\|}) + D_{\lambda,\nu}(\boldsymbol{q}_{\|}) \\ &\quad + \sum_\beta \Big[J_{\lambda,\beta}\Delta N_{\beta,\nu}(\boldsymbol{q}_{\|}) - J_{\nu,\beta}\Delta N_{\lambda,\beta}(\boldsymbol{q}_{\|})\Big]\, A_{\mathrm{QW}}\,. \end{aligned} \tag{13}$$

We see that the THz field generates a source for the ΔN dynamics which contains the transition-matrix element between two exciton states, $J_{\alpha,\beta} \equiv \sum_{\boldsymbol{k}_{\|}} \phi_\alpha^{L}(\boldsymbol{k}_{\|})\left[j_{\mathrm{c}}(\boldsymbol{k}_{\|}) - j_{\mathrm{v}}(\boldsymbol{k}_{\|})\right]\phi_\beta^{R}(\boldsymbol{k}_{\|})$. The term $D_{\lambda,\nu}(\boldsymbol{q}_{\|})$ in (13) contains exchange and scattering contributions which, e.g., lead to a damping of the correlations and can be replaced by $-i\gamma\Delta N_{\lambda,\nu}(\boldsymbol{q}_{\|})$ for the analytic investigation.

The THz current, J_{THz}, can also be studied in the exciton basis

$$\frac{\partial}{\partial t}J_{\mathrm{THz}} = -\frac{\gamma_J}{\hbar}J_{\mathrm{THz}} + \frac{1}{\hbar}\mathrm{Im}\left[\frac{1}{\mathcal{S}}\sum_{\mathbf{q}_{\|},\lambda,\nu}(E_\nu - E_\lambda)\,J_{\lambda,\nu}\Delta N_{\lambda,\nu}(\mathbf{q}_{\|})\right]. \tag{14}$$

The influence of the intraband correlations, $c_{c,c,c,c}$ and $c_{v,v,v,v}$, has been combined into the phenomenological decay constant, γ_J, because these terms essentially lead to a damping of J_{THz}.

We next evaluate analytically the linear response of a quasi-particle state to a weak THz probe. By solving (13) and (14) for a weak A_{QW}, we can determine the linear susceptibility and in turn the linear absorption which follows from $\alpha_{\mathrm{THz}}(\omega) = \mathrm{Im}\left[\omega/(n\,c)\chi_{\mathrm{THz}}(\omega)\right]$ provided that $|\chi_{\mathrm{THz}}|$ is sufficiently small. We thus obtain for the absorption the *THz-Elliott Formula* [12, 16]

$$\alpha_{\mathrm{THz}}(\omega) = \mathrm{Im}\left[\sum_{\nu,\lambda}\frac{S^{\nu,\lambda}(\omega)\Delta n^{(0)}_{\nu,\lambda} - \left[S^{\nu,\lambda}(-\omega)\Delta n^{(0)}_{\nu,\lambda}\right]^\star}{\epsilon_0 n c\omega(\hbar\omega + i\gamma_J)}\right]. \tag{15}$$

We see that the absorption depends on the initial state of the incoherent quasi-particle excitations given by

$$\Delta n^{(0)}_{\lambda,\nu} \equiv \frac{1}{\mathcal{S}}\sum_{\boldsymbol{q}_{\|}}\Delta N_{\lambda,\nu}(\boldsymbol{q}_{\|})_{(0)}\,. \tag{16}$$

where the subscript $_{(0)}$ denotes the quasi-stationary exciton correlations. Furthermore, we introduced the abbreviation

$$S^{\nu,\lambda}(\omega) = \sum_{\beta} \frac{(E_\beta - E_\nu)\, J_{\nu,\beta} J_{\beta,\lambda}}{E_\beta - E_\nu - \hbar\omega - i\gamma} \,. \tag{17}$$

The denominator of this response function leads to resonances corresponding to transitions between different exciton states, whereas the product of the matrix elements $J_{\nu,\beta}J_{\beta,\lambda}$ provides the selection rules.

To gain some more detailed insights, we first analyze the THz absorption for the limiting case where only diagonal correlations exist, i.e., $\Delta n^{(0)}_{\nu,\lambda} = \delta_{\nu,\lambda}\Delta n^{(0)}_{\nu,\nu} \equiv \delta_{\nu,\lambda}\Delta n_\nu$. In this situation, (15) reduces to

$$\alpha_{\text{atom}}(\omega) = \frac{\omega}{\epsilon_0 n c}\text{Im}\left[\sum_{\nu} \left(S^{\nu}_{\text{atom}}(\omega) - [S^{\nu}_{\text{atom}}(-\omega)]^{\star}\right) \Delta n_\nu\right] , \tag{18}$$

$$S^{\nu}_{\text{atom}}(\omega) = \sum_{\beta} \frac{|D_{\nu,\beta}|^2}{E_\beta - E_\nu - \hbar\omega - i\gamma} \,. \tag{19}$$

Here, we defined the excitonic dipole-matrix element $D_{\lambda,\nu} \equiv \langle\phi^L_\lambda|e\boldsymbol{r}\cdot\boldsymbol{e}_P|\phi^R_\nu\rangle = i\hbar J_{\lambda,\nu}/(E_\nu - E_\lambda)$ using the general connection of dipole- and current-matrix elements [15]. We have also assumed that the $(E_\nu - E_\lambda)$ in the numerator of (17) can be replaced by $\hbar\omega$ assuming narrow Lorentzian resonances in $S^{\lambda,\nu}$. If all the conditions outlined in the derivation are satisfied, we find that the THz analysis produces an atom-like absorption spectrum for the case where different atomic levels are populated according to Δn_ν.

4 Terahertz Response after Optical Excitation

Recent investigations [12, 21, 22] have analyzed microscopically how various optical interband excitation scenarios lead to different quasi-particle states in semiconductor structures. In this section, we present an overview of characteristic results. The numerical computations are performed for InGaAs-type two-band semiconductor systems. Due to the numerical complexity of the coupled equations, a planar arrangement of identical quantum wires is considered. The distance between neighboring wires is taken to be so large that they are electronically uncoupled but much smaller than the relevant light wavelength so that no diffraction pattern arises. This way, the arrangement best mimics a quantum well while the numerical effort is greatly reduced due to the lower dimensionality.

To demonstrate the capabilities of THz spectroscopy, we evaluate the THz response for three different experimentally relevant conditions where the energetic center of the optical excitation coincides with the $1s$ resonance, the $2s$ resonance, and the continuum of the optical absorption spectrum (see Fig. 1a–c).

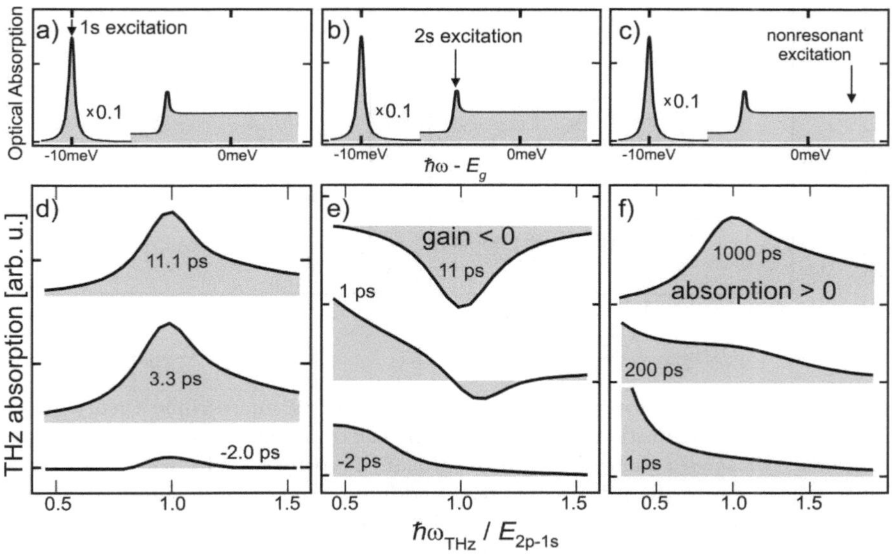

Fig. 1. THz response after optical excitation. For resonant excitation at the $1s$-exciton frequency (**a**), at the $2s$-exciton frequency (**b**) and for non-resonant excitation (**c**), the corresponding time-resolved spectra are shown in (**d**),(**e**) and (**f**), respectively

We start by assuming a resonant excitation at the lowest bound exciton energy, E_{1s}. It is shown in [12,21] that such an excitation can lead to the generation of a considerable concentration of incoherent $1s$-exciton populations for not too high intensities. The conversion of polarization into incoherent populations is mediated via the Coulomb and the phonon scattering such that it typically takes place within picoseconds. Figure 1d shows the induced absorption, probed via a weak THz pulse, for different times after the optical excitation. In general, the $1s$-$2p$ peak in the absorption can either originate from a polarization or a $1s$-exciton population [12, 21, 23]. Since the coherence time is at the range of 1 ps for the conditions assumed here, the $1s$-$2p$ resonance – remaining even 11 ps after the excitation – indicates that polarization is converted into true exciton populations. In Sect. 5, we present experimental data corresponding to this excitation scenario and demonstrate good qualitative agreement.

As a second example, we consider excitation at the $2s$ exciton resonance. Microscopic calculations for this excitation configuration reveal [12, 21] the remarkable result that the $2s$-polarization is converted into a mix of $2s$ and $2p$ populations. This is clearly possible from the energy conservation point of view, however, the direct optical generation of p-type excitons is unexpected at first sight since it involves a symmetry change of the optically induced $2s$-type polarization. As discussed in [12, 21], this symmetry breaking is di-

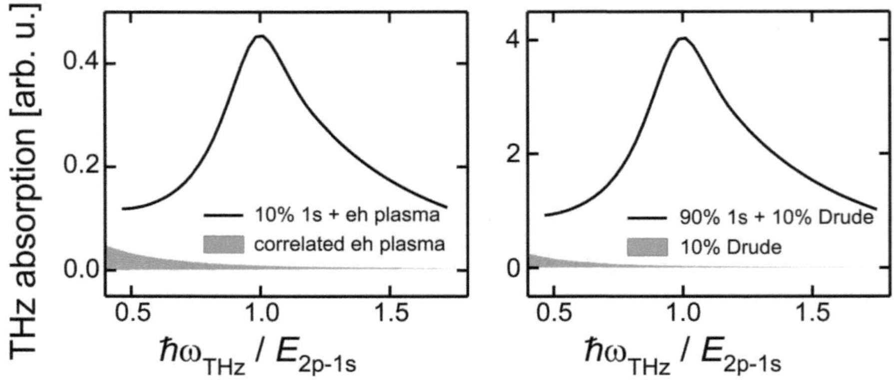

Fig. 2. *Left*: THz response computed from the full microscopic theory for a quantum-well configuration where 10 % of the electron-hole pairs are bound to 1s-excitons. The *shaded curve* shows the results of the correlated electron-hole plasma alone. *Right*: Fit to the microscopic result adding a pure Drude (*shaded area*) and a pure 1s-exciton contribution. Here, the Drude contribution had to be multiplied by 0.1 and the exciton contribution by 0.9, i.e., one has to assume 90 % 1s-excitons to fit the results of a calculation containing only 10 %

rectly related to the diffusive character of the Coulomb-induced scattering. Figure 1e shows a sequence of THz absorption spectra at different times after the optical excitation. As an important feature, we note that the THz absorption becomes negative around the $2p$-$1s$ transition energy. In other words, the generated quasi-particle state exhibits THz gain as a consequence of the population inversion between the $1s$ and $2p$ states. This gain opens up the possibility to generate THz amplifiers or lasers using the semiconductor as active THz-gain medium.

As a third example, we consider non-resonant excitation of a semiconductor above the bandgap but well below the LO-phonon frequency [12, 22]. In contrast to excitation resonant with the $1s$ and $2p$-exciton, here, the light is energetically resonant only with the ionized exciton states. Thus, it is understandable that, as the optical polarization decays, incoherent quasi-particle states with a significantly reduced amount of bound excitons are generated. In Fig. 1f, we notice that the THz response after 1 ps is qualitatively different compared to that obtained under resonant $1s$- and $2s$-excitation. The THz absorption roughly corresponds to a Drude response, which is characteristic for the plasma nature of the probed electron-hole-pair state. Only for later times – after roughly 1 ns – a clear $1s$-$2p$ transition emerges indicating the formation of incoherent $1s$-excitons. This behaviour has been verified in recent experiments [7, 9].

A somewhat controversial issue is the proper extraction procedure of excitonic populations from a given THz spectrum. Whereas the consistent microscopic theory unambiguously allows us to identify the exciton populations in the calculations, this is not at all trivial in the experiments. In [10] an ad hoc procedure was postulated where weighted Drude and excitonic THz spectra were simply added. Very good fits to the experimental results were obtained by adjusting the weight factors. Interpreting these factors as the respective population percentage, one has to postulate a relatively large percentage of excitonic populations. An example of such an additive spectrum is shown in the right frame of Fig. 2. Here, the THz absorption of an electron-hole system is constructed by adding Drude and exciton spectra assuming 90 % 1*s*-exciton and 10 % plasma population. The left frame of Fig. 2 shows the results of a microscopic calculation yielding virtually the same spectral shape, however for a configuration where only 10 % of the electron-hole pairs are bound to 1s-excitons. Hence, the ad hoc additive fit overestimates the excitonic population by a factor of nine showing that it is not an acceptable alternative to the full microscopic analysis.

5 Resonant Optical Pump THz Probe Experiments on (GaIn)As MQWs

In this section, we present examples of recent experiments using optical excitation and weak THz probe pulses.

5.1 Experimental Detail

The measurements are performed in a standard THz time-domain spectrometer setup. A 40 fs Ti:Sapphire oscillator at 80 MHz repetition rate is used for generation and detection of the THz radiation, a ps-Ti:Sapphire oscillator for optical excitation of the sample. The repetition rate of both lasers is actively locked with a jitter better than 2 ps thus limiting the time resolution to that value. The lasers are individually tunable and cover a spectral range from 1.72 eV to 1.45 eV. The THz radiation is generated with a photoconductive antenna (PCA) and focussed to a spot of about 300 μm diameter full width at half maximum (FWHM). After the sample, the transmitted THz radiation is recollimated and focussed on a 1 mm thick (110)-cut ZnTe detection crystal. Electro-optical sampling yields a spectral bandwidth of roughly 3 THz. The sample is mounted on a 200 μm pinhole; this reduces the low-energy, i.e., longer wavelength, bandwidth to about 0.6 THz, cf., the right hand side of Fig. 3. A helium cold finger cryostat with TPX windows allows for both THz and optical transmission and may be cooled down to 5 K. The complete THz path is encapsulated and purged with dry nitrogen to eliminate THz absorption from water vapor.

The sample (DBR42) contains twenty 8 nm $Ga_{0.94}In_{0.06}As$ QWs between 130 nm GaAs barriers and is grown by molecular-beam epitaxy (MBE) on an undoped 500 μm GaAs substrate. A linear absorption spectrum is shown in

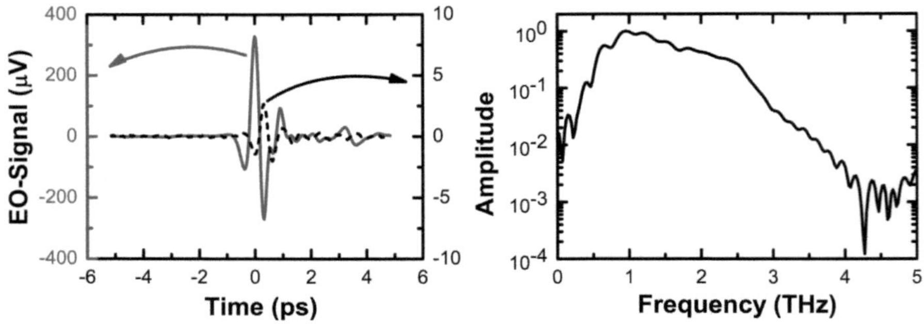

Fig. 3. *Left*: Timetrace of the THz waveform $E_{THz}(t)$. The transmitted THz field $E_0(t)$ without optical pump is shown as a *solid line*, the pump induced change $\Delta E_p(t)$ *dashed*. Note that the corresponding y-scale on the right hand side is magnified by a factor of 40. *Right*: Spectrum of the transmitted THz pulse without optical excitation

Fig. 4. The pronounced $1s$ heavy-hole (hh) resonance is observed at 1.471 eV, the $2s$ hh at 1.478 eV. The light holes (lh) are shifted to higher energies, the $1s$ lh at 1.494 eV, due to tensile strain. The dotted line in Fig. 4 shows the optical absorption of the substrate taken at the edge of the same wafer.

In order to deduce the spectral response of the sample after optical excitation, we measure both the transmitted THz field $E_0(t)$ without optical pump as well as the pump induced THz change, $\Delta E_p(t)$. The THz absorption α is then obtained from

$$\alpha(\omega) = \frac{2}{d} \ln \left(\frac{|E_0(\omega)|}{|E_p(\omega)|} \right), \tag{20}$$

where d denotes the sample thickness. $E_0(\omega)$ is obtained by Fourier transforming $E_0(t)$; similarly, $E_p(\omega)$ is the Fourier transform of $(E_0(t) + \Delta E_p(t))$.

Examples of the THz time-domain waveforms are shown in the left part of Fig. 3. The solid black line gives the THz probe field (left y axis), the pump-induced change of the latter is shown as dashed line (right y axis, note the expanded scale). Typically, a signal to noise ratio of about $2 \cdot 10^4$ in the electric field is achieved when using the pinhole. The relatively thick sample structure of roughly 0.5 mm allows for temporal separation of Fabry–Perot reflections from the sample surfaces. The spectral shape of the transmitted THz field through the sample without excitation is shown on the right hand side of Fig. 3.

5.2 Experimental Results

Typical induced THz absorption as a function of time delay is shown in Fig. 4, In this example, a resonant ps-pump pulse with a photon flux of 6.7×10^{11} photons cm^{-2} is incident on the sample.

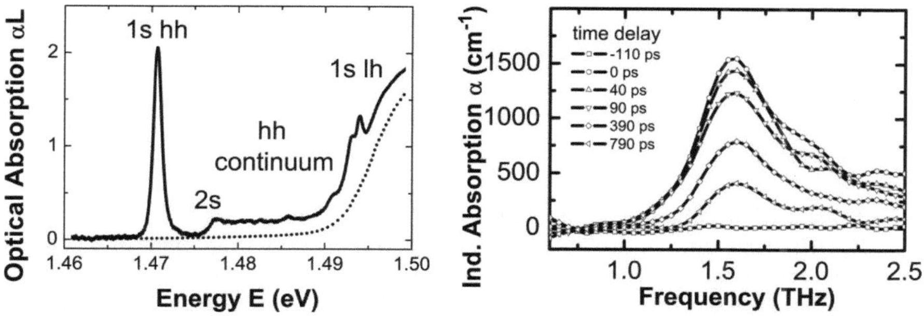

Fig. 4. *Left*: Linear absorption spectrum of the MQW (DBR42). The substrate absorption is shown *dotted*. *Right*: Induced THz absorption for various time delays under resonant ps excitation at the 1s exciton resonance

For negative time delays (-110 fs, squares) the sample is transparent, i.e., the transmission is spectrally flat. As the ps pump pulse temporally overlaps with the THz probe, a distinct resonance develops at 1.65 THz, about 7 meV. This corresponds to the $1s$-$2s$ hh spacing, shown in the left hand side of Fig. 4. The line shape is asymmetric, showing good agreement with the theoretical prediction Fig. 1d. Thus, we monitor directly the $1s$ hh exciton population by measuring the induced $1s$ to $2p$ absorption. At these densities absolute values of up to $1500\,\text{cm}^{-1}$ are achieved.

The amplitude of the resonance at 1.65 THz is reduced with increasing time delay. The decay time is on the order of several hundreds of ps to 1 ns, in contrast to the radiative lifetime of only about 10 ps. This long THz decay time is a consequence of the broad exciton momentum distribution generated in the semi-classical exciton generation process, where the optical field induces an interband polarization which is then converted into excitonic states predominantly by acoustic phonon scattering. This leads to an exciton population with a wide distribution of center-of-mass momenta [12]. Through subsequent scattering events, the excitons slowly relax on a timescale of hundreds of ps towards the light cone where they can radiate.

The results presented here on (GaIn)As MQWs are qualitatively similar to the pioneering observations reported in [7] for a GaAs/AlGaAs MQW system. However, the much larger heavy hole – light hole splitting in our InGaAs/GaAs sample allows us to focus on a nearly ideal two-band situation. These features should make it possible in the future to study also 2s- and continuum excitation conditions with the hope to test for the appearance of the theoretically predicted THz gain [12].

Acknowledgements

The authors want to thank W. W. Rühle, A. Leitenstorfer, and M. Koch for help with the experiment. This work is supported by the Deutsche

Forschungsgemeinschaft through the Quantum Optics in Semiconductors Research Group.

References

[1] J. Faist et al.: Quantum Cascade Laser, Science **264** (1994)

[2] E. Bründermann, D. R. Chamberlin, E. E. Haller: High duty cycle and continuous terahertz emission from germanium, Appl. Phys. Lett. **76**, 2991 (2000)

[3] S. Hoffmann et al.: Four-wave mixing and direct terahertz emission with two-color semiconductor lasers, Appl. Phys. Lett. **84**, 3585–3587 (2004)

[4] C. Brenner, S. Hoffmann M. R. Hofmann: Interaction of semiconductor laser dynamics with thz radiation (this volume)

[5] J. Černe et al.: Terahertz dynamics of excitons in GaAs/AlGaAs quantum wells, Phys. Rev. Lett. **77**, 1131–1134 (1996)

[6] R. Huber et al.: How many-particle interactions develop after ultrafast excitation of an electron-hole plasma, Nature **414**, 286–289 (2001)

[7] R. A. Kaindl et al: Ultrafast terahertz probes of transient conducting and insulating phases in an electron-hole gas, Nature **423**, 734–738 (2003)

[8] I. Galbraith et al.: Excitonic signatures in the photoluminescence and terahertz absorption of a GaAs/$Al_xGa_{1-x}As$ mqw, Phys. Rev. B **71**, 073302 (2005)

[9] W. Hoyer et al.: Many-body dynamics and exciton formation studied by time-resolved photoluminescence, Phys. Rev. B **72**, 075324 (2005)

[10] R. Huber et al.: Broadband terahertz study of excitonic resonances in the high-density regime in GaAs/$Al_xGa_{1-x}As$ quantum wells, Phys. Rev. B **72**, 161314(R) (2005)

[11] R. Huber et al.: Stimulated terahertz emission from intraexcitonic transitions in Cu_2O, Phys. Rev. Lett. **96**, 017402 (2006)

[12] M. Kira, S. W. Koch: Many-body correlations and excitonic effects in semiconductors, Prog. Quantum Electron. **30**, 155–196 (2006)

[13] C. Cohen-Tannoudji, J. Dupont-Roc, G. Grynberg: *Photons and Atoms – Introduction to Quantum Electrodynamics* (Wiley, 1989)

[14] M. Kira, F. Jahnke, W. Hoyer, S. W. Koch: Quantum theory of spontaneous emission and coherent effects in semiconductor microstructures, Prog. Quantum Electron. **23**, 189–279 (1999)

[15] H. Haug, S. W. Koch: *Quantum Theory of the Optical and Electronic Properties of Semiconductors* (World Scientific Publ., Singapore, 2004)

[16] M. Kira, W. Hoyer, S. W. Koch: Microscopic theory of the semiconductor terahertz response, Phys. Stat. Sol. (b) **238**, 443–450 (2003)

[17] H. W. Wyld, Jr., B. D. Fried: Quantum mechanical kinetic equations, Ann. Phys. **23**, 374–389 (1963)

[18] J. Fricke: Transport equations including many-particle correlations for an arbitrary quantum system: a general formalism, Ann. Phys. **252**, 479–498 (1996)

[19] R. Elliott: Theory of excitons, in *Polarons and Excitons*, C. Kuper(Ed.) (Oliver and Boyd, Edinburgh, 1963) pp. 269–293

[20] M. Kira, S. W. Koch: Quantum-optical spectroscopy of semiconductors, Phys. Rev. A **73**, 013813 (2006)

[21] M. Kira, S. W. Koch: Exciton-population inversion and terahertz gain in semiconductors excited to resonance, Phys. Rev. Lett. **93**, 076402 (2004)
[22] M. Kira, W. Hoyer, S. W. Koch: Terahertz signatures of the exciton formation dynamics in non-resonantly excited semiconductors, Solid State Commun. **129**, 733–736 (2004)
[23] S. W. Koch, M. Kira, G. Khitrova, H. M. Gibbs: Semiconductor excitons in new light, Nat. Mat. **5**, 523 (2006)

Nonlinear Terahertz and Midinfrared Response of n-Type GaAs

Michael Wörner[1], Peter Gaal[1], Wilhelm Kuehn[1], Klaus Reimann[1], Thomas Elsaesser[1], Rudolf Hey[2], and Klaus H. Ploog[2]

[1] Max-Born-Institut für Nichtlineare Optik und Kurzzeitspektroskopie, 12489 Berlin, Germany
[2] Paul-Drude-Institut für Festkörperelektronik, 10117 Berlin, Germany
`woerner@mbi-berlin.de`

Abstract. Our recent development of a simple and reliable method to generate THz pulses with high electric field amplitudes has paved the way for nonlinear optics in the THz regime. After a brief description of our source we present two experiments on n-type GaAs, which are in strong contrast to the predictions of Drude theory. (i) Nonlinear propagation of intense THz pulses through a thin n-type GaAs layer shows a coherent emission at 2 THz with a picosecond decay of the emitted field, despite the ultrafast carrier-carrier scattering at room temperature. While the linear THz response is in agreement with the Drude response of free electrons, the nonlinear response is dominated by the super-radiant decay of optically inverted impurity transitions. (ii) A nonlinear THz-pump–midinfrared-probe experiment shows a quantum kinetic phenomenon of the electron–LO-phonon dynamics. Ultrafast acceleration of free carriers in n-type GaAs in a strong THz field results in an oscillatory occurrence of midinfrared gain/absorption with the LO phonon frequency.

In most THz experiments the THz radiation is used as a linear probe. Using THz radiation for nonlinear excitation requires the ability to generate high enough THz intensities. The generation and field-resolved detection of high-field transients in the mid-infrared spectral range [1] has allowed for field-resolved nonlinear experiments on intersubband transitions in n-type modulation-doped GaAs/AlGaAs quantum wells providing valuable information on both intersubband Rabi oscillations [2–4] and nonlinear radiative coupling phenomena between quantum wells [5, 6]. Recently, we developed a simple and reliable method to generate THz pulses with high electric field amplitudes in the spectral range below 5 THz [7]. In this article we present our THz source and discuss two nonlinear THz experiments on n-type GaAs, which are in strong contrast to the predictions of Drude theory, which is only valid in the linear regime.

R. Haug (Ed.): Advances in Solid State Physics,
Adv. in Solid State Phys. **47**, 237–249 (2008)

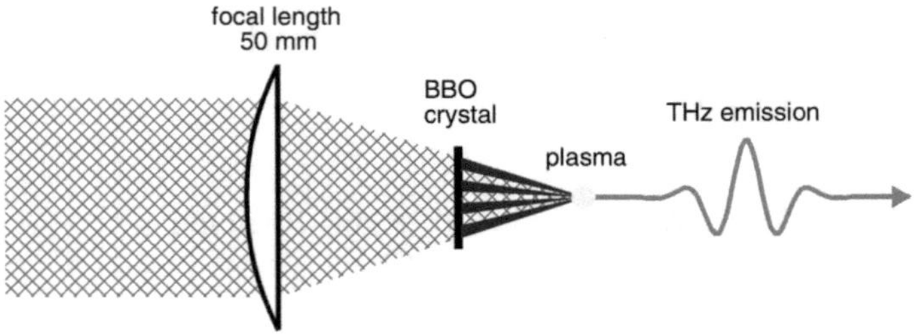

Fig. 1. Schematic of the setup for THz generation. Incident on the lens is the output of a Ti:sapphire amplifier with pulse energies of up to 500 μJ and pulse lengths down to 25 fs. The BBO crystal of 0.1 mm thickness is cut for type-I phasematched second-harmonic generation. In the focal region the intensity is high enough to generate a plasma in nitrogen gas, which then acts as the source of THz radiation

1 Generation of Single-Cycle THz Transients with High Electric-Field Amplitudes

Despite many advances in recent years, the generation, detection, and use of electromagnetic radiation [8, 9] in the frequency range of 0.1 to 10 THz are still far less developed than in other frequency ranges. Most THz generation schemes (see [10, 11]) provide quite low electric field amplitudes, severely restricting the possibilities to use THz radiation for nonlinear interactions. The highest amplitudes (150 kV/cm and 350 kV/cm) reported so far [12, 13] have been obtained using large-aperture photoconductors with bias voltages up to 45 kV. Here, we present a simple method to generate electric field amplitudes of more than 400 kV/cm in the THz range. Using electro-optic sampling, we directly measure the electric field as a function of time. This way we fully characterize the electric fields in amplitude and phase, in contrast to interferometric methods requiring additional assumptions for field reconstruction.

To generate THz radiation we use the nonlinear interaction of the fundamental (frequency ω) and the second harmonic (2ω) in a laser-generated plasma in nitrogen gas. Compared to previous uses of this method [11, 14, 15], we achieve much higher electric-field amplitudes using tighter focusing and shorter pulses (both leading to higher intensities). Further, we show that the spectrum of the THz pulses generated extends to 7 THz, considerably higher than previously demonstrated.

Our setup for THz generation is shown schematically in Fig. 1. A Ti:sapphire oscillator amplifier system generates pulses with a spectral width of 40 nm (corresponding to a bandwidth-limited pulse length of 22 fs) with pulse energies of up to 0.5 mJ at a repetition rate of 1 kHz. Both the pulse energy and the chirp of these pulses (and thus the actual pulse length) can be varied by an acousto-optic pulse shaper [16, 17] between the oscillator and the amplifier. These pulses are focused by a fused-silica lens with 50 mm focal length.

A 0.1 mm-thin BBO crystal cut for type-I SHG is inserted in the convergent beam about 5 mm before the focus. The THz radiation is generated in the plasma in the focal region. Both the focal length of the lens and the position of the BBO crystal are results of an optimization aimed at high electric field amplitudes. If one moves the BBO crystal closer to the focus, the intensity of the second harmonic and–as a result–the THz amplitudes become higher. This approach is, however, limited by damage in the crystal [18].

An undoped Si plate under Brewster's angle serves to separate the generated THz radiation from the remaining pump beam. Apart from its high transmission, Si has the further advantage of very low dispersion in the THz range, so that it does not distort the electric field transients.

Using off-axis parabolic mirrors the THz radiation is focused onto either a (110) ZnTe crystal or onto a (110) GaP crystal for electro-optic sampling [1]. To combine the THz radiation and the probe pulses, we use a second Si plate under Brewster's angle, which has high reflection for the s-polarized probe beam and high transmission for the p-polarized THz beam. The whole setup is enclosed and purged with dry nitrogen gas to prevent absorption from the rotational lines of water molecules.

Transients measured with our setup using a 0.4-mm thick ZnTe electro-optic crystal are shown in Fig. 2a. The THz detection range of ZnTe is limited [19, 20] by its optical phonon resonance (5.3 THz) to frequencies below about 4 THz. To check whether the spectrum (Fig. 2c) of the THz transient generated with the 25-fs input pulse extends to higher frequencies, we have measured this transient again with GaP (optical phonon frequency of 11 THz) as the electro-optic crystal (Fig. 2b). The measurement with GaP gives a much broader spectrum and an even higher amplitude of the electric field of more than 400 kV/cm, corresponding to a THz pulse energy of about 30 nJ.

Apart from providing very high electric field amplitudes, the present method of THz generation has the additional advantage of being easily tunable by changing the pulse length of the input pulse, i.e., by changing the chirp with the acousto-optic pulse shaper. Transients measured with the ZnTe electro-optic crystal for different lengths of the input pulse are shown in Fig. 2a and the corresponding spectra in Fig. 2c.

2 Nonlinear THz-Propagation through n-Type GaAs

In a first application of our high-field transients we study the nonlinear THz response of n-type bulk GaAs at room temperature. We observe a long-lived coherent THz emission excited by ultrashort THz pulses with electric field amplitudes of up to 70 kV/cm. In n-type GaAs, shallow donors (e.g., Si) with an electron binding energy $E_B = 6\,\text{meV}$ partially ionize at room temperature, thus providing free electrons in the conduction band. In the linear regime the frequency dependent conductivity is well described by the *Drude*

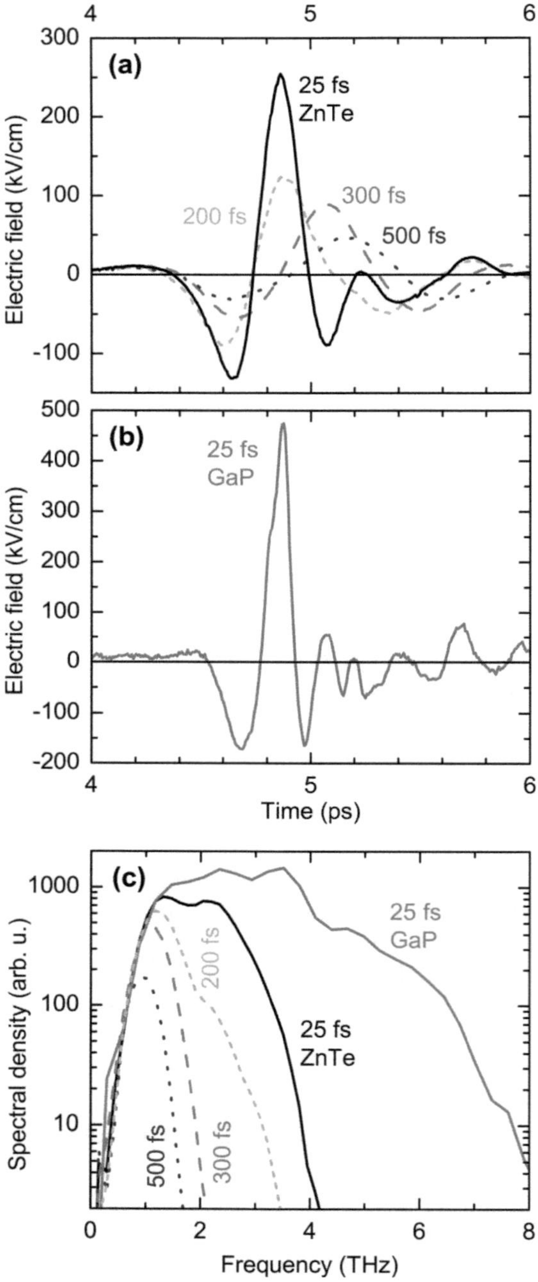

Fig. 2. (**a**) Electric-field transients measured by electro-optic sampling in ZnTe for different pulse lengths of the incoming pulse (pulse energy held constant at 0.5 mJ), varied by changing the amount of chirp. (**b**) Electric-field transient for a 25-fs pulse measured by electro-optic sampling in a 0.1-mm thick GaP crystal. (**c**) Spectra obtained by Fourier transform of the transients. For the shortest pulse length of 25 fs the spectrum measured with ZnTe as electro-optic crystal shows a high-frequency cut-off around 4 THz

theory [21], which considers a quasi-free electron gas with a mean momentum relaxation time τ_{Drude}, resulting in a *homogeneously* broadened Lorentzian absorption band centered at the frequency $\nu = 0$ [22]. In the intermediate doping range (we studied a donor concentration of $N_D = 10^{17}\,\text{cm}^{-3}$) a considerable portion of the electrons ($\approx 30\,\%$) remains even at room temperature in localized impurity ground states. Above a THz field strength of ≈ 40 kV/cm the nonlinear population dynamics between localized impurity levels and delocalized continuum states results in a coherent THz emission from the sample, which strongly resembles far-infrared laser action based on donor transitions in silicon [23]. The emission is centered at 2 THz and displays a picosecond decoherence time, in sharp contrast to the Drude scattering time of $\tau_{\text{Drude}} = 150\,\text{fs}$ derived from the linear THz response of the sample. The emitted THz field is due to phase-coherent stimulated emission from impurities having a populated excited state and an unpopulated ground state (inversion).

In our experiment, which is shown schematically in Fig. 3b, we use electro-optic sampling of the THz radiation in ZnTe, which yields the full response of the sample in a one-beam propagation experiment. The sample with a thickness of 500 nm, which is much smaller than the wavelength of the THz radiation, is placed in the focus of the incoming THz beam between two parabolic mirrors. The nonlinear response of the sample is extracted by taking the difference between the electric field transients of the transmitted pulse and of the incident pulse. An example is shown in Figs. 3a and c together with the spectrum of the incident pulse (Fig. 3b). A comparison of the expected linear response, which has a typical Drude-like shape, and the nonlinear response showing a pronounced peak at 2 THz is shown in Fig. 3d.

At low incident electric field amplitudes the difference signal occurs only during the incident pulse (see, e.g., the curve for $E_{\text{max}} = 32$ kV/cm in Fig. 4a). Above a certain amplitude the sample shows additionally a coherent oscillation after the incident pulse (curves for $E_{\text{max}} \geq 51$ kV/cm in Fig. 4a). This oscillation with a frequency around 2 THz (see Fig. 4b) lasts for 1.5 ps, ten times longer than the scattering time of Drude theory. For moderate intensities, the spectral power at this frequency increases superlinearly with the intensity (the intensity is proportional to the square of the field amplitude) of the incident field (see Fig. 4c). At very high intensities the spectral power saturates.

Samples with lower doping concentrations ($2\times10^{16}\text{cm}^{-3}$ and $1\times10^{16}\text{cm}^{-3}$) do not show the coherent oscillation around 2 THz. This indicates that the coherent polarization is caused by transitions which involve states of un-ionized donors.

Before excitation, the equilibrium electron distribution follows a Boltzmann statistics with a significant ($\approx 30\,\%$ in the sample with $N_D = 10^{17}\,\text{cm}^{-3}$) fraction of donors neutral at room temperature. Thus, photoinduced ionization and subsequent recombination of impurities contribute to the THz response of the sample. The driving pulse has sufficient field strength

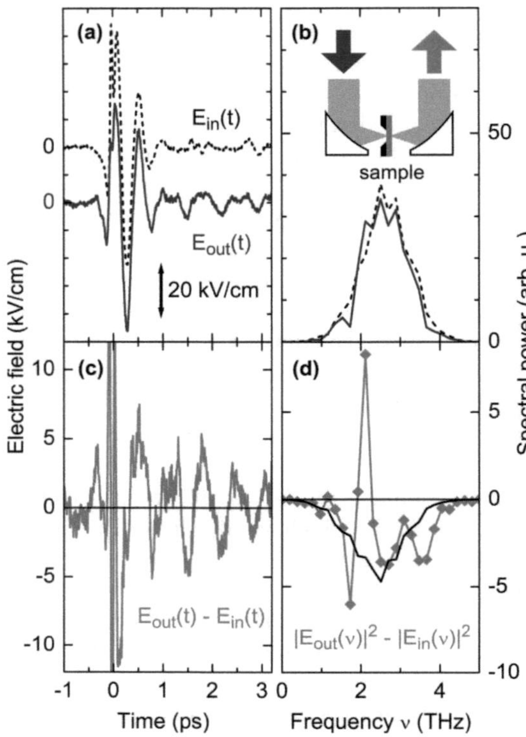

Fig. 3. Nonlinear propagation of THz pulses through the n-type GaAs layer without the semi-insulating GaAs substrate, which has been removed by wet etching. *Inset* in (b): the sample is placed in the focus between the parabolic mirrors. (**a**) *Dashed line*: incident pulse $E_{\text{in}}(t)$, *blue solid line*: transmitted pulse $E_{\text{out}}(t)$. (**b**) Corresponding power spectra. (**c**) The difference signal $E_{\text{out}}(t) - E_{\text{in}}(t)$ (*red curve*) shows for $t > 0$ weakly damped oscillations with a frequency of $\nu = 2\,\text{THz}$. (**d**) The difference of the power spectra (*red diamonds*) exhibits a pronounced, *positive* peak at 2 THz–in distinct contrast to the expected result from the Drude theory (*black solid line*)

to ionize neutral donors by transferring initially bound electrons from the S-like ground state to continuum states, in this way generating additional impurities with an unpopulated ground state. Under such nonequilibrium conditions, the electron gas thermalizes by carrier-carrier scattering on a sub-picosecond time scale and populates both high-lying impurity states and the conduction band. In impurities with a populated P-like excited state and an unpopulated ground state, stimulated THz emission can occur on the corresponding dipole-allowed optical transition, which overlaps with the spectrum of the THz driving pulse. As a result, the THz pulse saturates the optical gain within a few picoseconds, leading to the emission of a super-radiant coherent electromagnetic wave. This model also explains the observed dependence on

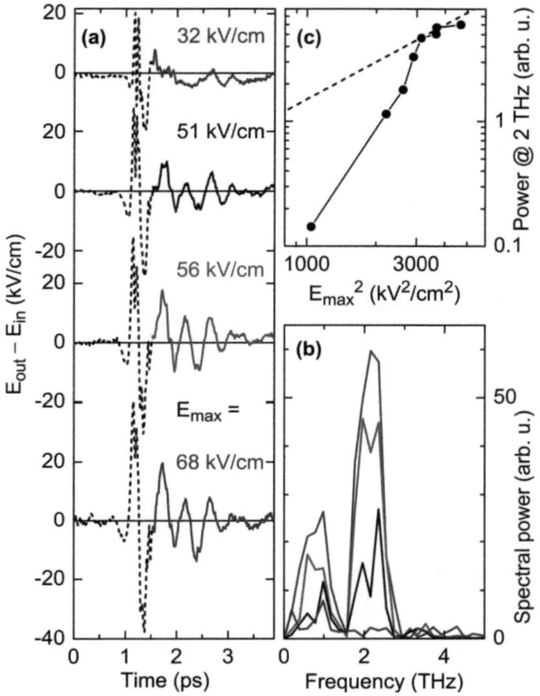

Fig. 4. (**a**) Difference signals $E_{\mathrm{out}}(t) - E_{\mathrm{in}}(t)$ for different amplitudes $E_{\max}$ of the incident THz pulse as indicated. The coherent oscillations after the main pulse (*colored solid lines*) occur above a certain threshold of $E_{\max} \approx 40$ kV/cm. (**b**) Fourier transforms of the respective colored part of $E_{\mathrm{out}}(t) - E_{\mathrm{in}}(t)$. (**c**) *Circles*: Spectral power of the oscillation around 2 THz as a function of $E^2_{\max}$ shown in a double-logarithmic diagram. The *dashed line* shows the linear relation for comparison

$E_{\max}$. For low $E_{\max}$, the ionization probability increases exponentially with $E_{\max}$, resulting in the observed superlinear increase of the coherent emission. For high $E_{\max}$, the coherent emission saturates, since at most all electrons can emit coherently.

The experimental results are confirmed by model calculations of electronic states in a disordered impurity potential. The single particle Schrödinger equation was numerically solved with the Coulomb potentials of 30 randomly distributed positively charged donors in a $67 \times 67 \times 67\,\mathrm{nm}^3$ box with periodic boundary conditions. The static dielectric constant $\epsilon_s = 12.4$ and effective electron mass $m_{\mathrm{eff}} = 0.067 m_0$ of GaAs correspond to an effective Bohr radius of $a_B = 10\,\mathrm{nm}$ and to an effective Rydberg energy of $E_{\mathrm{Ryd}} = 6\,\mathrm{meV}$.

For such parameters, an isolated hydrogen-like impurity displays an 1S $\leftrightarrow$ 2P transition frequency of $\approx 1\,\mathrm{THz}$ ($4\,\mathrm{meV}$), as experimentally observed in far-infrared luminescence experiments [24]. For the doping density in our sample, the average distance between donors is around 20 nm, i.e., of the

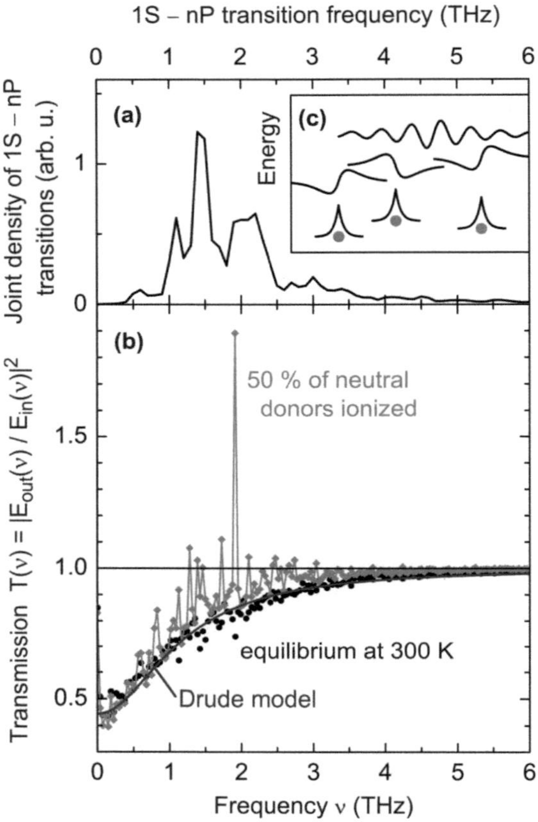

Fig. 5. Results of solving the single particle Schrödinger equation with the Coulomb potentials of randomly distributed impurities. (**a**) The *solid line* shows the joint density of states as a function of the energy difference between the ground state and excited states. (**b**) *Dots*, calculated transmission spectrum of the n-type GaAs layer at room temperature from our quantum mechanical model, which reproduces very well the result from the Drude model (*solid line*). Diamonds, ionization of half the neutral donors results in a pronounced gain around 2 THz. (**c**) Sketch of typical wave functions in the potential of disordered impurities

same order of magnitude as the Bohr radius. Thus, wave functions of neighboring impurities partially overlap, resulting in a delocalization of electron states [25]. This tendency towards delocalization is partially compensated for by the *Anderson* localization [26] caused by the random distribution of impurities. The effect of these two opposing tendencies is that more than 80 % of the donors have a 1S ground state wave function strongly localized on the impurity site. The first excited states are much less localized. This results in a spread of their energy, depending on the number and distance of neighboring impurities, and in a distribution of 1S↔2P transition energies, forming a broad band around 2 THz (8 meV) (Fig. 5a). All other states of the disordered impurity potential are delocalized among all impurities and form a continuum.

This model gives the THz transmission spectrum $T(\nu)$ of a 500 nm thick n-type GaAs layer shown in Fig. 5b. The dots were calculated for an equilibrium electron population at room temperature. The numerical result fits almost perfectly the analytical result of the Drude model (solid line) and the measured linear optical response of the system.

The diamonds in Fig. 5b show the numerical result of the discrete state model for a non-equilibrium situation, in which half of the initially neutral donors have been ionized by transferring electrons to continuum states. The most prominent feature of this transmission spectrum $T(\nu)$ is the optical gain around 2 THz, which is due to the population inversion on donors with one electron in a P-like excited state.

Our data show for the first time a nonlinear THz response beyond the Drude theory. Excitation with strong THz pulses results in coherent emission on impurity transitions with surprisingly long picosecond decoherence times. Ultrafast electron redistribution in higher lying impurity states and the conduction band continuum is essential for establishing a population inversion in impurity atoms with an unpopulated ground state. The nonlinear phenomena demonstrated here may lead to novel THz emitters and optical switches.

3 Terahertz Field-Induced Midinfrared Gain and Absorption in n-Type GaAs

As a second application of our high-field THz transients we present the first nonlinear THz-pump–midinfrared-probe experiment, which shows an interesting quantum kinetic effect in the electron–LO-phonon dynamics of rapidly accelerated carriers in n-type GaAs. The experimental setup is sketched in Fig. 6. Both a high-field THz transient [7] and a synchronized midinfrared (MIR) transient [1] are focused collinearly onto the same sample as in the previous section [27]. Subsequently, the time-dependent electric field of the THz and MIR pulses is measured with electro-optic sampling in a thin ZnTe crystal. Both the THz-pump and the MIR-probe beam are chopped with different frequencies allowing for independent measurements of $E_{\mathrm{THz}}(t)$, $E_{\mathrm{MIR}}(t,\tau)$, and $E_{\mathrm{Both}}(t,\tau)$. The latter transient is measured when both pulses are applied. τ is the delay between the THz and the MIR pulse and t is real time. Figure 6 shows such transients for $\tau = 77$ fs. The nonlinear signal of interest is obtained by subtracting from $E_{\mathrm{Both}}(t,\tau)$ the two single color measurements: $E_{\mathrm{NL}}(t,\tau) = E_{\mathrm{Both}}(t,\tau) - E_{\mathrm{THz}}(t) - E_{\mathrm{MIR}}(t,\tau)$, shown as the solid line in Fig. 7a on an expanded time scale. The sample shows a coherent, nonlinear emission, which is for this particular τ in phase with the MIR pulse, demon-

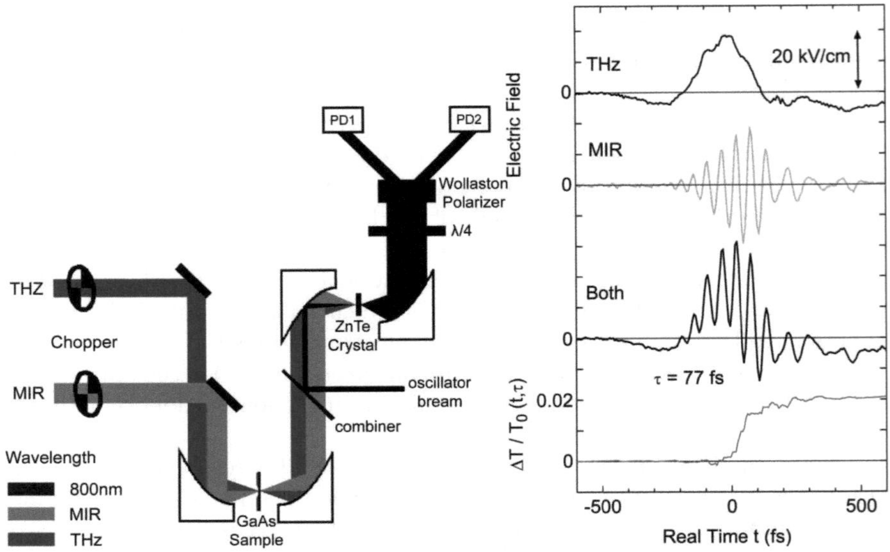

Fig. 6. *Left*: THz-pump–MIR-probe setup: both THz $E_{\mathrm{THz}}(t)$ and midinfrared transients $E_{\mathrm{MIR}}(t,\tau)$ propagate collinearly through a 500-nm thick n-type GaAs sample and are measured subsequently by electro-optic sampling in a thin ZnTe crystal. τ is the delay between the THz and the MIR field. Dual-frequency chopping of both incoming beams allow for independent measurements of E_{THz}, E_{MIR}, and E_{Both} (both pulses are transmitted through the sample). *Right*: Measured transients for $\tau = 77$ fs. The red curve shows the buildup of the transmission change according to (1)

strating a THz-field induced MIR gain of the sample. A quantitative analysis of the nonlinear transmission change can be gained from

$$\frac{\Delta T}{T_0}(t,\tau) = \int_{-\infty}^{t} E_{\mathrm{NL}}(t',\tau)\, E_{\mathrm{MIR}}(t',\tau)\, dt' \Big/ \int_{-\infty}^{\infty} \left|E_{\mathrm{MIR}}(t',\tau)\right|^2 dt' \,, \quad (1)$$

which is for $t \to \infty$ identical to the usual pump-probe signal. The red curve in Fig. 6 shows the buildup of the transmission change $\Delta T/T_0(t,\tau)$ in real time t demonstrating that the nonlinear increase of the MIR pulse energy occurs in a narrow time window. In Fig. 7b we plot $\Delta T/T_0(\infty,\tau)$ as a function of the delay τ. The nonlinear signal shows an oscillatory behavior changing with the frequency of the GaAs LO phonon between gain and absorption.

The free-carrier absorption of n-type GaAs [28] in the MIR occurs by the simultaneous absorption of a MIR photon and the absorption or emission of a LO phonon. MIR pump-probe experiments on n-type InAs [29] have already shown that photo-excited incoherent hot phonon populations can drastically enhance the free carrier absorption in the MIR.

We performed model calculations based on the Heisenberg equations of motion of classical observables of the Fröhlich polaron and on quantum ki-

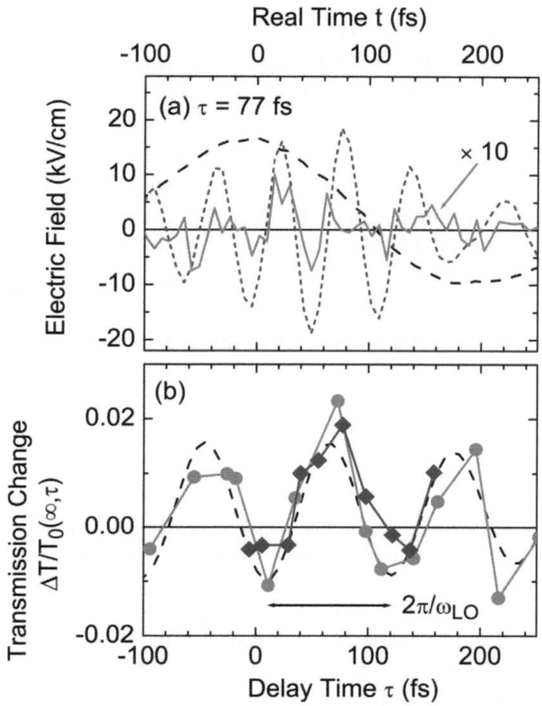

Fig. 7. (**a**) $E_{\mathrm{THz}}(t)$ (*dashed line*), $E_{\mathrm{MIR}}(t,\tau)$ (*dotted line*), and $E_{\mathrm{NL}}(t,\tau)$ (*solid line*) for $\tau = 77$ fs on an expanded time scale. (**b**) Transmission change $\Delta T/T_0(\infty,\tau)$ of the MIR pulse as a function of τ obtained from two independent measurement series (*circles* and *diamonds*). *Dashed line*: sine wave with LO phonon frequency for comparison

netic theories similar to that of [30] (details of these calculations will be published elsewhere). It strongly supports the following interpretation of the experimental data. Acceleration of free carriers in n-type GaAs in a strong THz field leads to an ultrafast stripping of the LO phonon cloud from the electron in the Fröhlich polaron of GaAs. The transient nonlinear modulation of the free carrier absorption above the LO phonon frequency gives direct information on the transient phonon cloud around the electron. Our model calculations predict an oscillatory occurrence of midinfrared gain/absorption with the longitudinal optical (LO) phonon frequency in excellent agreement with the experiment.

4 Conclusion

In contrast to most THz experiments, where the THz radiation is used as a linear probe, our novel ultrafast THz source provides electric fields high enough for nonlinear optics in the THz range. We have used this THz source for two experiments on n-type GaAs: (i) Nonlinear propagation of intense THz pulses through a thin n-type GaAs layer shows a coherent emission at 2 THz with a picosecond decay of the emitted field, despite the ultrafast carrier-carrier scattering at room temperature [2]. While the linear THz response is in agreement with the Drude response of free electrons, the nonlinear response is dominated by the super-radiant decay of optically inverted impurity transitions. (ii) A nonlinear THz pump–midinfrared probe experiment shows a quantum kinetic effect in the electron–LO phonon dynamics. Ultrafast acceleration of free carriers in n-type GaAs in a strong THz field results in an oscillatory occurrence of midinfrared gain/absorption with the LO phonon frequency.

References

[1] K. Reimann, R. P. Smith, A. M. Weiner, T. Elsaesser, M. Wörner: Direct field-resolved detection of terahertz transients with amplitudes of megavolts per centimeter, Opt. Lett. **28**, 471–473 (2003)

[2] C. W. Luo, K. Reimann, M. Wörner, T. Elsaesser, R. Hey, K. H. Ploog: Phase-resolved nonlinear response of a two-dimensional electron gas under femtosecond intersubband excitation, Phys. Rev. Lett. **92**, 047402–1–4 (2004)

[3] C. Luo, K. Reimann, M. Wörner, T. Elsaesser: Nonlinear terahertz spectroscopy of semiconductor nanostructures, Appl. Phys. A **78**, 435–440 (2004)

[4] C. W. Luo, K. Reimann, M. Wörner, T. Elsaesser, R. Hey, K. H. Ploog: Rabi oscillations of intersubband transitions in GaAs/AlGaAs MQWs, Semicond. Sci. Technol. **19**, S285–286 (2004)

[5] T. Shih, K. Reimann, M. Wörner, T. Elsaesser, I. Waldmüller, A. Knorr, R. Hey, K. H. Ploog: Nonlinear response of radiatively coupled intersubband transitions of quasi–two-dimensional electrons, Phys. Rev. B **72**, 195338–1–8 (2005)

[6] T. Shih, K. Reimann, M. Wörner, T. Elsaesser, I. Waldmüller, A. Knorr, R. Hey, K. H. Ploog: Radiative coupling of intersubband transitions in GaAs/AlGaAs multiple quantum wells, Physica E **32**, 262–265 (2006)

[7] T. Bartel, P. Gaal, K. Reimann, M. Wörner, T. Elsaesser: Generation of single-cycle THz transients with high electric-field amplitudes, Opt. Lett. **30**, 2805–2807 (2005)

[8] D. Dragoman, M. Dragoman: Terahertz fields and applications, Prog. Quantum Electron. **28**, 1–66 (2004)

[9] C. A. Schmuttenmaer: Exploring dynamics in the far-infrared with terahertz spectroscopy, Chem. Rev. **104**, 1759–1780 (2004)

[10] T. Löffler, M. Kreß, M. Thomson, T. Hahn, N. Hasegawa, H. G. Roskos: Comparative performance of terahertz emitters in amplifier-laser-based systems, Semicond. Sci. Technol. **20**, S134–141 (2005)

[11] M. Kress, T. Löffler, S. Eden, M. Thomson, H. G. Roskos: Terahertz-pulse generation by photoionization of air with laser pulses composed of both fundamental and second-harmonic waves, Opt. Lett. **29**, 1120–1122 (2004)

[12] D. You, R. R. Jones, P. H. Bucksbaum, D. R. Dykaar: Generation of high-power sub-single-cycle 500-fs electromagnetic pulses, Opt. Lett. **18**, 290–292 (1993)
[13] E. Budiarto, J. Margolies, S. Jeong, J. Son, J. Bokor: High-intensity terahertz pulses at 1-kHz repetition rate, IEEE J. Quantum Electron. **32**, 1839–1846 (1996)
[14] H. Hamster, A. Sullivan, S. Gordon, W. White, R. W. Falcone: Subpicosecond, electromagnetic pulses from intense laser-plasma interaction, Phys. Rev. Lett. **71**, 2725–2728 (1993)
[15] D. J. Cook, R. M. Hochstrasser: Intense terahertz pulses by four-wave rectification in air, Opt. Lett. **25**, 1210–1212 (2000)
[16] P. Tournois: Acousto-optic programmable dispersive filter for adaptive compensation of group delay time dispersion in laser systems, Opt. Commun. **140**, 245–249 (1997)
[17] F. Verluise, V. Laude, Z. Cheng, C. Spielmann, P. Tournois: Amplitude and phase control of ultrashort pulses by use of an acousto-optic programmable dispersive filter: pulse compression and shaping, Opt. Lett. **25**, 575–577 (2000)
[18] P. Allenspacher, R. Baehnisch, W. Riede: Multiple ultrashort pulse damage of AR-coated beta-barium borate, Proc. SPIE **5273**, 17–22 (2004)
[19] A. Leitenstorfer, S. Hunsche, J. Shah, M. C. Nuss, W. H. Knox: Detectors and sources for ultrabroadband electro-optic sampling: Experiment and theory, Appl. Phys. Lett. **74**, 1516–1518 (1999)
[20] Q. Wu, X.-C. Zhang: 7 terahertz broadband GaP electro-optic sensor, Appl. Phys. Lett. **70**, 1784–1786 (1997)
[21] P. Drude: Zur Elektronentheorie I, Ann. Phys. (Leipzig) **1**, 566–613 (1900)
[22] N. Katzenellenbogen, D. Grischkowsky: Electrical characterization to 4 THz of n- and p-type GaAs using THz time-domain spectroscopy, Appl. Phys. Lett. **61**, 840–842 (1992)
[23] S. G. Pavlov, R. K. Zhukavin, E. E. Orlova, V. N. Shastin, A. V. Kirsanov, H.-W. Hübers, K. Auen, H. Riemann: Stimulated emission from donor transitions in silicon, Phys. Rev. Lett. **84**, 5220–5223 (2000)
[24] I. Melngailis, G. E. Stillman, J. O. Dimmock, C. M. Wolfe: Far-infrared recombination radiation from impact-ionized shallow donors in GaAs, Phys. Rev. Lett. **23**, 1111–1114 (1969)
[25] F. Stern, R. M. Talley: Impurity band in semiconductors with small effective mass, Phys. Rev. **100**, 1638–1643 (1955)
[26] P. W. Anderson: Absence of diffusion in certain random lattices, Phys. Rev. **109**, 1492–1505 (1958)
[27] P. Gaal, K. Reimann, M. Wörner, T. Elsaesser, R. Hey, K. H. Ploog: Nonlinear terahertz response of n-type GaAs, Phys. Rev. Lett. **96**, 187402–1–4 (2006)
[28] W. G. Spitzer, J. M. Whelan: Infrared absorption and electron effective mass in n-type gallium arsenide, Phys. Rev. **114**, 59–63 (1959)
[29] T. Elsaesser, R. J. Bäuerle, W. Kaiser: Hot phonons in InAs observed via picosecond free-carrier absorption, Phys. Rev. B **40**, 2976–2979 (1989)
[30] W. Magnus, W. Schoenmaker: Dissipative motion of an electron-phonon system in a uniform electric field: An exact solution, Phys. Rev. B **47**, 1276–1281 (1993)

Part VI

Defects, Dislocations and Strain

Effect of Hydrogen and Grain Boundaries on Dislocation Nucleation and Multiplication Examined with a NI-AFM

Afrooz Barnoush, Bo Yang, and Horst Vehoff

Department of Materials science, Saarland University,
Bldg. D22 P.O. Box 151150, D-66041 Saarbruecken, Germany
afrooz.barnoush@gmail.com

Abstract. A nanoindenting AFM (NI-AFM) with an environment chamber was constructed to study the effect of hydrogen on decohesion and dislocation nucleation and the effect of grain boundaries on dislocation nucleation and multiplication. Ultra fine grained Ni single crystals were examined. It could be clearly shown that hydrogen influences the pop in width and length. Testing single grains with grain sizes below one micron at different rates inside a NI-AFM showed that the rate dependence of ultra fine grained Ni is a result of the interaction of the growing dislocation loops with the boundary.

1 Introduction

Over the last few years great advances have been achieved in the development of both destructive and non-destructive techniques for the accurate characterization of materials on a sub-micron (or nanometer) scale. One of these modern techniques is instrumented nanoindentation (NI). During the NI process, a penetration vs. applied load curve is recorded. These curves can be used to extract local mechanical properties of the material. The more comprehensive form of instrumented NI is "nanoindentation atomic force microscopy" (NI-AFM) which is capable of imaging the surface like a scanning force microscope, positioning the tip on a specific point of the imaged area, and then performing the indentation [1]. This technique allows to study the mechanical properties of second phases, thin films and dislocation-boundary interactions with great precision [2–6]. This is partly because of the unique ability of a NI-AFM to detect dislocation nucleation events on well-prepared surfaces [5,7]. During the initial elastic loading sequence of a nanoindentation experiment, the lateral dimensions of the volume of material that deforms are significantly smaller than the mean dislocation spacing for annealed metals. As a result, the indenter produces shear stresses underneath the tip contact region that approaches the theoretical shear strength [8–10]. A pop-in occurs in the load displacement curves at the onset of plasticity, which correlates to homogeneous dislocation nucleation. Low dislocation densities are required; otherwise the indenter mainly activates existing sources such as Frank–Read sources, and no pop-ins occur. In addition the in situ imaging capability of

R. Haug (Ed.): Advances in Solid State Physics,
Adv. in Solid State Phys. **47**, 253–269 (2008)

a NI-AFM allows to test areas that are either free of defects, i.e., second phases and grain boundaries, or testing intentionally near these defects by in situ imaging of the surface and precisely positioning the tip before indentation [1,11,12]. On the basis of these unique abilities we decided to revisit two extensively studied and discussed topics in materials science, namely "hydrogen embrittlement" (HE) and "strain rate sensitivity of nanocrystalline (nc) and ultra fine grained (UFG) materials". Both topics seems to be quite diverse, however, in fact, microscopically they are both controlled by dislocation nucleation and rate laws. Since strain rate sensitivity depends on the interaction of the dislocation core with segregates like hydrogen and on the probability of dislocation to overcome local barriers like forest dislocations. The first part of the paper will discuss the strain rate sensitivity observed in Ni below a grain size of 1 μm and the second part will be devoted to HE in Ni.

1.1 Strain Rate Sensitivity

The strain rate sensitivity of typical fcc metals is usually benign. However, many groups observed strong strain rate dependence in nanocrystalline fcc materials [13, 14], even at low temperatures. Depending on grain size and temperature additional deformation mechanisms, such as emission of partial dislocations from grain boundaries [15], micro twinning, grain rotation [13] and conventional creep mechanisms [16] which should operate even at room temperature are discussed, simulated and shown experimentally under special conditions. Global observation of these mechanisms using conventional mechanical tests at different temperatures and loads, like creep and tensile tests, only yields results averaged over the whole microstructure. Whereas in simulations [13], relatively small model systems consisting of typical one million atoms are considered. Therefore we revisited this topic with specifically designed NI-AFM tests.

1.2 Hydrogen Embrittlement

The deleterious effect of hydrogen on mechanical properties was first documented in 1875 by *Johnson* [17], who reported that hydrogen in iron and steel causes a reduction in ductility and fracture stress. Since then it has been shown that HE is not restricted to iron and steel but occurs in many materials. In addition to the reduction in ductility and fracture stress, it has been shown that hydrogen can alters the fracture mode from ductile to brittle [18], increases slip localization [19], and increases slip planarity [20]. In Nb, V, Zr, Ti, and alloys based on them, the presence of the hydrogen can result in the formation of the brittle hydride phase [21, 22]. Numerous mechanisms have been proposed to explain the effect of hydrogen on the mechanical properties of materials, but their relative importance still remains unclear, and it is likely that a combination of these processes may simultaneously contribute to embrittlement [18–23]. Moreover, it is widely observed that hydrogen dislocation interaction is a prelude to embrittlement. The effect of hydrogen on deformation has been explored using conventional tensile and compression tests on relatively large specimens [19, 20], and also by in situ transmission electron microscopy (TEM) [23]. Although the qualitative effects of hydrogen can be observed from conventional mechanical experiments, the results are difficult to interpret quantitatively. This is mainly due to difficulties in exploring the details of plastic deformation locally. In addition, environmental TEM has been used to study in detail the effect of hydrogen on dislocations (plastic deformation). These in situ TEM experiments on thin films (plain stress) are difficult to interpret, since the electron beam produces high hydrogen fugacities and local heating [24]. Mainly based on these two different experimental approaches hydrogen enhanced decohesion (HEDE) and/or hydrogen enhanced local plasticity (HELP) are discussed. In the HEDE modelit is assumed that hydrogen accumulates at trapping sites such as cracks or interfaces, and reduces the bond strength there. Supporting experiments show directly that, for example, the crack tip opening angle (CTOA) in stressed Fe-3wt.% Si and nickel single crystal decreases progressively with increasing hydrogen pressure. This can only be explained by a mixture of slip and bond breaking, whereby the latter increases with increasing pressure.

For HELP two important mechanisms have been established: (1) hydrogen facilitates dislocation motion(2) hydrogen promotes planar slip [23].

To reduce the hydrogen-dislocation interaction volume we design a novel environmental NI-AFM for controlled electrochemical testing [25, 26].

2 Experimental

2.1 Materials and Sample Preparation

Ni plates were produced by pulse electrodeposition (PED) with rectangular dimensions of 65 mm×35 mm and thicknesses of 2 to 4 mm. Initial grain sizes of 20 nm, 30 nm, 40 nm and 70 nm were obtained depending on the PED conditions. The obtained PED material had high purity (>99.9 %) and no porosity. Experimental details of the PED are given by *Hempelmann* et al. [27]. Single crystal Ni was produced using a modified Bridgman technique. Misalignment of the single crystal plates were reduced to less than 2°, checked by Electron Back-Scatter Diffraction (EBSD). The samples were mechanically polished up to 0.25 µm. The Ni crystal were annealed at 1200°C in a vacuum better than 10^{-6} mbar for 24 hours. To obtain grain sizes up to 1 µm from a 20 nm starting nc-Ni material, the final heat treatments in vacuum (10^{-6} mbar) are summarised in Table 1.

Figure 1 gives examples of the resulting microstructures. For some grain sizes bimodal distributions could not be avoided (Fig. 1c, h, i). Since the NI experiment is very sensitive to surface condition and roughness, the samples

Table 1. Heat treatments to produce grain sizes from nc to UFG

Heat treatement	Mean Grain Size (nm)	Figure 3
as-deposited	21.2	–
150°C for 0.5 h	24	a
170°C for 0.5 h	37	b
170°C for 4 h	bimodal	c
190°C for 2 h	80	–
210°C for 2 h	250	d
250°C for 2 h	450	e
280°C for 2 h	720	f
300°C for 2 h	900	g
300°C for 4 h	bimodal	h
440°C for 2.5 h	bimodal	i
800°C for 2 h	40,000	–

were electropolished in a Methanolic 1 M H_2SO_4 solution [25]. To diminish oxidation effects which strongly affect the reproducibility of the experiments as demonstrated in [14] the specimens were immediately indented after electropolishing.

2.2 NI-AFM Method

The indentation tests were performed using a Hysitron Triboscope® in conjunction with a Digital Instruments (DI) Nanoscope II® taking the control of the AFM unit. For all experiments a tip with a Berkovich geometry has been used.

2.2.1 Strain Rate Sensitivity Tests with the NI-AFM

The load-displacement curves were evaluated as a function of the loading rate and the distance between indent and the grain boundary. During the first testing series grain sizes from micro to nano scale were investigated. The loading and unloading rates of the indentation were changed over a wide range in order to reproduce the findings of the tensile tests on bulk samples as reported elsewhere [14]. In the second testing series single grains were examined. The indenter was centered within single grains in order to study the interaction between dislocations and grain boundaries. For a given heat treatment that means a given grain size distribution, we always indented in the centers of the largest grains in order to guarantee that only the middle sections of grains were examined.

2.2.2 HE Tests with the NI-AFM

For the nanoindentation system, a three-electrode electrochemical setup was developed, as shown in Fig. 2. A platinum wire was used as a counter electrode

Fig. 1. Atomic force microscope (AFM)(SEM, (i)) – images of the microstructure of nc Ni obtained with the heat treatments given in Table 1

and an Ag/AgCl as a reference electrode. All electrochemical potentials in this study are therefore reported against an Ag/AgCl reference electrode. The cell was made from Teflon™ to be fitted to the DI Multimode sample holder. Nanoindentation was conducted inside this electrochemical cell while the sample was covered with approximately 2 mm of electrolyte. To eliminate the oxygen effect on the electrochemical reaction, the whole system rested in a chamber containing a protective atmosphere of nitrogen and helium. The solution was injected from outside the chamber through a polyethylene tube connected to a MicroFil™ pipette.

3 Results

3.1 Strain Rate Sensitivity in nc-Ni and UFG Ni

For grain sizes from micro to nano the loading and unloading rates were changed over a wide range. Figure 3 shows the effect of different loading rates on the load displacement curves. The maximal load was kept constant

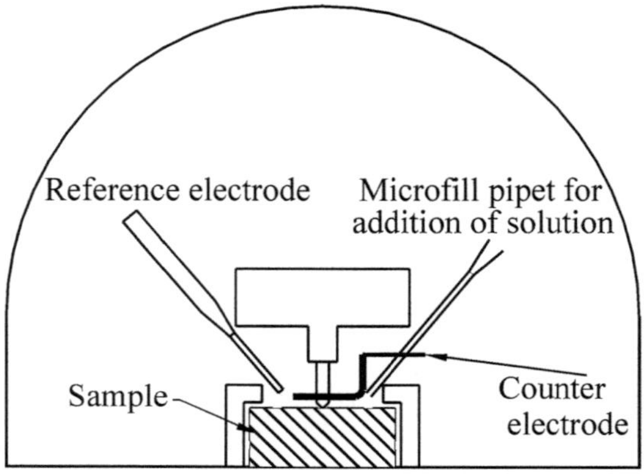

Fig. 2. Experimental setup used in HE tests

at 4000 μN and the loading rates were changed from 10 μN/s over 100 μN/s up to 1000 μN/s.

The material shows strain rate effects by changing the loading rate during NI. Clearly the observed effects are similar to the results of the tensile tests [14]: no strain rate sensitivity for large grains (Fig. 3a), high sensitivity for smaller grains (Fig. 3b,c). Therefore, not only experiments where the grain size is in the order of the tip size are interesting for NI, but also tests in materials where the grain size is much smaller than the tip. This is especially interesting for examination of strain rate sensitivity of thin films.

In addition we examined the interaction of the indenter with single grain boundaries. For this purpose we choose the grain size just a little bit larger than the indent size, that means grain sizes of above 200 nm.

Figure 4 shows such a typical experiment [14]. The influence of grain size on the pop-in width and on the displacement depth can be seen in the load displacement curves recorded for different grain sizes (shown in the AFM-pictures, Fig. 4). In order to interpret our results we have to clarify that the effect of the interaction between the dislocations and grain boundaries is not overridden by the indentation size effect. Figure 5 demonstrates the effect of boundaries on the indentation size effect. Below a grain size of 900 nm the decrease in the penetration depth is not due to the indentation size effect since the effect of lateral boundaries clearly overrides this effect (Fig. 5).

From Fig. 4 it can be directly deduced that when the plastic zone reaches the boundary the penetration depth decreases. In our view the decrease in pop-in width is an indication that the dislocations nucleated below the indenter interact directly with the surrounding grain boundaries. Above a grain size of 900 nm the pop-in width became independent of grain size (Fig. 6), which means that the dislocations only interact with each other but do not interact with the boundary here.

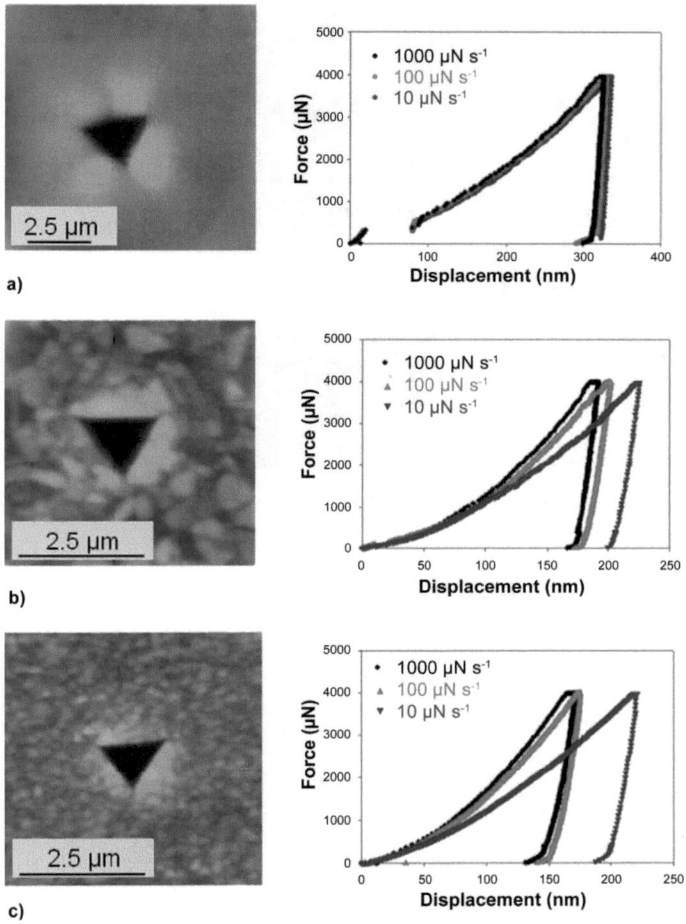

Fig. 3. Effect of loading rate on the load-displacement curve for Ni with different grain sizes (grain sizes: (**a**) 40000 nm, (**b**) 250 nm, (**c**) 21 nm) at room temperature

Below and above this grain size (900 nm) we studied the effect of the indentation rate on the load-displacement curve. The maximal load was kept constant at 300 µN and the loading rates were changed from 10 µN/s over 20 µN/s up to 40 µN/s. An example of a typical measurement is given in Fig. 7. It shows that in the case where the pop-in width is independent of grain size (above 900 nm in Fig. 6) no rate effects occurred. However, in the second case when the boundaries restrict the pop-in width a strong rate dependence and a decreasing penetration depth at a constant load were found (grain size below 900 nm; Fig. 7b).

This indicates that dislocation pinning at grain boundaries is the rate controlling mechanism.

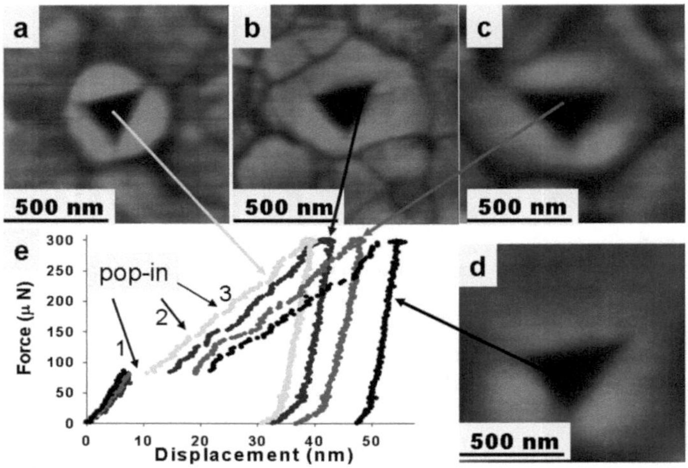

Fig. 4. Influence of grain size on the pop-in width

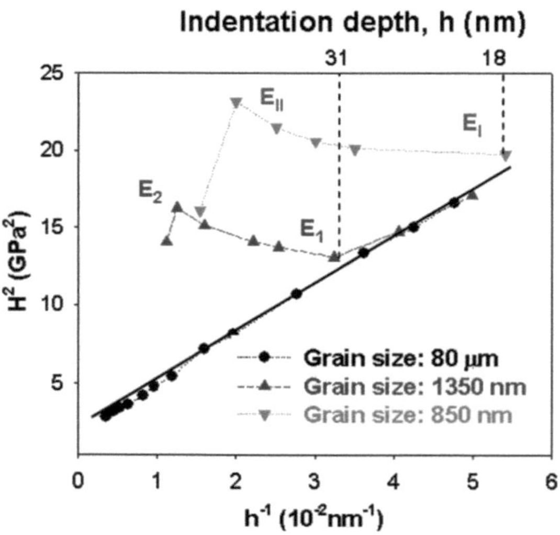

Fig. 5. The effect of lateral grain boundaries on the Nix Gao plot [5]

3.2 HE in Ni

The in situ electrochemical nanoindentation of nickel (111) surfaces was carried out in a 0.05 M Na_2SO_4 electrolyte prepared from an analytical grade Na_2SO_4 and double distilled water. To reduce the air formed oxide layer of nickel, the pH of the solution was set to 6 by adding trace amounts of analytical grade H_2SO_4 [28]. A new passive layer forms after subsequent passivation, which is stable under cathodic potentials, i.e., this passive layer is not vul-

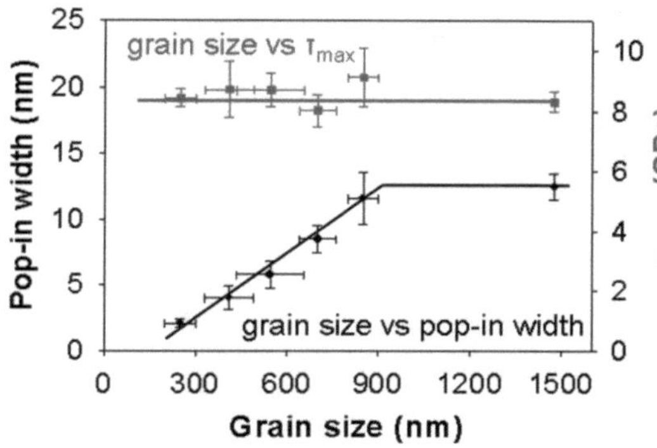

Fig. 6. Pop-in width and maximum shear stress vs. individual grain size

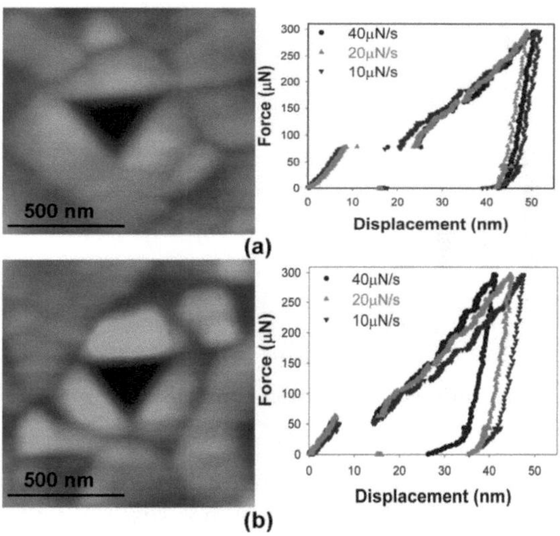

Fig. 7. Rate dependence of the force-penetration curve: (**a**) for larger grains (900 nm) no rate effects were found; (**b**) for smaller grains (450 nm): pop-in width dependent on grain size, strong rate effects were observed (experiments at RT)

nerable to cathodic reduction [28]. This stable surface film guarantees that all results reported below are not due to film effects [26]. The freshly electropolished nickel sample was installed in the nanoindentation setup (Fig. 2) and the chamber was then filled with a mixture of nitrogen and helium. The specimen was kept at a cathodic potential of −1000 mV while the electrochemical cell was being filled with the electrolyte. The air-formed oxide layer was removed by keeping the sample at this potential for 15 minutes as recommended by *MacDougall* et al. [28]. The sample was then swept with a scan rate of 10 mV/s to an anodic potential of 500 mV to passivate the surface and

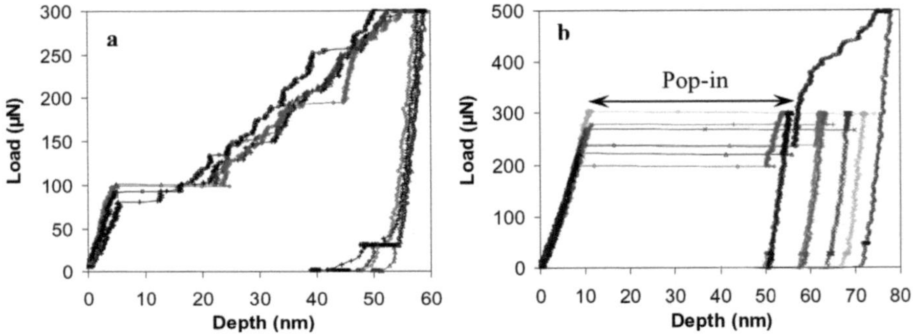

Fig. 8. Load displacement curves for nickel at cathodic (**a**) and anodic (**b**) potentials

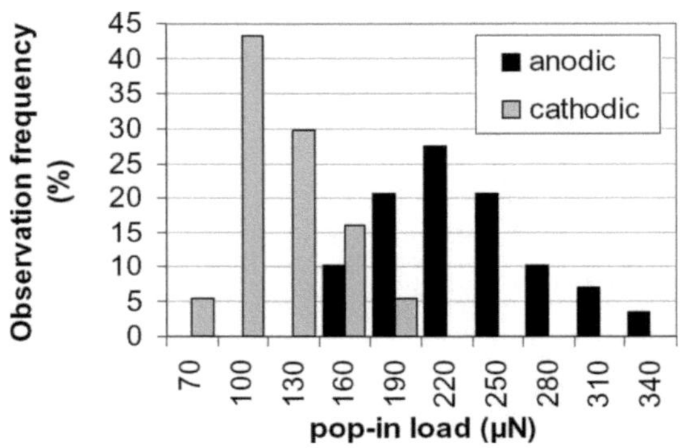

Fig. 9. Pop-in load distribution under various polarization conditions for nickel

produce the stable passive oxide layer. Under these conditions, the passive layer of nickel has a thickness of about 1.2–1.4 nm and is not susceptible to cathodic reduction in pHs between 4 and 8 [28], i.e., at the pH used. This stable thin surface film guarantees that all results reported below are not due to film effects as has been shown previously [26]. The sample was kept at anodic potential for 15 min before switching to cathodic potential. The nanoindentation tests were started after approximately one minute of hydrogen charging at cathodic potential. Typical load displacement curves obtained under cathodic potential are shown in Fig. 8a. After several indentations, the potential was again switched to anodic and maintained for several seconds before indentation. Figure 8b shows typical anodic load displacement curves with pop-in loads that are considerably higher than those at cathodic potential. This sequence was performed several times to check the reproducibility of the observations for each polarization condition. The mean values of the pop-in load distribution under each condition are shown in Fig. 9.

For an indentation test, the applied shear stress that nucleates a dislocation can be assumed to be the maximum shear stress beneath the indenter at the onset of a pop-in. According to continuum mechanics the maximum shear stress is reached at a point below the sample surface of approximately 0.48 times the contact radius. Computer simulations support this approximation [29]. The position and value of the maximum shear stress $z_{(\tau_{max})}$, and τ_{max}, are given by [30].

$$z_{(\tau_{max})} = 0.48 \left(\frac{3PR}{4E_r}\right)^{\frac{1}{3}} \tag{1}$$

$$\tau_{max} = 0.31 \left(\frac{6E_r^2}{\pi^3 R^2} P\right)^{\frac{1}{3}} \tag{2}$$

where P is the applied load, R is the radius of the tip curvature, and E_r is the reduced modulus, given by [30].

$$\frac{1}{E_r} = \frac{1-\nu_1^2}{E_1} + \frac{1-\nu_2^2}{E_2} \tag{3}$$

where E is the elastic modulus and ν is the Poisson's ratio. The subscripts (1) and (2) indicate the tip and the sample respectively. For nickel and copper and a diamond tip, (3) gives E_r=191 GPa and 111 GPa respectively. The tip radius was obtained by fitting a Hertzian model to the elastic loading part of the load displacement curves [12]. If we insert this tip radius into (2), we obtain a maximum shear stress for each pop-in load. This maximum shear stress is responsible for the homogeneous dislocation nucleation at $z_{(\tau_{max})}$ below the tip. Classic dislocation nucleation theory as summarized by *Hirth* and *Lothe* [31] suggests that the shear stress depends on the energy required to generate a dislocation loop. The free energy of a circular dislocation loop of radius r is given by

$$\Delta G = 2\pi r W + \pi r^2 \gamma - \pi r^2 b \tau \tag{4}$$

where W is the line energy of the dislocation loop, b is the Burgers vector for a partial dislocation in an FCC lattice, τ is the external shear stress acting on the loop, and γ is the stacking fault energy (SFE). The first term on the right-hand side of (4) describes the energy required to create a dislocation loop in a defect free lattice. The second term indicates the increase in energy due to the creation of a stacking fault. The sum is equal to the total increase in lattice energy due to the formation of a dislocation loop. The last term gives the work done by the applied stress τ as a result of the Burgers vector displacement and indicates the work done on the system to expand the dislocation loop. The line energy W for the loop, which results from the lattice strain in the vicinity of the dislocation for $r > \rho$, is given by [31]

$$W = \frac{2-\nu}{1-\nu}\frac{\mu b^2}{8\pi}\left(\ln\frac{4r}{\rho} - 2\right) \tag{5}$$

where μ is the shear modulus and ρ is the dislocation core radius. Using (5), we can rewrite (4) as follows:

$$\Delta G = \frac{2-\nu}{1-\nu}\frac{\mu b^2 r}{4}\left(\ln\frac{4r}{\rho}-2\right) + \pi r^2\gamma - \pi r^2 b\tau\,. \tag{6}$$

Figure 10 shows a plot of (6) calculated for nickel. For a given shear stress, at small radii, the total increase in the lattice energy as a result of the dislocation loop formation is larger than the external work done by the applied shear stress, and the total free energy is positive. This situation changes, however, as the radius grows in size, so that with larger radii, the free energy becomes negative and the free energy of formation has a critical maximum value ΔG_C at a loop size r_C, which could be found by setting $\partial\Delta G/\partial r = 0$. This means that the growth of any nucleus with a radius smaller then r_C requires an activation energy to overcome the ΔG_C energy barrier for forming a stable dislocation loop. The available thermal energy at room temperature is 0.026 eV. Therefore, where the free energy maximum is lower than 0.026 eV, spontaneous dislocation nucleation is possible. This is only the case, if the applied shear stress is lower than a critical shear stress, τ_C ($\approx$ 3 GPa for nickel). At stresses higher than this value, the free energy curve is always negative. This is exactly what happens during a pop-in in nanoindentation. In Fig. 11, the free energy curves for all measured pop-in loads of nickel (Fig. 9) are plotted using its physical properties [32, 33] and a tip radius of 1 (μm). A very good agreement between the homogeneous dislocation nucleation model and experimental results in nickel was obtained at anodic potentials. This means pop-in only observed for loads which their free energy maximum is lower than available thermal energy at room temperature. The pop-ins at cathodic potentials, however, seem to overcome the existing energy barrier for dislocation nucleation at lower shear stresses (loads). Since the thermal energy available during nanoindentation at room temperature is very low (0.026 eV), the only possible external contribution to overcome the activation energy for dislocation nucleation is hydrogen, charged electrochemically into the metal. Hydrogen is therefore obviously reducing the activation energy required for dislocation nucleation in nickel. For a first approximation, we can rewrite (6) introducing the effect of hydrogen as follows:

$$\Delta G = \frac{2-\nu}{1-\nu}\frac{\mu_H b^2 r}{4}\left(\ln\frac{4r}{\rho}-2\right) + \pi r^2\gamma_H - \pi r^2 b\tau \tag{7}$$

where μ_H is the altered modulus due to hydrogen, and γ_H is the hydrogen affected stacking fault energy. The effects of hydrogen on the other parameters involved in (7) are either neglected or inapplicable. We can now use the experimental data to calculate the effect of hydrogen on these constants. To do this, we apply the rule of spontaneous homogeneous dislocation nucleation to (7), i.e., $G_C = 0$. This results in the following equation:

$$\tau_H = \frac{2-\nu}{1-\nu}\frac{\mu_H b}{4r_C}\left(\ln\frac{4r_C}{\rho}-2\right) + \frac{\gamma_H}{b} \tag{8}$$

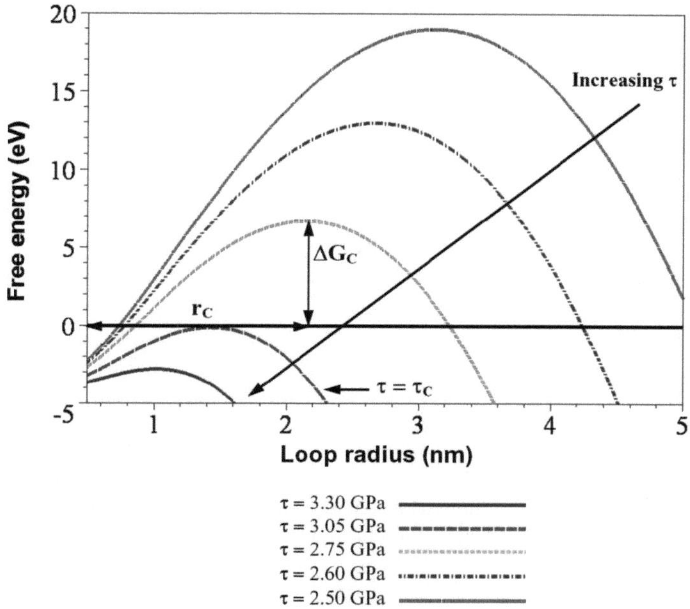

Fig. 10. Change in free energy of homogeneous dislocation nucleation for nickel as a function of dislocation loop radius (r) for various applied shear stresses (τ) on the crystal lattice. The activation energy for dislocation nucleation (ΔG_C) and critical loop radius (r_C) are shown for one of the curves

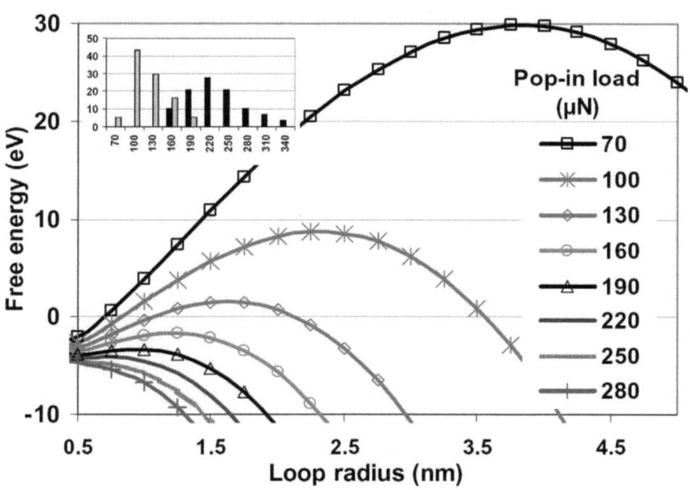

Fig. 11. Change in free energy of homogeneous dislocation nucleation for nickel as a function of dislocation loop radius (r) for various experimentally observed pop-in loads in Fig. 9 (Pop-in load distribution plot, *upper left corner*)

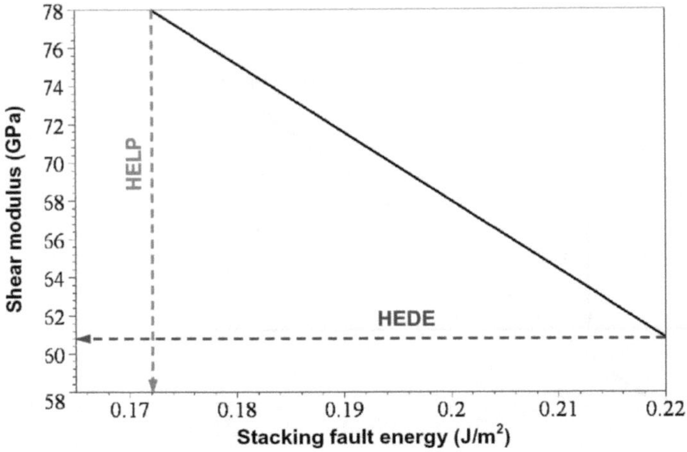

Fig. 12. Possible changes in shear modulus and SFE due to hydrogen for overcoming the homogeneous dislocation nucleation energy barrier. Exclusive change in shear modulus (μ) results in HEDE. Exclusive change in SFE (γ) results in HELP

where τ_H is the shear stress required for the homogeneous dislocation nucleation in hydrogen charged crystals. Using the most frequently observed pop-in load at a cathodic potential of 100 μN (Fig. 9), we can calculate τ_H using (2) and insert it into (8). The resulting relation between μ_H and γ_H is plotted in Fig. 12.

The plot shows the values for shear modulus and stacking fault energy which fulfills (8) for the most frequently observed pop-in load in the hydrogen charged nickel sample (100 μN). Assuming a constant stacking fault energy for nickel ($\gamma_H = \gamma = 0.22\,\mathrm{J/m^2}$), hydrogen reduces the shear modulus from 78 GPa to 61 GPa according to Fig. 12. A 22 % reduction of the shear modulus is therefore obtained. Fundamental property which determine the elastic constants is the binding energy [32]. This have a simple physical meaning: the reduction in the elastic modulus is equal to the reduction of the strength of the interatomic bonds assumed in the HEDE model.

On the other hand, assuming a constant shear modulus ($\mu_H = \mu =$ 78 GPa) results in the reduction of the stacking fault energy from $0.22\,\mathrm{J/m^2}$ [33] to $0.172\,\mathrm{J/m^2}$, which alters the dissociation of partial dislocation. This induces an increase of the equilibrium distance and of the recombination work due to the increase in the separation of partials. This means an increase in the activation energy required for cross-slip of the dislocations, which increase the tendency of the dislocations to stay on the same slip plane. This localizes slip and supports the HELP mechanism. This is in agreement with previously reported observation of slip planarity in nickel both experimentally [20] and through simulations [34, 35]. According to the results presented here, it seems that both the HEDE and the HELP models are working in parallel to each other in nickel. This explains the discrepancy in reported mechanisms [18, 20, 36]. Depending on the experimental procedure used to investigate the HE, either HEDE or HELP has been observed and reported. In our previous experiments, for example, in which cyclically hardened notched nickel single crystals were tested in low-cycle fatigue, HEDE was observed [18, 36]. Where in the compression experiments on perfectly annealed nickel samples, *McInteer* et al. observed slip planarity and reported HELP [20]. This means that, depending on the condition of the sample (purity, defect density, heat treatment, cold work, etc.) and the observation technique (mechanical testing method, sample geometry, hydrogen charging method, etc.), one of the two mechanisms surpasses the other. In this study, however, by selecting a proper local measurements technique that is able to probe a small, fully-characterized volume of the material, we see that hydrogen may result in both HEDE and HELP in nickel.

4 Conclusion

The effect of strain rate and hydrogen on dislocation nucleation is demonstrated in tests with ultra fine grained and single crystalline nickel in air and in situ electrochemical nanoindentation experiments. The test with grains sizes just a little bit larger than the indent size clearly demonstrated that the strain rate sensitivity at room temperature results from the preferential interaction of dislocations with grain boundaries. In these tests the pop-in load was not influenced by the grain size. In hydrogen, however, the pop-in load was reduced drastically. In the thermodynamic frame work of nucleation theory these results can be only explained if it is assumed that hydrogen either reduces the stacking fold energy or the interatomic bond strength. The former effect is in agreement with hydrogen enhanced local plasticity (HELP), while the latter is evidence for hydrogen enhanced decohesion (HEDE). It is shown that both the HEDE and the HELP models are working in parallel to each other, and by selecting an appropriate local measurements technique it is possible to see that hydrogen may result in both HEDE and HELP in nickel. To support these assumptions similar measurements on other crystal systems are under way.

References

[1] B. Yang, H. Vehoff: The effect of grain size on the mechanical properties of nanonickel examined by nanoindentation, Zeitschrift für Metallkunde/Materials Research and Advanced Techniques **95**, 499–504 (2004)

[2] Z. Zhang, H. Kristiansen, J. Liu: A method for determining elastic properties of micron-sized polymer particles by using flat punch test, Computational Materials Science **39** (2), 305–314 (2007)

[3] M. D. Uchic, D. M. Dimiduk: A methodology to investigate size scale effects in crystalline plasticity using uniaxial compression testing, Materials Science and Engineering A **400–401**, 268–278 (2005)
[4] C. Volkert, E. Lilleodden: Size effects in the deformation of sub-micron Au columns, Philosophical Magazine **86** (33–35 SPEC. ISSUE), 5567–5579 (2006)
[5] B. Yang, H. Vehoff: Dependence of nanohardness upon indentation size and grain size – a local examination of the interaction between dislocations and grain boundaries, Acta Materialia **55**, 849–856 (2007)
[6] H. Guo, B. Lu, J. Luo: Response of surface mechanical properties to electrochemical dissolution determined by in situ nanoindentation technique, Electrochemistry Communications **8**, 1092–1098 (2006)
[7] M. Oden, H. Ljungcrantz, L. Hultman: Characterization of the induced plastic zone in a single crystal TiN(001) film by nanoindentation and transmission electron microscopy, Journal of Materials Research **12**, 2134–2142 (1997)
[8] H. Bei, E. P. George, J. L. Hay, G. M. Pharr: Influence of indenter tip geometry on elastic deformation during nanoindentation, Physical Review Letters **95**, 1–4 (2005)
[9] M. Göken, M. Kempf, M. Bordenet, H. Vehoff: Nanomechanical characterizations of metals and thin films, Surface and Interface Analysis **27**, 302–306 (1999)
[10] K. Durst, M. Göken, H. Vehoff: Finite element study for nanoindentation measurements on two-phase materials, Journal of Materials Research **19**, 85–93 (2004)
[11] M. Kempf, M. Göken, H. Vehoff: Nanohardness measurements for studying local mechanical properties of metals, Applied Physics A Materials Science & Processing **66**, S843–S846 (1998)
[12] B. Yang, H. Vehoff: Grain size effects on the mechanical properties of nanonickel examined by nanoindentation, Materials Science and Engineering A **400–401**, 467–470 (2005)
[13] R. J. Asaro, S. Suresh: Mechanistic models for the activation volume and rate sensitivity in metals with nanocrystalline grains and nano-scale twins, Acta Materialia **53**, 3369–3382 (2005) cited by (since 1996): 48
[14] H. Vehoff, D. Lemaire, K. Schüler, T. Waschkies, B. Yang: The effect of grain size on strain rate sensitivity and activation volume – from nano to ufg nickel, Int. J. Mat. Res. **98**, 4 (2007)
[15] H. Van Swygenhoven, A. Caro: Plastic behavior of nanophase metals studied by molecular dynamics, Phys. Rev. B **58**, 11246–11251 (1998)
[16] E. Ma: Watching the nanograins roll, Science **305**, 623–624 (2004)
[17] W. H. Johnson: On some remarkable change produced in iron and steel by the action of hydrogen and acids, Proceedings of the Royal Society of London **23**, 168–179 (1875)
[18] H. Vehoff, H. K. Klameth: Hydrogen embrittlement and trapping at crack tips in Ni single crystals., Acta Metallurgica **33**, 955–962 (1985) cited by (since 1996): 13
[19] D. G. Ulmer, C. J. Alstetter: Hydrogen-induced strain localization and failure of austenitic stainless steels at high hydrogen concentrations, Acta Metallurgica et Materialia **39**, 1237 (1991)
[20] W. McInteer, A. W. Thompson, I. M. Bernstein: The effect of hydrogen on the slip character of nickel, Acta Metallurgica **28**, 887 (1980)

[21] A. Pundt, R. Kirchheim: Hydrogen in metals: Microstructural aspects, Annual Review of Materials Research **36**, 555–608 (2006)
[22] H. Vehoff: *Hydrogen in Metals III* (Springer Berlin / Heidelberg 1997)
[23] I. M. Robertson, H. K. Birnbaum: HVEM study of hydrogen effects on the deformation and fracture of nickel., Acta Metallurgica **34**, 353–366 (1986) cited by (since 1996): 75
[24] G. M. Bond, I. M. Robertson, H. K. Birnbaum: On the determination of the hydrogen fugacity in an environmental cell tem facility., Scripta metallurgica **20**, 653–658 (1986) cited By (since 1996): 14
[25] A. Barnoush, H. Vehoff: Electrochemical nanoindentation: A new approach to probe hydrogen/deformation interaction, Scripta Materialia **55**, 195–198 (2006)
[26] A. Barnoush, H. Vehoff: In situ electrochemical nanoindentation of a nickel (111) single crystal: Hydrogen effect on pop-in behaviour, International Journal of Materials Research **97**, 1224–1229 (2006)
[27] H. Natter, R. Hempelmann: Tailor-made nanomaterials designed by electrochemical methods, Electrochimica Acta **49**, 51–61 (2003)
[28] B. MacDougall, M. Cohen: Anodic oxidation of nickel in natural sulfate solution, J. Electrochem. Soc. **121**, 1152 (1974)
[29] K. J. Kim, J. H. Yoon, M. H. Cho, H. Jang: Molecular dynamics simulation of dislocation behavior during nanoindentation on a bicrystal with a Σ=5(210) grain boundary, Materials Letters **60**, 3367–3372 (2006)
[30] K. Johnson: *Contact mechanics* (Cambridge University Press 2003)
[31] J. P. Hirth, J. Lothe: *Theory of dislocations* (Wiley, New York 1981)
[32] C. Kittel: *Introduction to Solid State Physics*, 4th edition ed. (John Wiley, New York 1971)
[33] R. P. Reed, R. E. Schramm: Relationship between stacking-fault energy and x-ray measurements of stacking-fault probability and microstrain, J. Appl. Phys **45**, 4705 (1974)
[34] M. Wen, S. Fukuyama, K. Yokogawa: Atomistic simulations of hydrogen effect on dissociation of screw dislocations in nickel, Scripta Materialia **52**, 959–962 (2005)
[35] J. Chateau, D. Delafosse, T. Magnin: Numerical simulations of hydrogen-dislocation interactions in fcc stainless steels. Part I: hydrogen-dislocation interactions in bulk crystals, Acta Materialia **50**, 1507–1522 (2002)
[36] H. Vehoff, W. Rothe: Gaseous hydrogen embrittlement in FeSi and Ni single crystals, Acta Metallurgica **31**, 1781 (1983)

Tuning the Strain in $LaCoO_3$ Thin Films by the Heteroepitaxial Growth on Single Crystal Substrates

Dirk Fuchs[1], Erhan Arac[1,2], Thorsten Schwarz[1,2], and Rudolf Schneider[1]

[1] Forschungszentrum Karlsruhe, Institut für Festkörperphysik, Hermann-von-Helmholtz-Platz 1, 76344 Eggenstein-Leopoldshafen, Germany
dirk.fuchs@ifp.fzk.de

[2] Physikalisches Institut, Universität Karlsruhe, 76128 Karlsruhe, Germany

Abstract. Epitaxial strain in $LaCoO_3$ thin films has been changed by the growth of thin films on different substrate materials, i.e., (001) oriented $SrLaAlO_4$, $LaAlO_3$, $SrLaGaO_4$, $(LaAlO_3)_{0.3}(Sr_2AlTaO_6)_{0.7}$, and $SrTiO_3$. The films were deposited by pulsed laser deposition. The lattice mismatch of the in-plane lattice parameters between the substrate, a_s, and $LaCoO_3$, a_b, $\epsilon = (a_s\text{-}a_b)/a_b$, ranges from $\epsilon = -1.32\,\%$ for the growth on $SrLaAlO_4$ to $\epsilon = +2.76\,\%$ for films on $SrTiO_3$. The characterization of the structural properties such as film lattice parameters, unit cell volume and epitaxial strain was carried out by $\theta/2\theta$ scans and reciprocal space mapping on a two-circle and four-circle x-ray diffractometer, respectively. We succeeded in the growth of single phased (001) oriented LCO films on all the substrate materials. For the lattice parameters of the $LaCoO_3$ films we observed a quasi-elastic behaviour in the range of $-1\,\% < \epsilon < +2\,\%$. The out of plane lattice parameter c increases and the in-plane lattice parameters a and b decrease nearly linearly with decreasing in-plane lattice parameter a_s of the substrate material. This indicates an elastic coupling of the film to the substrate and therefore the possibility to tune the Co–O bond-length and thus the crystal field splitting, Δ_{CF}, by epitaxial strain.

1 Introduction

The different possible spin states of the perovskite $LaCoO_3$ (LCO) [1,2], i.e., a nonmagnetic low spin state with a spin-value of $S = 0$ below $T = 35\,\mathrm{K}$, a high spin state with $S = 2$ above $T = 100\,\mathrm{K}$ and a possible intermediate spin-state with $S = 1$ for $35\,\mathrm{K} < T < 100\,\mathrm{K}$, are very sensitive to variations of the Co–O bond length and the Co–O–Co bonding angle because of the subtle balance between the crystal field splitting, Δ_{CF}, and the interatomic exchange energy, J_{ex} [3]. A decrease of the Co–O bond length leads for example to a decrease of Δ_{CF} and can thus cause an increased population of higher spin states. The balance between Δ_{CF} and J_{ex} can also easily be influenced by, e.g., hole or electron doping [4], and chemical or external pressure [5].

A simple avenue to change the Co–O bonding length is the heteroepitaxial growth of LCO thin films on different single crystal substrate materials. Dif-

R. Haug (Ed.): Advances in Solid State Physics,
Adv. in Solid State Phys. **47**, 271–275 (2008)

Table 1. In-plane lattice parameters, a_s and b_s, and the corresponding lattice mismatch, $\epsilon = (a_s - a_b)/a_b$, for the different substrate materials assuming a cube on cube growth of $LaCoO_3$ ($a_b \approx 3.80$ Å) on the substrate material

substrate	orientation	abbreviation	$a_s = b_s$ (Å)	ϵ(%)
$SrLaAlO_4$	(001)	SLAO	3.75	−1.32
$LaAlO_3$	(001)	LAO	3.78	−0.53
$SrLaGaO_4$	(001)	SLGO	3.84	+1.05
$(LaAlO_3)_{0.3}(Sr_2AlTaO_6)_{0.7}$	(001)	LSAT	3.87	+1.84
$SrTiO_3$	(001)	STO	3.90	+2.76

ferent in-plane lattice parameters of the substrates may influence the epitaxial strain and thus, the bonding lengths and bonding angles of the film material. For that reason we have investigated the possibility of the epitaxial growth of LCO on various substrates. Beside the prerequisites for epitaxial growth such as chemical stability, the near lattice matching and also the matching of the thermal expansion coefficients we additionally required the feasibility for a 'Cube on Cube growth' of LCO on the substrate material, i.e., *a*-axes and *b*-axes of the substrate- and LCO film material are parallel to each other. Furthermore, because of the pseudo cubic crystal structure of perovskites the substrate material should also have a square shaped surface cell. We have figured out the following suitable and commercially available substrate materials: (001) oriented $SrLaAlO_4$ (SLAO), $LaAlO_3$ (LAO), $SrLaGaO_4$ (SLGO), $(LaAlO_3)_{0.3}(Sr_2AlTaO_6)_{0.7}$ (LSAT), and $SrTiO_3$ (STO). The in-plane lattice parameters, a_s and b_s, of the substrate materials and the corresponding lattice mismatch to LCO, $\epsilon = (a_s - a_b)/a_b$, where a_b is the pseudo cubic bulk lattice parameter of LCO, are listed in Table 1. The lattice mismatch is ranging from $\epsilon = -1.32\,\%$ for SLAO to $\epsilon = +2.76\,\%$ for STO.

2 Experimental

The film deposition was carried out by pulsed laser deposition. The deposition parameters such as substrate temperature and oxygen partial pressure were optimized with respect to the crystalline quality of the films. The fluence and repetition rate of the laser light as well as the target – substrate distance were adopted from a previous optimization of the growth of LCO on LSAT [6]. For the characterization of the structural properties of the epitaxial films such as the out-of plane *c*-axis lattice parameter and mosaicity we carried out $\theta/2\theta$ scans and rocking curves on a two-circle x-ray diffractometer. We also carried out a reciprocal space mapping on asymmetric Bragg reflections on a four-circle x-ray diffractometer in order to characterize the in-plane lattice parameters, a and b, and the epitaxial strain. The composition of the films were checked by Rutherford backscattering spectrometry and energy dispersive x-ray analysis.

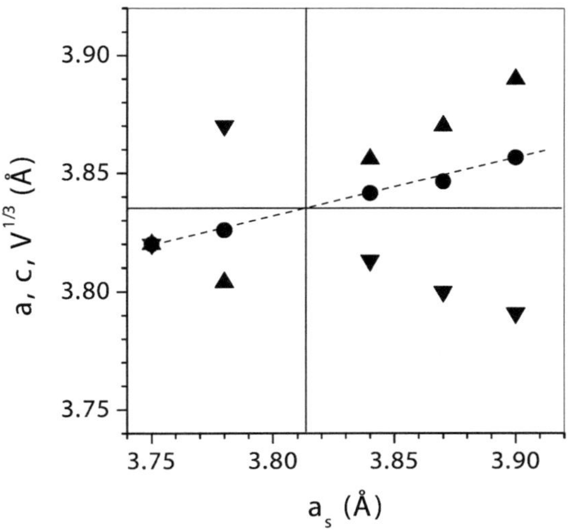

Fig. 1. Lattice parameters a (*up triangles*) and c (*down triangles*) and the unit cell volume $V^{1/3}$ (*circles*) of the LCO films as a function of the lattice parameter a_s of the corresponding substrate material. The *dotted line* is a guide to the eye

3 Results and Discussion

We succeeded in the growth of single phased (001) oriented LCO films on all the substrates. The mosaicity of the films which we determined from the full width at half maximum of the rocking curve at the 002 reflection was comparable for all the films and amounts to about 0.1°. In Fig. 1 we have plotted the lattice parameters and the unit cell volume of the LCO films as a function of the out-of-plane lattice parameter, a_s, of the substrate material. With increasing a_s the in-plane film lattice parameter a also increases and the out-of-plane lattice parameter c deceases resulting in a pseudo tetragonal structure. The tetragonal distortion, $\Delta_{TD} = |a - c|/(a + c)$, is largest for the films on STO and LAO, i.e., 2.6 % and 2.3 %, respectively, and about zero for the films on SLAO. However, deviations from the nearly linear behavior of a and c vs. a_s are clearly present for the LCO films grown on SLAO and STO. On the one hand, films grown on SLAO seem to be nearly fully relaxed with a pseudo cubic structure of $a = 3.82$ Å$= b \approx c$. Therefore, the maximum of compressive strain which we were able to apply to the films was restricted to about −1 %. On the other hand, the growth on STO substrates results in a strong formation of macroscopic cracks in the films after the deposition process. Because of the high tensile strain, i.e., +2.76 %, the cracking is most likely caused by a structural relaxation and thus the relief of tensile strain in the film. Therefore, the maximum of tensile strain which could be applied to the LCO films was limited somewhat below the expected value of $\epsilon =$ +2.76 %. We want to point out that the unit cell volume, V, is not constant

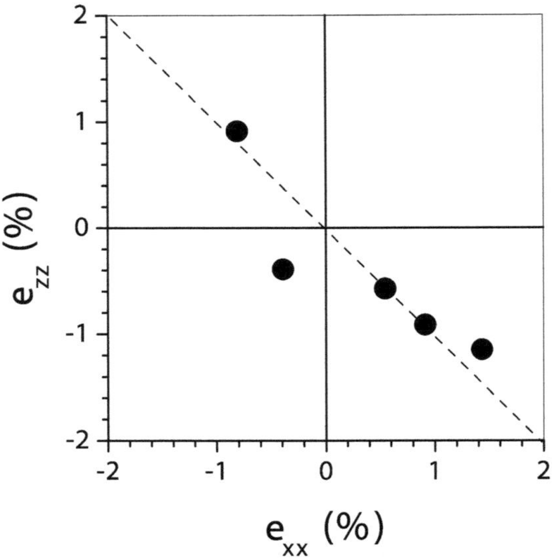

Fig. 2. Out-of-plane lattice strain, ϵ_{zz} as a function of the in-plane lattice strain ϵ_{xx} assuming a mean pseudo cubic lattice parameter of $a = 3.835$ Å, as suggested from Fig. 1. The *dotted line* reflects $\epsilon_{zz} = -\epsilon_{xx}$ which results in a poison ratio of $\nu = 1/3$

and also increases nearly linearly by about +3 % with increasing a_s within the measured range. Figure 1 suggests that a pseudo cubic structure with $a \approx c \approx 3.835$ Å will be obtained for films grown on substrates with $a_s \approx 3.815$ Å. This can be deduced from the crossing point of a versus a_s and c versus a_s at $a_s \approx 3.815$ Å. If we assume a mean pseudo cubic lattice parameter of $a = 3.835$ Å for fully relaxed LCO films we can calculate the resulting in-plane-,ϵ_{xx}, and out-of-plane lattice strain, ϵ_{zz}, of the LCO films. In Fig. 2. we have plotted ϵ_{zz} as a function of ϵ_{xx}. The dotted line represents $\epsilon_{zz} = -\epsilon_{xx}$ which results in a poison ratio of $\nu = 1/(1 - 2\epsilon_{xx}/\epsilon_{zz}) = 1/3$. A poison ratio of $\nu = 1/3$ has been also observed for epitaxially strained films of the hole doped cobaltates, i.e., $La_{0.7}A_{0.3}CoO_3$– A = Ca, Sr or Ba, and seems to be also typical for the manganates.

4 Summary

Epitaxialy strained thin films of $LaCoO_3$ have been prepared successfully by pulsed laser deposition on different substrate materials. Because of the square shaped surface cell of the chosen substrate materials the films could be grown in a cube-on-cube growth mode onto the substrates. The lattice mismatch of the in-plane lattice parameters between the substrate, a_s, and $LaCoO_3$, a_b, ranges from $\epsilon = -1.32$ % to +2.76 %. The nearly linear behavior of a and c as a function of the substrate lattice parameter a_s within the range of $-1\,\% < \epsilon < 2\,\%$, indicates an elastic coupling of the film to the substrate and therefore the possibility to tune the Co–O bond-length and thus Δ_{CF} by the epitaxial strain. However, beside the elastic behavior of the films in that range of ϵ we do also observe an increase of the unit cell volume, V, by about $+3\,\%$ which does also possibly contribute to a change in Δ_{CF}. In order to find out a possible relationship between the structural and magnetic properties of LCO we are planning ongoing experiments, where we will characterize the magnetic properties of epitaxial strained LCO films in detail.

References

[1] J. Q. Yan, J. S. Zhou, and J. B. Goodenough: Phys. Rev. B **69**, 134409 (2003)
[2] M. A. Korotin, S. Yu. Ezhov, I. V. Solovyev, V. I. Anisimov, D. I. Khomskii, and G. A. Sawatzky: Phys. Rev. B **54**, 5309 (1996)
[3] D. M. Sherman: *Advances in Physical Geochemistry* edited by S. K. Saxena (Springer, Berlin 1988)
[4] D. Fuchs, P. Schweiss, P. Adelmann, T. Schwarz, and R. Schneider: Phys. Rev. B **72**, 014466 (2005)
[5] J. S. Zhou, J. Q. Yan, and J. B. Goodenough: Phys. Rev. B **71**, 220103 (2005)
[6] D. Fuchs, C. Pinta, T. Schwarz, P. Schweiss, P. Nagel, S. Schuppler, R. Schneider, M. Merz, G. Roth, and H. v. Löhneysen: Phys. Rev. B **75**, 144402 (2007)

Atomic Migration Phenomena in Intermetallics with High Superstructure Stability

Rafa Kozubski[1], Andrzej Biborski[1], Mirosaw Kozowski[1], Véronique Pierron-Bohnes[2], Christine Goyhenex[2], Wolfgang Pfeiler[3], Marcus Rennhofer[3], and Bogdan Sepiol[3]

[1] Interdisciplinary Centre for Materials Modelling, M. Smoluchowski Institute of Physics, Jagellonian University,
Reymonta 4, 30-059 Kraków, Poland
rafal.kozubski@uj.edu.pl

[2] Institut de Physique et Chimie des Matériaux de Strasbourg, Groupe d'Etudes des Matériaux Métalliques CNRS-ULP,
23 rue du Loess BP 43, 67034 Strasbourg, France
vero@ipcms.u-strasbg.fr

[3] Fakultät für Physik der Universität Wien,
Strudlhofg. 4, A-1090 Wien, Austria
wolfgang.pfeiler@univie.ac.at

Abstract. Most of the contemporary materials based on intermetallic phases are either multiple bulk phases, or nanostructured layers deposited on appropriate substrates. In each case, the desired properties of the materials are due to chemical order and the preparation technology consists of a generation of specific processes mediated by atomic migration. It is shown how a nanoscopic (atomistic) image of the atomic migration phenomena results from an indirect experimental technique in combination with Monte Carlo (MC) and Molecular Dynamics (MD) simulations. "Order-order" relaxations were observed in phases representing three typical cubic superstructures of high stability: $L1_2$ (Ni_3Al), B2 (NiAl), and $L1_0$ (FePt).

Detailed analysis of the atomic-jump statistics yielded by MC and MD simulations elucidated: (i) the origin of the multi-time-scale character of the process, (ii) the effect of triple-defect formation on the kinetics of the "order-order" relaxation in B2 binaries, (iii) the effect of free surfaces on the superstructure stability in $L1_0$ nano-layers.

1 Introduction

Intermetallic compounds are still of interest in materials science as being the basis for the technology of many promising engineering and functional materials (see, e.g., [1, 2] and numerous references therein). Most of the properties which make these materials attractive and promising (e.g., high strength at high temperatures, favorable magnetic properties etc.) stem from the chemical (atomic) long-range order (LRO) observed in wide temperature ranges. This indicates the importance of basic investigations of the phenomenon of

R. Haug (Ed.): Advances in Solid State Physics,
Adv. in Solid State Phys. **47**, 277–288 (2008)

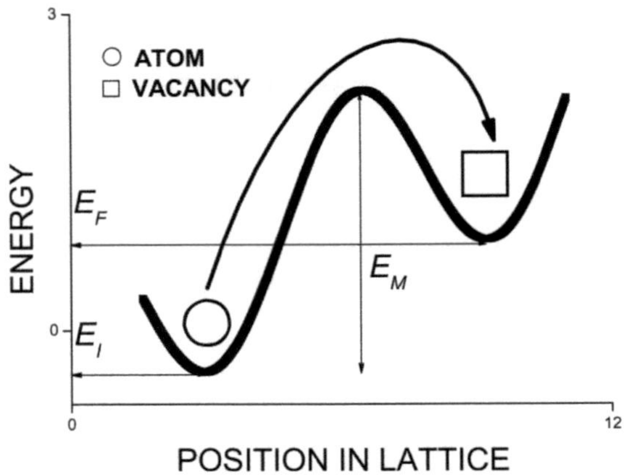

Fig. 1. System energy during an atomic jump to nn vacancy

atomic ordering focused especially on the mechanism and kinetics of the superstructure formation in intermetallics. In particular, the kinetic aspect of the process is crucial when intelligently designing the related contemporary advanced technologies.

Physically, atomic ordering is a specific structural transformation consisting of the generation/elimination of antisite defects and thus involving atomic migration, whose predominant mechanism in metals and alloys consists of elementary jumps of atoms to nearest-neighboring (nn) vacancies. The features of the kinetics of chemical ordering originate, therefore, from the physics of these elementary jumps. Stability of a crystalline structure of a solid means that potential energies of atoms show minima at the lattice sites and that, accordingly, the particular crystalline structure of the whole system means a minimum of its total energy. A possible jump of an atom from the initial lattice site to a nn vacant one always means surmounting an energy barrier (actually, a saddle point of the potential energy) separating the two positions (Fig. 1), which (in a classical approach) is possible due to thermal energy.

Within the "activated-state-rate theory" (see, e.g,. [3]), the energetics of an atomic jump is described by means of three energy contributions of the whole system: the energy E_M of the system with a jumping atom at the saddle point and the energies E_I and E_F corresponding the atom occupying the initial and the final lattice sites, respectively. These energies determine the probability $p_{I \to F}$ of the jump to occur at temperature T:

$$p_{I \to F} = C \times \exp\left[-\frac{E_M - E_I}{k_B T} \right] \times p_v^{(F)} \qquad (1)$$

where k_B denotes the Boltzmann constant and $p_v^{(F)}$ is a probability for a vacancy residing on the final lattice site.

When occurring in a superstructure, the jumps are classified from an "atomistic" (what kind of atom jumps) and a structural (between what lattice

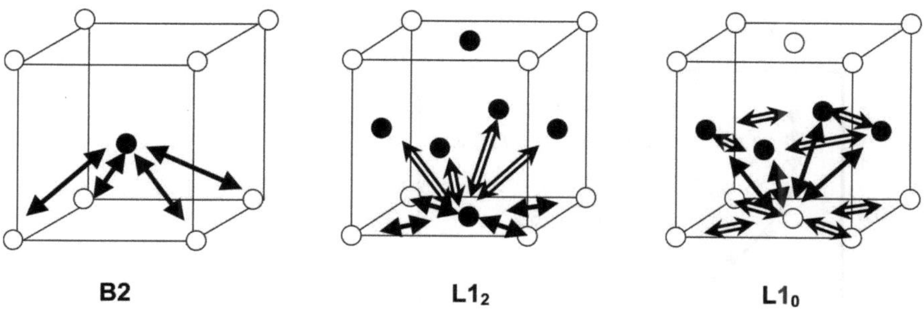

Fig. 2. Unit cells of B2, $L1_2$, and $L1_0$ superstructures. *Solid* and *open arrows* mark ordering and LRO-neutral atomic jumps, respectively

sites the jump is executed) point of view. If the initial and final lattice sites belong to different sublattices, the particular atomic jump affects the degree of LRO in the crystal by generating/eliminating antisite defects. The degree of LRO is usually quantified by a parameter η depending on the antisite concentration and equal to 1 and 0 in the case of perfect LRO and perfect disorder, respectively.

The geometry of a particular superlattice determines the character of possible atomic jumps. Three most common bcc and fcc-based superstructures are shown in Fig. 2.

While any atomic jump to a nn vacancy will affect the degree of LRO (i.e., the concentration of antisite defects) in the B2 superstructure, the $L1_2$ and $L1_0$ superlattices offer possibilities for LRO-neutral atomic jumps within one superlattice.

Investigation of ordering kinetics in intermetallic phases is, therefore, interesting and useful both with respect to technological application and due to its basic impact addressing non-equilibrium thermodynamics combined with quantum theory of solids focused at these materials.

As will be shown in this paper, the superlattice geometry definitely affects ordering kinetics in intermetallic compounds. The phenomena were studied experimentally and modeled on an atomistic scale by Monte Carlo (MC) and Molecular Dynamics (MD) simulations. Three examined intermetallic compounds: Ni_3Al, NiAl and FePt represented the $L1_2$, B2, and $L1_0$ superstructures, respectively. In parallel to the effect of superlattice geometry, the influence of nanostructuring on chemical ordering processes is addressed.

2 Experimental Studies

Although the intermetallic compounds examined within the project show very high superstructure stability and either disorder at very high temperature (FePt, T_t = 1600 K), or maintain very high degree of LRO up to the melting point (NiAl, T_m = 1910 K; Ni_3Al, T_m = 1660 K), their degree of LRO

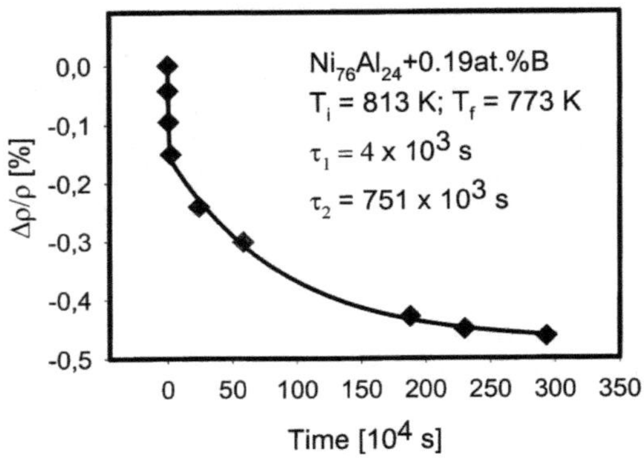

Fig. 3. "Order-order" relaxation in $Ni_{76}Al_{24}$+B showing the operation of two time scales [6]

is, of course, temperature dependent. A sudden change of temperature of a system in equilibrium is, therefore, followed by an "order-order" relaxation towards the new degree of LRO, a process controlled by atomic migration in non-steady-state conditions. As measurable relaxations are observed at temperatures not higher than ca. 1100 K, where the temperature dependence of the degree of LRO is very weak, experimental monitoring of the process requires highly sensitive techniques, among which resistometry proved to be very much adequate. Since the first direct evidence of "order-order" relaxation in Ni_3Al [4], the studies have been extensively continued and developed [5–12].

The results showed:

- specific and superstructure-geometry-dependent relationships between the activation energies for "order-order" relaxations and self-diffusion in the same materials (both processes controlled by atomic migration) [6, 8, 10];
- correlation between the superstructure stability and the activation energy for the "order-order" process [5,9] (the predominant effect on the vacancy formation energy has been definitely found);
- discontinuous change of the relaxation dynamics in FePt at about 800 K [12] (the effect correlated with a similar one observed in Fe diffusion in FePt [13]);
- multi-time scale character of the relaxations occurring in $L1_2$- and $L1_0$-ordered intermetallics [6, 10] (Fig. 3);
- surprisingly low rate of "order-order" relaxations in the B2-ordered NiAl [8, 10] in view of very high vacancy concentration observed in this system [14] (Fig. 4):

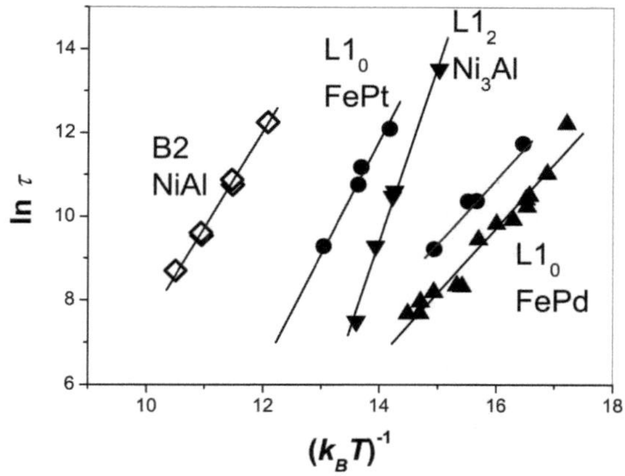

Fig. 4. Arrhenius plots of the relaxation times of order-order kinetics in Ni_3Al [6], NiAl [8], FePd [10], and FePt [12]

3 Modelling

"Order-order" relaxations as controlled by atomic migration proceeding via vacancy mechanism have been modeled by MC and MD atomistic simulations realized in two variants involving either fixed, i.e., temperature independent [15, 16], or temperature dependent number of vacancies.

3.1 Thermal Vacancy Thermodynamics

The two variants of atomistic simulations follow from the thermodynamics of thermal vacancies in ordered intermetallics which have been studied within a model based on a concept of *Schapink* [17]. A bcc or fcc lattice gas composed of two kinds of atoms and corresponding vacancies was considered within the Bragg–Williams approximation [18–20]. The ratio of the atomic concentrations was in correspondence to the composition of the system in question and the pair interactions between atoms favored long-range ordering in the related superstructure. Zero-energy interactions were assumed between vacancies. Such a lattice gas shows two properties:

- long-range ordering and a continuous (bcc) or discontinuous (fcc) "order-disorder" transition at temperature T_c (bcc) or T_t (fcc);
- tendency for decomposition into a vacancy-poor and a vacancy-rich phase.

The latter tendency is observed for a wide range of temperature (usually comprising the temperature T_c (T_t)) and a wide range of the overall vacancy concentration in the gas. The vacancy-poor phase, whose vacancy concentration C_v and the degree of LRO (i.e., antisite concentrations C_a) follow

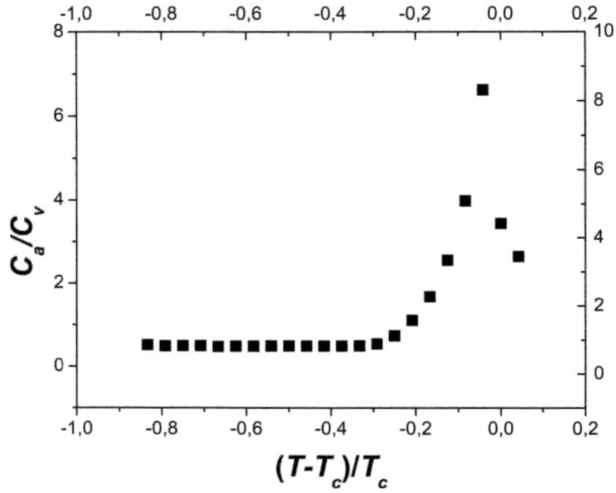

Fig. 5. Reduced-temperature dependence of the ratio C_a/C_v in B2-ordered AB binary with pair-interaction energies favoring the creation of A-antisites

from the solution of the Bragg–Williams model, is identified with the intermetallic crystal in question. In general, in fcc intermetallics the value of C_v results much lower than in the bcc intermetallics. Moreover, although both C_v and C_a increase with temperature, in fcc intermetallics C_a increases much faster than C_v and this fact justifies the approximation of fixed (temperature-independent) vacancy concentration commonly applied in atomistic simulations.

It appeared of special interest to analyze the temperature dependence of the C_a/C_v ratio obtained for B2-ordering AB binaries .

C_a/C_v (T) plots corresponding to a B2 AB intermetallics with pair-interaction energies yielding high asymmetry between the formation energies of particular types of vacancies and antisites (i.e., promoting the generation of, e.g., A-antisites and A-vacancies) show plateau meaning a proportionality between the equilibrium concentrations of both defects [20]. An extreme case, where the level of the plateau values equals exactly 0.5 is shown in Fig. 5). The effect is associated with the formation of so-called triple defects [21] widely discussed in literature for several decades (see, e.g., [8] and references therein). The process is schematically presented in Fig. 6.

Initially perfectly B2-ordered binary intermetallic disorders via atomic jumps to nn vacancies generated in pairs, which guarantees the preservation of bcc geometry. In the case of comparable probabilities for the generation of A- and B-types of antisites and vacancies, the antisites are created in pairs and the process may continue even via a single pair of vacancies. This corresponds to the case of C_a increasing with temperature much faster than C_v. In contrary, in the case of the above probabilities substantially differing one from each other, only one sort of antisites is produced and, as a result, most of the vacancies shift to the counterpart sublattice forming specific

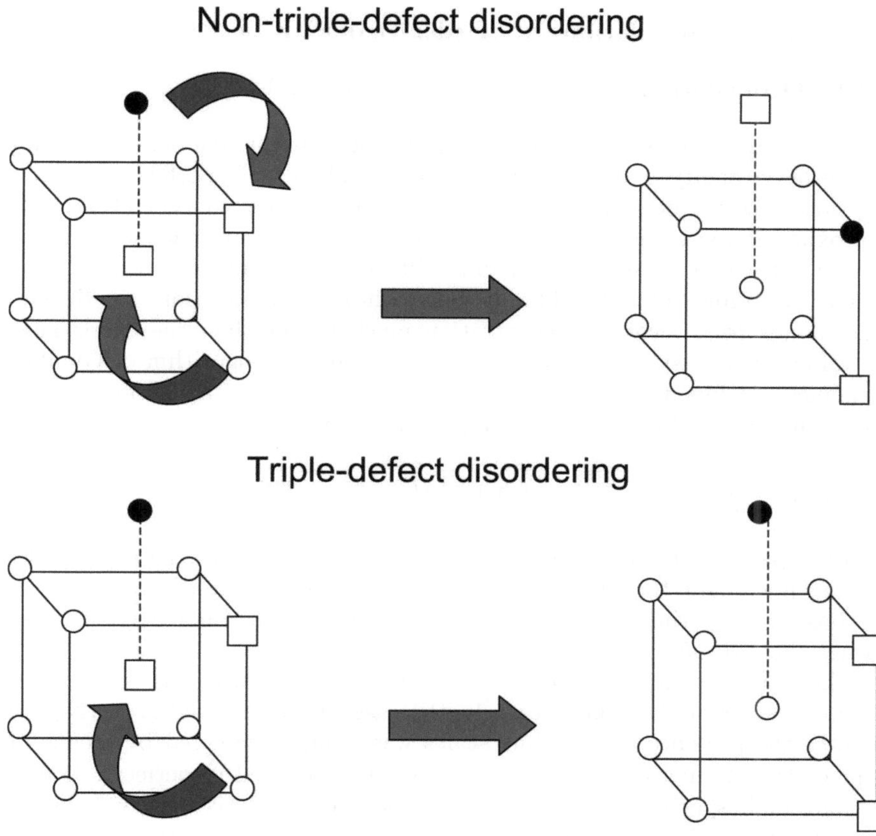

Fig. 6. Schemes of non-triple defect and triple defect disordering of a B2-ordered AB binary

defect complexes (triple defects) composed of two A-(B-)vacancies and one A-(B-)antisite. Such complexes are energetically stable and make the vacancies effectively immobile. The proposed scheme implies that (i) a triple-defect-disordering system contains comparable numbers of vacancies and antisites (in an extreme proportion of 2:1) – the effect reflected by the graph in Fig. 5; (ii) most of vacancies (as trapped in triple defects) do not take part in the process of disordering, which therefore, proceeds rather slowly. It is clear that when simulating the triple-defect ordering/disordering the assumption of fixed C_v is no longer justified.

It should be stressed that, in contrary to most of the related publications, the result of Fig. 5 has been obtained without any a priori assumptions of the existence/absence of particular types of vacancies and antisites in the system.

3.2 Atomistic Simulations of "Order-Order" Kinetics

3.2.1 "Order-Order" Kinetics in Bulk Material

Extensive simulations of the process occurring in non-triple-defect $L1_2$ and $L1_0$ ordered binary and quasi-binary intermetallics were performed within the variant of temperature-independent (fixed) vacancy concentration. The approximation was additionally acceptable due to very low vacancy concentration in the real systems (Ni_3Al and FePt): even one vacancy present in the simulated sample (40 x 40 x 40 unit cells with periodic boundary conditions) corresponds to a vacancy concentration much higher than the experimental values. The basic simulations involved a standard algorithm of Glauber dynamics combined with vacancy mechanism of atomic migration with a probability $\Pi_{I\to F}$ for an atomic jump to nn vacancy given by:

$$\Pi_{I\to F} = \frac{\exp\left[-\frac{\Delta E}{k_B T}\right]}{1+\exp\left[-\frac{\Delta E}{k_B T}\right]} \quad \text{where} \quad \Delta E = E_F - E_I \tag{2}$$

with the parameters defined in Fig. 1 and (1). Initially, an Ising-type model was assumed and the atomic pair-interaction energies were evaluated on the basis of experimental and theoretical results concerning particular intermetallics [10, 15].

The studies resulted in the elucidation of the origin of the multi-time-scale character of the relaxations. In the case of $L1_2$-ordered A_3B (Ni_3Al) binaries the presence of two time scales was definitely proved by means of Laplace transformations of the $\eta(t)$ isotherms simulated in perfectly homogeneous bulk specimens. By means of a detailed analysis of atomic jump statistics it was shown that the effect originated from the specific correlation of atomic jumps induced by the superstructure-geometry [15, 16]. A single time scale appeared in turn in the $\eta(t)$ isotherms simulated in homogeneous $L1_0$-ordering AB systems[1]. The latter result corroborates recent theoretical findings of *Ohno* and *Mohri* [22] indicating that only one of the three time scales experimentally observed in $L1_0$ ordering kinetics is related to homogeneous ordering within the antiphase domains. The results concerning $L1_2$-ordered A_3B binaries obtained in the above way were then perfectly reproduced by more sophisticated "residence-time" kinetic MC simulations dedicated to Ni_3Al with embedded atom method (EAM) energetics including saddle-point energies [16]. An additional confirmation of the standard MC results followed from MD simulations [23].

Most recently, MC simulations of "order-order" kinetics in triple-defect B2-ordering AB binaries have been taken up with vacancy thermodynamics (i.e., temperature-dependent vacancy concentration) involved.

[1] Very weak contribution of a second time scale in "order-order" relaxations in $L1_0$-ordered AB binaries reported in previous publications [10, 12] has recently been found an artefact of the simulation procedure.

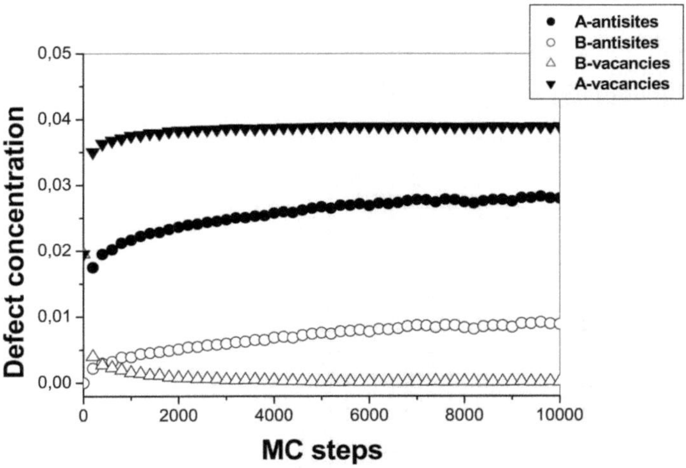

Fig. 7. Simulated "order-order" isotherms of defect concentrations in triple-defect B2-ordered AB binary at $T = 0.5 \times T_c$

The preliminary test consisted of the simulation of disordering $\eta(t)$ isotherms starting from a perfectly B2-ordered system with atomic pair-interaction energies yielding purely triple-defect-type disordering within the Bragg–Williams approximation (Fig. 5). The vacancies were initially distributed at random with C_v equal to the equilibrium value according to Bragg–Williams thermodynamics. The resulting $C_a(t)$ and $C_v(t)$ isotherms are plotted in Fig. 7.

It is clearly visible that the disordering relaxation proceeds in two stages: (i) an extremely fast (covering several MC steps) shift of almost all vacancies to the A- sublattice accompanied by almost exclusive generation of A-antisites with $C_a/C_v \approx 0.5$; (ii) very slow further increase of C_a. This preliminary result shows that the simulated B2 "order-order" kinetics undoubtedly reflects the features of the experimental one and confirms the postulated model of triple-defect disordering [8]. At present, the methodology is developed towards a unification of the approximation levels connected with vacancy thermodynamics and MC simulation of ordering.

3.2.2 "Order-Order" Kinetics in $L1_0$-Ordered Binaries with Free Surfaces

The increasing interest in nano-layered $L1_0$-ordered intermetallics showing magnetic anisotropy [2] motivated the authors to adapt the developed MC simulation methodology to systems limited by free surfaces. Up to now, free (001)-surface-limited layers of $L1_0$-ordering AB binary systems were simulated by removing the periodic boundary conditions in z crystallographic direction [24]. The pair-interaction energies between nn and nnn atoms were

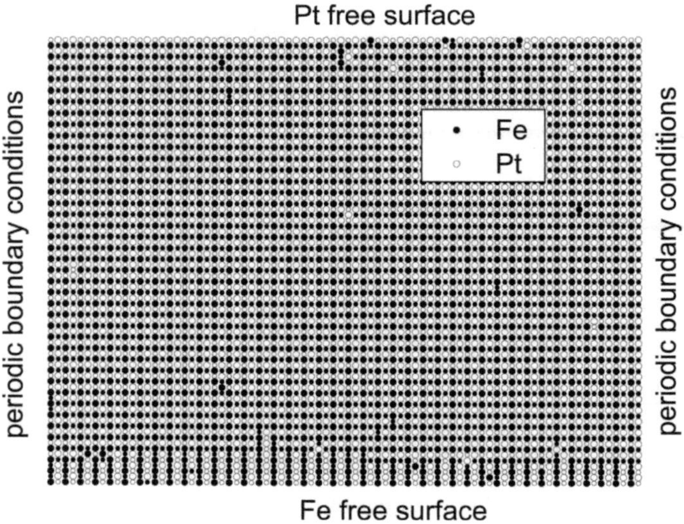

Fig. 8. (010) section of the simulated $L1_0$-ordered FePt layer showing the $L1_0$ z-variant to x-variant re-orientation

evaluated on the basis of "ab-initio" calculations performed for FePt [25]; no interactions between atoms and vacancies, as well as between vacancies were assumed. The problem of $L1_0$ ordering in such systems involves two categories of $L1_0$ variants with the monoatomic A and B planes parallel (z-variant) or perpendicular (x- and y-variants) to the (001) surface (see Fig. 2). The simulations based on the Glauber dynamics combined with vacancy mechanism of atomic jumps showed that the z-variant $L1_0$ superstructure (yielding the technologically desired magnetic anisotropy) appears unstable in the layers and the monoatomic planes spontaneously re-orient creating x- and y-variant domains (Fig. 8) [24].

The process starts preferentially on the Fe-monoatomic free surface and a discontinuous growth of the x- and y-$L1_0$-variant domains inward the layer is observed. The rate of this growth decreases while the process advances and thus, percolation takes place only in the case of sufficiently thin layers. The process is explicable in terms of atomic-bond structure changing due to the re-orientation process. The reorientation of the z-variant to the x-(y-) variant brings about a decrease of the configurational energy proportional to $k_B \times T_t$ per unit cell located at the (001) surface. This decrease generates driving force for the process, whose value related to the entire layer is, however, inversely proportional to the layer thickness.

The values of the pair interaction energies applied in the simulations yield substantial difference between the system energy increases caused by the jumps of Fe and Pt surface atoms to antisite positions below the respective surfaces. This explains the preferential nucleation of the process on the Fe free surface.

The $L1_0$ z-variant instability in (001)-oriented intermetallic nano layers, though first previewed by MC simulations, has recently been observed experimentally in FePd [11] and FePt [26] layers. The picture arising from these results corroborates that in FePd and FePt relatively low temperatures (773 K to 873 K) cause a partly re-orientation, most probably starting from the free surface.

The simulation studies are continued and it is planned to test the MC results by MD simulation involving "ab-initio" calculated energetics.

Acknowledgements

The presented studies are pursued within research projects realized in the framework of COST 535 and COST P19 actions.

References

[1] J.-C. Zhao, J. H. Westbrook: MRS Bull. **28**, 622 (2003)
[2] H. Coufal, L. Dhar, C. Denis Mee: MRS Bull. **31**, 374 (2006)
[3] R. Kozubski: Prog. Mater. Sci. **41**, 1 (1997)
[4] R. Kozubski, M. C. Cadeville: J. Phys. F.: Met. Phys. **18**, 2569 (1988)
[5] R. Kozubski, J. Sotys, M. C. Cadeville, V. Pierron-Bohnes, T. H. Kim, P. Schwander, J. P. Hahn, G. Kostorz, J. Morgiel: Intermetallics **1**, 139 (1993)
[6] R. Kozubski, W. Pfeiler: Acta Mater. **44**, 1573 (1996)
[7] H. Lang, K. Rohrhofer, P. Rosenkranz, R. Kozubski, W. Pšchl, W. Pfeiler: Intermetallics **10**, 283 (2002)
[8] R.Kozubski, D. Kmieć, E. Partyka, and M. Danielewski: Intermetallics **11**, 897 (2003)
[9] E. Partyka and R. Kozubski: Intermetallics **12**, 213 (2004)
[10] R. Kozubski, M. Kozowski, V. Pierron-Bohnes, W. Pfeiler: Z. Metallkde. **95**, 880 (2004)
[11] Ch. Issro, W. Püschl, W. Pfeiler, P. Rogl, W. Soffa, R. Kozubski, V. Pierron-Bohnes: Scr. Mater. **53**, 447 (2005)
[12] R. Kozubski, C. Issro, K. Zapaa, M. Kozowski, M. Rennhofer, E. Partyka, V. Pierron-Bohnes, W. Pfeiler: Z. Metallkde. **97**, 273 (2006)
[13] M. Rennhofer, B. Sepiol, M. Sladecek, D. Kmieć, S. Stankov, G. Vogl, M. Kozlowski, R. Kozubski, A. Vantomme, J. Meersschaut, R. Rffer, A. Gupta: Phys. Rev. B **74**, 104301 (2006)
[14] H.-E. Schaefer, K. Frenner, R. Wurschum: Intermetallics **7**, 277 (1999)
[15] P. Oramus, R. Kozubski, V. Pierron-Bohnes, M. C. Cadeville, W. Pfeiler: Phys. Rev. B **63**, 174109 (2001)
[16] P. Oramus, R. Kozubski, V. Pierron-Bohnes, M. C. Cadeville, C. Massobrio, W. Pfeiler: Defect and Diffusion Forum **194–199**, 453 (2001)
[17] F. W. Schapink: Scr. Metall. **3**, 113 (1969)
[18] S. H. Lim, G. E. Murch, W. A. Oates: J. Phys. Chem. Solids **53**, 181 (1992)

[19] R. Kozubski: Acta Metall. Mater. **41**, 2565 (1993)
[20] A. Biborski, R. Kozubski: Triple deffect and constitutional vacancies in B2 binaries. Direct Bragg–Williams thermodynamics., Proceedings of the Third International Conference on Multiscale Materials Modeling MMM2006, Freiburg, Germany, P. Gumbsch (Ed.), (Fraunhofer IRB Verlag 2006) pp. 854–857
[21] R. J. Wasilewski: J. Phys. Chem. Solids **29**, 39 (1968)
[22] M. Ohno, T. Mohri: Philos. Mag. **83**, 315 (2003)
[23] P. Oramus, C. Massobrio, M. Kozowski, R. Kozubski, V. Pierron-Bohnes, M. C. Cadeville, W. Pfeiler: Comput. Mater. Sci. **27**, 186 (2003)
[24] M. Kozowski, R. Kozubski, V. Pierron-Bohnes, W. Pfeiler: Comput. Mater. Sci. **33**, 287 (2005)
[25] T. Mohri, Y. Chen: Mater. Trans. **43**, 2104 (2002)
[26] M. Rennhofer: private communication, to be published

Material Science with Positrons: From Doppler Spectroscopy to Failure Prediction

Matz Haaks, Patrik Eich, Judith Fingerhuth, and Ingo Müller

Helmholtz Institut für Strahlen- und Kernphysik, Universität Bonn,
Nußallee 14–16, 53115 Bonn, Germany
haaks@iskp.uni-bonn.de

Abstract. We describe an alternative approach for a reliable lifetime prediction employing the local concentration of lattice defects as a precursor for fatigue failure. We present positron annihilation spectroscopy (PAS) as a non-destructive technique sensitive for defect concentrations in the range relevant to plasticity in metals.

The Bonn Positron Microprobe (BPM), a currently unique device, provides a fine focused positron beam with a selectable beam diameter from 5 to 200 μm assisted by an inbuilt fully functional scanning electron microscope (SEM). Using the BPM, plasticity and fatigue can be measured with a lateral resolution from some microns up to the range of millimeters.

Employing laterally resolved PAS and the empirical supposition of a linear relation between the defect concentration and the logarithm of the number of fatigue cycles, the point of failure was successfully predicted on the common carbon steel AISI 1045. For a generalization of the precursor method, a minimal model of fatigue based on a cellular automaton was developed. First results from a one-dimensional implementation are presented.

1 Introduction

Fatigue of repeatedly loaded components is one of the major reasons for the decay of material properties and finally for failure. In the case of severe accidents of airplanes, trains and cars this is a widely discussed topic in the news. From our daily experience we know that the lifetime of components exposed to repeated load is finite. Generally, a component will fail if the cyclic load amplitude exceeds 50 % of the yield strength, which is still within the 'reversible' elastic region (Hooke's law applies). Since almost 150 years the lifetime of components subjected to cyclic load is estimated employing the Wöhler-test, where the stress amplitude is plotted versus the logarithm of the number of cycles to failure. This method employs idealized conditions, especially a sinusoidal or triangular control wave [1]. More realistic conditions concerning a representative stochastic distribution of load amplitudes are implemented in the *Gassner*-test [2]. The lifetime of a component under operational conditions is extrapolated from these test series by the means of statistics. However, these tests are destructive and to complete a full diagram

R. Haug (Ed.): Advances in Solid State Physics,
Adv. in Solid State Phys. **47**, 289–300 (2008)

a huge number of samples has to be fatigued until failure which may take up to 10^9 load cycles for each sample.

Here we describe a new approach for a reliable prediction of fatigue failure. On the atomic scale fatigue is due to the accumulation of dislocations and other lattice defects followed by the formation of dislocation structures, e.g., persistent slip bands. During tensile testing or cyclic fatigue the dislocation density rises by several orders of magnitude from the well annealed state until fatigue failure occurs. A crack will initiate when a critical density of defects has been accumulated locally. Employing the defect density as precursor, failure can be extrapolated from the very beginning of a fatigue test.

Positron annihilation spectroscopy (PAS) is an technique having a high sensitivity for these defects and a large dynamic range at the same time [3–5], reaching from the well annealed state of an alloy to the critical defect density when the material fails [6]. The method doesn't require an advanced sample preparation and can be applied in-situ during tensile or fatigue experiments [7, 8]. Since PAS is non-destructive, the results gained can be verified by conventional tests on the same sample. In comparison the classical experimental methods are far from non-destructive, like transmission electron microscopy, or not sensitive enough like hardness testing, flow stress measurement, or the measurement of the internal strains with X-rays or neutrons.

2 Positrons as Probes for Atomic Defects

Already in the 1960s it was realized, that Positron Annihilation Spectroscopy (PAS) is a versatile tool to detect vacancy-like defects and dislocations in metals [9, 10] and semiconductors [11] nondestructively with an outstanding sensitivity. These defects constitute an open volume, which acts as an attractive potential for positively charged particles. Due to its diffusive motion the positron acts as a probe on the atomic scale scanning about 10^6 lattice positions. Hence, from a defect concentration of 10^{-6} per atom the annihilation signal is significantly changed.

The interaction of positrons with matter can be divided into four sections: thermalization, diffusion, trapping, and finally annihilation with an electron from its vicinity. Implanted in condensed matter, a positron looses all its kinetic energy within a few picoseconds, which is rather short compared to its lifetime in matter (from about 100 ps in closed packed metals to several ns in polymers). For monoenergetic positrons from a slow positron beam, having typical energies from 10 to 50 keV, the implantation profile reaches its maximum a few μm below the surface and can be described by a *Makhov*-distribution [15]. For instance, at a positron energy of 30 keV the mean implantation depth is 1.6 μm for ferrous alloys. At the end of the energy loss the positron is in thermal equilibrium with the lattice ($E_{\mathrm{kin}} = 3/2\,kT \approx 0.04\,\mathrm{eV}$ at room temperature (RT)) [12–14].

Once thermalized, the positron diffuses through the lattice and behaves like a free particle. Repelled from the positively charged nuclei, its probability of presence has a maximum in the interstitial regions of the lattice [16], while its motion can be described as a 3-dimensional random-walk [17]. The positron is highly mobile, having a diffusion coefficient at RT in the order of $10^{-4}m^2s^{-1}$ [18].

Every open volume in the lattice, causing a local increase of the distance of the atomic positions, acts as an attractive potential for the diffusing positron, if not positively charged. The probability for trapping into different kinds of defects is described by trapping models [19, 20]. For instance, one atomic vacancy forms a deep positron trap with a binding energy around 1 eV due to the absent repulsion of the missing nucleus. Dislocations form shallow traps with a binding energy of 50–100 meV, which is comparable to the positrons kinetic energy at RT. Hence, an escape from this kind of trap is probable [21–23]. Experimentally measured positron lifetimes being characteristic for dislocations differ significantly from lifetimes obtained for defect-free materials and are almost equal to the characteristic lifetimes for vacancies measured in materials having a high vacancy concentration (prepared, e.g., by electron irradiation). Hence, trapping into dislocations obviously forms an intermediate state, from which the positron is trapped by the deep potential of the associated vacancy-like defect (e.g., jogs, vacancies bound by the elastic strain field of the dislocation) [7].

The creation of dislocation during fatigue is always accompanied by the production of vacancies and interstitial atoms. The most eminent processes for production of point-like defects are jog dragging and the annihilation of edge dislocations [24, 25]. At room temperature PAS is extremely sensitive for changes in the density of vacancy-like defects. For instance, the density of atomic vacancies in steels can be observed from 1.2×10^{-6} to 4.3×10^{-4} vacancies per dislocation density are detected by changes in the density of the associated vacancy-like defects.

When the positron finally annihilates with an electron, the rest mass of both particles is transformed into two γ-quanta of 511 keV emitted antiparallel in the center-of-mass-system. By transformation into the laboratory system the longitudinal component of the electron momentum causes a Doppler shift in the γ-energy, which results in a broadening of the annihilation line. The momentum of the thermalized positron ($\sim$40 meV) can be neglected compared to the electron's momentum (1–10 eV). The contribution of electrons with low momenta to the Doppler broadening is quantified by a shape parameter of the momentum distribution, the S-parameter, which is defined as the ratio of a central area of the annihilation peak and the integral over the whole peak [26]. Thus, the S-parameter extracts annihilation events with valence electrons having low momenta.

Since the electron density distribution in an open volume defect differs significantly from that in the ideal lattice, the S-parameter can be employed as a measure for the defect density, with respect to the sensitivity limits.

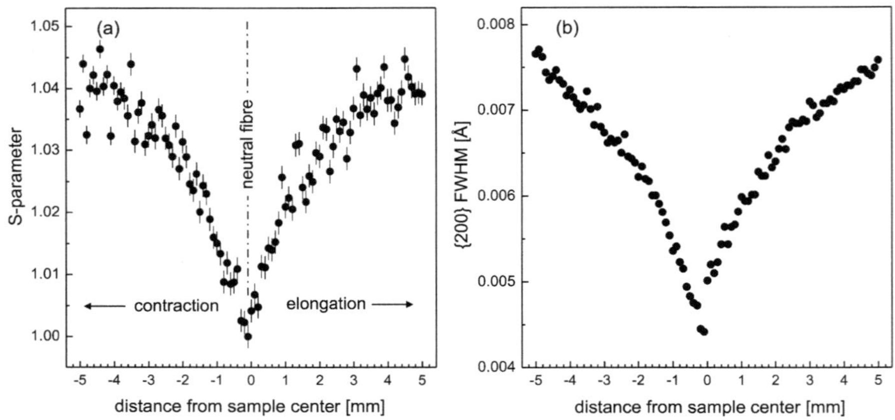

Fig. 1. Laterally resolved scans over the cross-section of a ferritic carbon steel (AISI 1045) sample deformed in three-point bending geometry. (**a**) S-parameter scan. (**b**) Broadening of the {200} reflections of the α-phase (statistical errors are similar to (a))

The S-parameter depends as well on the arbitrary choice of the borders used for determining the central area, as on the energy resolution of the gamma spectrometer. To make measurements comparable, the S-parameter should be normalized to an appropriate reference value. For an investigation on plastically deformed or fatigued metals this would be the S-parameter of the well annealed state of the same material.

The sensitivity of positrons to plasticity and fatigue in metals [8, 27, 28] and semiconductors [29] has been shown in several experiments. For instance, tensile tests on ferrous alloys show a linear correlation of the S-parameter on the true strain, beyond a sensitivity threshold which corresponds to the transition from elastic to plastic deformation [30, 32].

As a validation of the PAS method, laterally resolved studies on plasticity in the ferritic carbon steel AISI 1045 (equals to the European norm C45E) were performed after deformation in a three-point bending test. This geometry is characterized by a linear stress gradient along the bending radius and a neutral fiber between elongation and contraction.

After releasing the stress the remaining plastic deformation was analyzed employing two complementary methods: The local distribution of vacancy-like defects was measured by scanning the sample with positrons employing the S-parameter. Dislocations cause a distortion in the distances of the crystallographic planes which is expressed by a broadening of the X-ray reflections. Hence, the distribution of dislocations was investigated by X-rays diffraction in Debye–Scherrer geometry.

Scans with both methods were done at exactly the same positions on the same sample. Figure 1 shows the S-parameter (a) and the broadening of

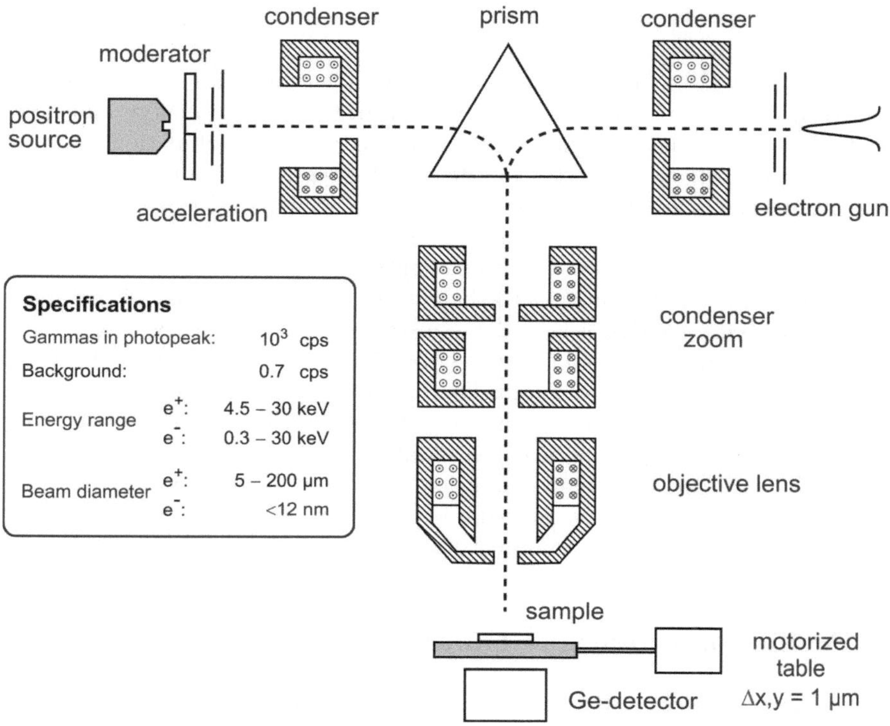

Fig. 2. Design of the Bonn Positron Microprobe with current specifications. The BPM provides a fine-focused positron beam having an adjustable beam diameter in the micron range and standard SEM features. The sample position is magnetically field-free, which allows defect studies on ferromagnetic alloys

the {200} reflections of the ferritic α-phase in scans over the cross-section of the bent sample. Comparing both methods, a linear correlation between the density of vacancy-like defects and dislocations was found [33]. This reflects the proportionality between the production rates of dislocations and associated vacancy-like defects during plastic deformation. Since Debye–Scherrer diffraction requires powder conditions (at least 10^4 grains in the irradiated volume), only adapted geometries can be studied using high intense X-ray sources. In contrast, PAS can be applied to virtually all kinds of samples.

3 The Bonn Positron Microprobe

Most effects of plastic deformation and material fatigue show a strongly inhomogeneous defect distribution over the sample volume where the region of maximum stress concentration is in the micron range. Hence, for an understanding of these processes by the means of positron annihilation a lateral resolution in the micron range is required.

The Bonn Positron Microprobe (BPM) [34] provides a fine focused positron beam in the micron range with adjustable beam energy and a selectable beam diameter from 5–200 μm. The BPM is a combination of a tiny moderated positron source (^{22}Na, 10 mCi) having a small phase space and a conventional scanning electron microscope (SEM). Positron and electron source are mounted on the opposite sites of a magnetic prism which bends both beams downward by 90° into the entrance plane of a SEM condenser zoom. An objective lens focuses the beams onto the sample, which is mounted on a motorized table that is laterally adjustable with an accuracy of 1 μm. The positron beam diameter can be set between 5 and 200 μm, which fits for most plasticity and fatigue experiments. There is no need for an additional focusing by a strongly inhomogeneous magnetic field behind the sample position what allows the study of ferromagnetic materials, e.g., steels. The annihilation radiation is recorded by a high resolution Germanium-detector ($\Delta E = 1\,\text{keV}$ @ 478 keV), mounted below the sample's position. The BPM is a laboratory instrument intended for automated measurement series similar to a SEM.

4 Prediction of Fatigue Failure

Preliminary studies on the prediction of failure from the early states of fatigue have shown that failure occurs when a critical defect density is reached locally. These studies where carried out on ferrous and titanium alloys using rotating bending geometry like in the classical Wöhler method [30, 32]. Based on these studies fatigue failure was predicted from a series of fatigued samples of the carbon steel AISI 1045 in an standard engineering geometry: A flat-bar samples having a central bore hole where the globally applied stress concentrates.

To identify the exact size and position of the maximum affected volume a finite element simulation (FEM) was carried out employing the Von-Mises stress σ_{VM}, which is a scalar measure on the stress tensor, independent of the stress orientation. Plasticity studies on deformation caused by indentation have shown the correlation of σ_{VM} with the S-parameter [35]. The FEM calculation was done by simulating an uni-axial elastic traction force on one end of the flat-bar sample. It revealed a strong localization of the distribution of σ_{VM} at the sides of the central bore hole.

The samples were fatigued using a servo-hydraulic testing machine applying a sinusoidal control wave with a maximum stress amplitude of 160 MPa, which is around 60 % of the materials yield strength. The defect distribution was measured on each sample by line scans with the BPM from the side of the bore hole to the edge of the sample. Additionally, the defect distribution in the area of maximum stress concentration in vicinity of the bore hole was imaged by a 2-dimensional scans with a resolution of 20 μm [36].

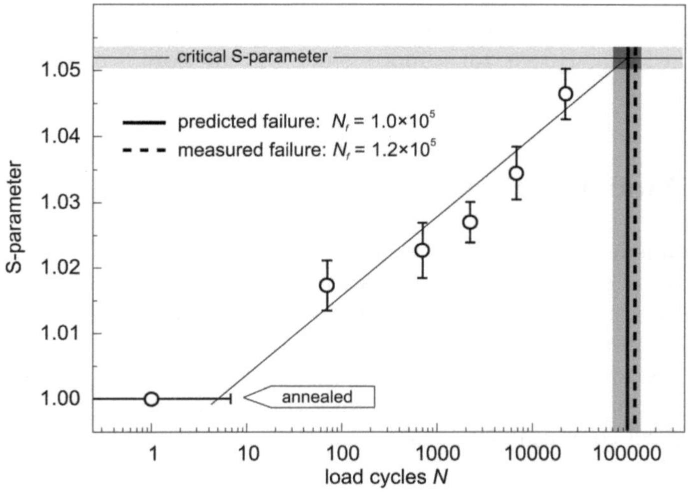

Fig. 3. Diagram for the prediction of fatigue failure on the carbon steel AISI 1045. Extrapolation is done by a regression of the maximum defect density $S(logN)$ (*diagonal line*). The critical defect density, where failure occurs, is obtained at the tip of a fatigue crack in the same alloy (*vertical line*). The predicted failure is given by a *full vertical line* and validated by measurement (*dashed vertical line*). The shaded areas indicate the estimated statistical error

The data obtained by the line scans fit to a model derived from linear fracture mechanics [37]. The local maximum of the defect density was obtained by the S-parameter directly beneath the side of the bore hole and plotted versus the logarithm of the cycle number N (hollow circles in the prediction diagram Fig. 3). The extrapolation of the point of failure was done by a regression of $S(\log N)$ assuming a linear dependence of the local defect density with the logarithm of the load cycle number, as indicated by preliminary studies on ferrous alloys [30].

The critical S-parameter S_{crit}, corresponding to the critical defect density, was obtained at the position of the cracktip in a fatigue crack produced in the same alloy by a 2-dimensional positron scan. Based on the supposition that a fatigue crack will initiate when the critical defect density is reached locally, it is assumed that failure occurs when the extrapolated defect density reaches the critical value S_{crit}: in our case at 1.0×10^5 fatigue cycles (full vertical line in Fig. 3). After prediction the result was validated by further fatiguing the samples to failure ($N_f = 1.2 \times 10^5$ cycles: dashed vertical line in Fig. 3). An excellent match between the predicted and the measured values was found.

But here a general problem in our method appears: The assumption that the defect density increases linearly with the logarithm of the number of fatigue cycles is purely empiric, and, albeit it works well with ferrous alloys, it cannot be generalized for other alloys. For instance, similar test series on titanium alloys reveal a more complicated behavior that cannot be extrapolated that straight forward [32]. Hence, for a more general prediction, a model of fatigue is required.

5 A Minimal Model of Fatigue

The physical phenomena important for fatigue reach over several scales from the macroscopic scale of component design or sample geometry over the mesoscopic scale of the crystallographic structure and dislocation networks to the atomic scale of the interaction between dislocations and obstacles. Despite the enormous increase of computer power within the past years, a complete description of fatigue in a multiscale model will still stay a project for the future. Hence, a model computable in reasonable times requires radical simplifications.

Here we propose a minimal model based on a cellular automaton. Each grain of the crystallographic structure is represented by one cell of the automaton. In this context all inner properties of a single grain are reduced to a simple set of scalar variables assigned to each cell. All grains are having the same fixed volume, but randomly seeded crystallographic orientations. The positions of the grains are denoted by coordinates which may vary during the execution of the model. The material properties are implemented in a simplified material law assuming fully elastic deformation below the yield stress and simple slip in the system activated depending on the orientation in the plastic region above. Hereby the initial critical shear stress $\tau_{\text{crit},0}$ is calculated from the yield stress via the *Sachs*-factor [38]. Cyclic deformation is represented by oscillation boundary conditions simulating an uni-axial fatigue test under total strain control. The deformation affecting each cell is calculated from the macroscopically applied strain regarding the material law and the conservation of volume.

During execution of the model, the accumulation of defects and hence, the increase in the dislocation density is stored for each cell in a scalar variable. Since the critical shear stess τ_{crit} depends on the square root of the dislocation density it will increase during calculation. The critical tensile stress σ_{crit} above that the slip systems are activated is calculated from τ_{crit} and the orientation dependent Schmid-factor μ (ρ_0 is the assumed initial dislocation density):

$$\tau_{\text{crit}} \sim \sqrt{\rho_{\text{disl}}}$$

$$\sigma_{\text{crit}} = \frac{\tau_{\text{crit}}}{\mu} = \frac{\tau_{\text{crit},0}}{\mu}\sqrt{\frac{\rho}{\rho_0}} .$$

The model terminates when either a critical defect density is reached in one cell or a critical distance between two neighbors is exceeded, which represents intragranular and intergranular failure, respectively.

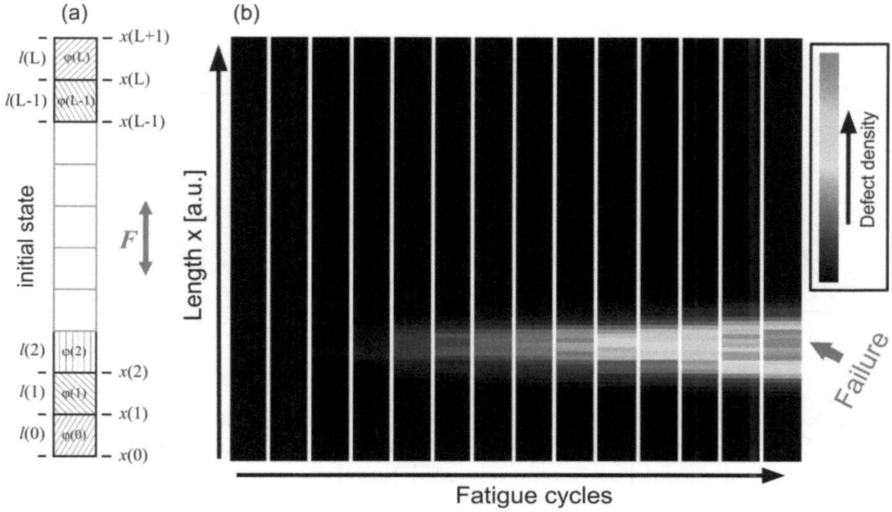

Fig. 4. (**a**) A first 1-dimension implementation of our model as a chain of grains. The randomly seeded orientations are symbolized by hatching. (**b**) Formation of lateral defect structures ('weak spot') during the progress of the simulation. Each *vertical bar* shows the defect density (*color coded*) as successive snap-shots of the simulation on a non-linear time scale *(online color)*. When the critical defect density is reached the model terminates

In a first approach, our model was implemented in a one-dimensional chain of cube shaped grains all having the same volume but random orientations. In the initial state an uniform low defect density was assigned to all cells. This corresponds to the bamboo-structure that can be found in annealed metallic wires (see Fig. 4a). Pure annealed nickel was chosen as model material, since it is a widely researched and well understood metal. For the description of the elastic-plastic behavior, the material law at room temperature was implemented. In ferromagnetic materials the Young's modulus depends on the magnetisation [31]. Regarding this, the Young's moduli of the cells were randomly variated around its mean value with a variance of 5 %. The external stress, driving the system, results in an elongation and compression of the chain in longitudinal direction. Under this prerequisite the stress field is being homogeneous over the chain's length.

Figure 4b shows a set of snap-shots of the simulation at several successive numbers of fatigue cycles, displaying the defect density coded in colors. Already in the early states of fatigue an inhomogeneous distribution of the defect density appears due to the initial random distribution of the orientations. Starting from a grain with an orientation favorable for plastic deformation, and hence, a faster cylic hardening compared to its neighbors, the defect distribution localizes in a 'weak spot' having a higher hardness than the surrounding cells.

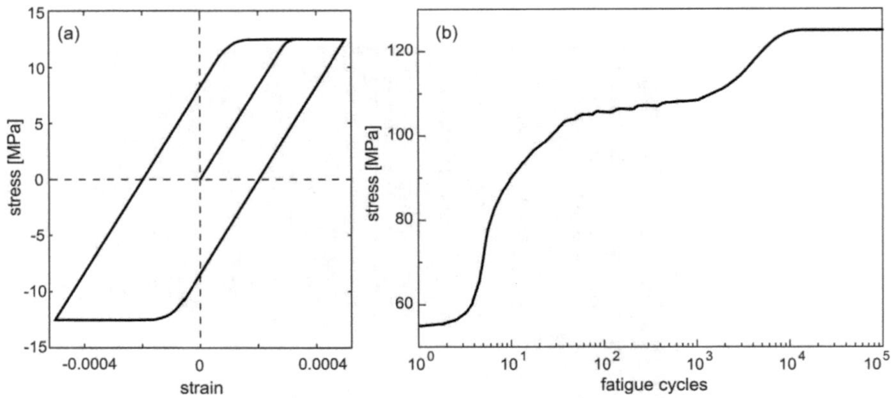

Fig. 5. Mechanical hysteresis loop of one fatigue cycle (**a**) and cyclic hardening curve (**b**) of the total system simulated by a one-dimensional implementation based on the material parameters of pure annealed Nickel

Outgoing from the localization in the beginning, the defect density increases locally during the further progress of the simulation and, by affecting neighboring cells, the 'weak spot' spreads over the chain. Finally the critical defect density is reached and the simulation terminates.

The simulation, even in its currently basic implementation, reproduces experimental results found by applying a nearly homogeneous stress field in a cyclic fatigue test. In 2-dimensional scans of the defect density, performed at several stages of fatigue, the formation and propagation of weak spots was observed [32].

Figure 5a shows a mechanical hysteresis loop of the whole chain monitored over a single fatigue cycle. The mechanical energy deposited in the system, equivalent to the area under the loop, is stored in the model as accumulated plastic deformation. During the progress of the simulation this results in cyclic hardening (see Fig. 5b).

In summary, our cellular automaton is fast enough to calculate the precursor for failure over at least 10^5 fatigue cycles. Despite to its radical simplifications, it reproduces essential features of fatigue like a mechanical hysteresis, the accumulation of defects, and cyclic hardening curves. However, this minimal model reproduces a formation of defect structures, their increase in intensity, and spread in space similar to the experimental data.

6 Conclusions

Positron annihilation spectroscopy (PAS) is a versatile and non-destructive tool for the study of open volume defects in the lattice. Due to its sensitivity to defect concentrations relevant for plasticity, changes in the microstructure can be detected already in the early stages of fatigue. Assuming an empiric linear relation between the defect density and the logarithm of the number of fatigue cycles, failure could be successfully extrapolated in a carbon steel. The result of the prediction was validated by a conventional destructive test series. Since this relation is not assumed as universal over the broad range of technical relevant alloys, a minimal model of fatigue was developed and tested in a one-dimensional implementation.

Both, the experimental results and the first outputs of the simulations show that the local defect density can be used as a precursor for fatigue failure prediction.

Acknowledgements

We like to acknowledge Ralf Sindelar for FEM simulations and fatigue testing, Gunter Schütz for sharing his knowledge in cellular automata and Karl Maier for his general support and fruitful discussions.

References

[1] A. Wöhler: Zeitschrift f. Bauwesen **8**, 642 (1858)
[2] E. Gassner: Luftwissen **6**, 61 (1939)
[3] R. West: Adv. Phys. **22**, 263 (1973)
[4] L. C. Smedskjaer, M. J. Fluss: Experimental Methods of Positron Annihilation for the Study of Defects in Metals, in: *Methods of Experimental Physics* Vol. 21, J. N. Mundy et al. (Eds.) (Academic Press, New York, London 1983)
[5] P. J. Schultz, K. G. Lynn: Rev. Mod. Phys. **60**, 701(1988)
[6] T. E. M. Staab, R. Krause-Rehberg, B. Kieback: J. Mater. Sci. **33**, 3833 (1999)
[7] T. Wider, K. Maier, U. Holzwarth: Phys. Rev. B **60**, 179 (1989)
[8] U. Holzwarth, P. Schaaff: Phys. Rev. B **69**, 094110 (2004)
[9] I. Y. Dekhtyar, D. A. Levina, V. S. Mikhalenkov: Sov. Phys. Dokl. **9**, 492 (1964)
[10] I. K. MacKenzie, T. L. Khoo, A. B. MacDonald, B. T. A. McKhee: Phys. Rev. Lett. **19**, 946(1967)
[11] I. Y. Dekhtyar, V. S. Mikhalenkov, S. G. Sakharova: Fiz. Tverd. Tela **11**, 3322 (1969)
[12] R. H. Ritchie: Phys. Rev. **114**, 644 (1959)
[13] A. Perkins, J. P. Carbotte: Phys. Rev. B **1**, 101(1970)
[14] K. Jensen, A. Walker: J. Phys. Condens. Matter **2**, 9757 (1990)
[15] A. F. Makhov: Sov. Phys. Solid State **2**, 1934, 1942, 1945 (1961)
[16] W. Brandt, R. Paulin: Phys. Rev. B **5**, 2430 (1972)
[17] M. Puska, R. Nieminen: Rev. Mod. Phys. **66**, 841 (1994)
[18] E. Soininen, H. Houmo, P. A. Huttunen, J. Mäkinen, A. Vehanen, P. Hautojärvi: Phys. Rev. B **41**, 6227 (1990)
[19] W. Frank, A. Seeger: Appl. Phys. **3**, 61 (1974)
[20] B. Bergensen, T. McMullen: Solid State Commun. **24**, 421 (1977)
[21] L. C. Smedskjaer, M. Manninen, M. J. Fluss: J. Phys. F **10**, 2237 (1980)
[22] C. Hidalgo, S. Linderoth: J. Phys.: Metal Phys. **18**, L263 (1988)

[23] K. Petersen, I. A. Repin, G. Trumpy: Condens. Matter. **8**, 2815 (1996)
[24] G. Saada: Acta Metal. **9**, 965 (1961)
[25] P. Hirsch, D. Warrington: Philos. Mag. **6**, 735 (1961)
[26] I. K. MacKenzie, J. A. Eady, R. R. Gingerich: Phys. Lett. A **33**, 279 (1970)
[27] C. Zamponi, S. Sonneberger, M. Haaks, I. Müller, T. Staab, G. Tempus, K. Maier: J. Mater. Sci. **39**, 6951 (2004)
[28] U. Holzwarth, A. Barbieri, S. Hansen-Ilzhöfer, P. Schaaff, M. Haaks: Appl. Phys. A **73**, 467 (2001)
[29] T. E. M. Staab, C. Zamponi, M. Haaks, I. Müller, S. Eichler, K. Maier: Mater. Sci. Forum, **445–446**, 510 (2004)
[30] K. Bennewitz, M. Haaks, T. Staab, S. Eisenberg, Th. Lampe, K. Maier: Z. f. Metallkd. **93**, 778 (2002)
[31] B. S. Berry. Elastic and Anelastic Behavior, in: *Metallic Glasses*, J. J. Gilman, H. J. Leamy (Eds.) (American Society for Metals, Metals Park - Ohio 1978), pp. 161
[32] M. Haaks, K. Maier: Predicting the Lifetime of Steels, in *Extreme events in nature and society*, S. Albeverio, V. Jentsch, H. Kantz (Eds.) (Springer, Berlin 2005), pp. 209
[33] M. Haaks, I. Müller, A. Schöps, H. Franz: Phys. Stat. Sol. a **203**, R31 (2006)
[34] H. Greif, M. Haaks, U. Holzwarth, U. Männig, M. Tongbhoyai, T. Wider, K. Maier, H. Bihr, B. Huber: Appl. Phys. Lett. **71**, 2115 (1997)
[35] J. Dryzek, E. Dryzek: Tribol. Lett. **13**, 309 (2003)
[36] P. Eich, M. Haaks, R. Sindelar, K. Maier: Proceedings of the 14th International Conference on Positron Annihilation, Hamilton 2006, Phys. Stat. Sol. c (in press, 2007)
[37] A. Liu: Summary of Stress-Intensity Factors, in: *ASM Handbook* Vol. 19, N. D. DiMatteo, S. R. Lampman (Eds.) (1997) pp. 981
[38] G. Sachs: Z. Ver. Dtsch. Ing. **72**, 734 (1928)

X-Ray Diffraction Residual Stress Analysis: One of the Few Advanced Physical Measuring Techniques That Have Established Themselves for Routine Application in Industry

Wolfgang Nierlich[1] and Jürgen Gegner[1,2]

[1] SKF GmbH, Department of Material Physics, Ernst-Sachs-Strasse 5, D-97424 Schweinfurt, Germany
[2] Institute of Material Engineering, University of Siegen, Paul-Bonatz-Strasse 9–11, D-57068 Siegen, Germany
juergen.gegner@skf.com

Abstract. The conventional procedure of X-ray diffraction (XRD) residual stress measurement is improved by means of a modification of the beam path of the diffractometer and an iterative technique that includes a pre-analysis of the near-peak line-profile. The achieved short measuring times of 5 and around 10 min per residual stress value and retained austenite content, respectively, serve as precondition for routine industrial applications over the last three decades within SKF. The line width represents a measure of material ageing within the lifetime cycle of a rolling bearing: calibration curves for the (near-) surface and the sub-surface failure mode are presented. Material response analysis permits differentiation of these failure modes and between low- and high-cycle fatigue.

1 Economic Physical Measuring

X-ray diffraction (XRD) residual stress analyses, even today, mainly occur at university institutes or established research institutions and few commercial service laboratories. Only at rather few places, they are applied as tool for routine investigations in the industry. The reason is, first of all, the required experience in interpreting the results. In the past, however, also the extensive expenditure of time for the complex measuring process must be mentioned as an obstacle.

An optimized sample throughput in terms of high quantity with corresponding series of measurements of residual stress depth profiles from the edge to the core likewise represents the precondition for both economic measuring and gaining the necessary practical experience. In the present paper, it shall be shown how this pioneering task is solved in the SKF Department of Material Physics in Schweinfurt by modifying simple conventional devices and rationalized design of a measurement process control. The proven method has been permanently established in the service operation since about 30 years and is only now replaced by the installation of new up-to-date equipment primarily due to problems with the spare part supply.

R. Haug (Ed.): Advances in Solid State Physics,
Adv. in Solid State Phys. **47**, 301–314 (2008)

2 Accelerated Measurement Technique

The residual stress measurements discussed in this paper are performed applying the XRD diffraction line shift-based $\sin^2 \psi$ method [1]. The mean penetration depth of the used Cr Ka radiation for iron materials equals around 5 μm. Therefore, this represents the distance, over which each value is essentially averaged. Residual stress profiles into the depth of the workpiece are obtained by repeated XRD measurements after stepwise material removal. The etching technique of electro-polishing is applied. This mild method of material removal does not generate new processing residual stresses. Appropriate mathematical methods for correcting the measuring error caused by the removed layer are given in the literature [2]. As long as the depth of electrolytic thinning is low compared to the wall thickness of the sample, the relaxation effect can be neglected.

The evaluation of each individual residual stress measuring value requires the determination of the peak position of a series of diffraction lines, depending on the varying orientation of the specimen. Thus, the fastest possible identification of line positions is in demand.

Already for about 30 years, position-sensitive detectors (PSD) are known for the determination of the position of X-ray diffraction lines. However, complicated handling and necessary intricate repeated calibration reflect the times of commencement of these devices. The consistent manufacturer recommendation of employing a physicist or engineer for the XRD measurements loses track of business reality due to the 95 % use of manpower resources for certainly advanced but still routine work. The then made decision for the well proven scintillation detector thus becomes comprehensible.

2.1 Reduction of Measuring Time by Modified Diffractometer Beam Path

XRD equipment is usually designed to achieve high resolution with minimum instrumental broadening. To gain intensity in the beam path, however, it is necessary to modify the system with the courage to simplification contrary to its original purpose. Major changes stem from recommendations made by *U. Wolfstieg* [3]. It concerns steps that, for the large line broadenings of the recorded α-Fe (211) diffraction reflex (excited by Cr Kα X-ray radiation) of about 5 to 7.5° of hardened rolling bearing steels, result in negligible effects on the practical determination of peak position and line width. The dispersion, which defines the line shift relevant to residual stress measurements, is not influenced. Figure 1 shows the beam path of a standard goniometer with conventional focusing Bragg–Brentano geometry. Items 1 to 4 indicate the interventions made:

Position 1: a square instead of a line focal spot is used.

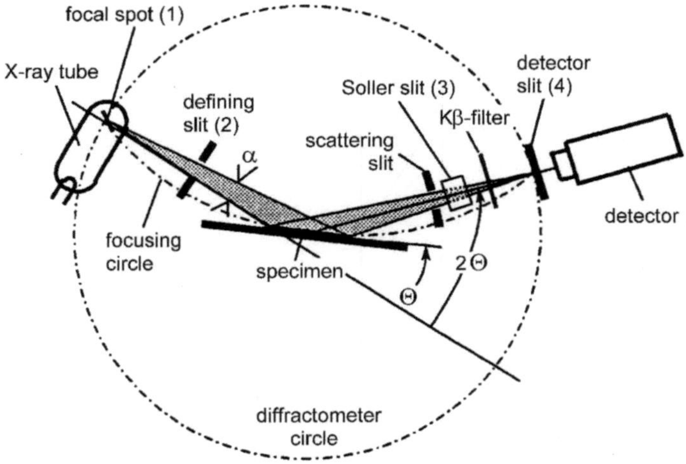

Fig. 1. Schematic representation of the beam path of a standard diffractometer. α and Θ respectively denote the aperture and the glancing Bragg angle (diffraction angle 2Θ)

Position 2: the horizontally and vertically adjustable beam-defining slit is placed at a distance of two-third of the measuring circle radius from the focal spot.

Position 3: the parallelization slits are removed.

Position 4: a detector slit with 2° aperture is used.

These simplifying measures provide an increase of the detected X-ray intensity by a factor of ten.

2.2 Reduction of Measuring Time by Automated Self-Adjusting XRD Line Analysis

A further shortening of the recording time is achieved by increasing the efficiency of the measuring process control by means of a self-developed strategy for accelerated XRD line evaluation [4]. For residual stress measurements according to the XRD $\sin^2\psi$ method [1], the diffraction angles of the X-ray interferences of the a-Fe (211) lattice planes are to be determined for a certain number of sample orientations. For this purpose, commonly the complete diffraction line is registered with appropriate accuracy regarding counting statistics, i.e., with considerable time expenditure for scanning with the scintillation detector.

To avoid the time consuming standard procedure, a suitable number of evaluation points (e.g., 10) is placed symmetrically around the peak maximum in an angular range of 3°. The measurements are performed using the pulse-controlled recording system and the line position is then determined

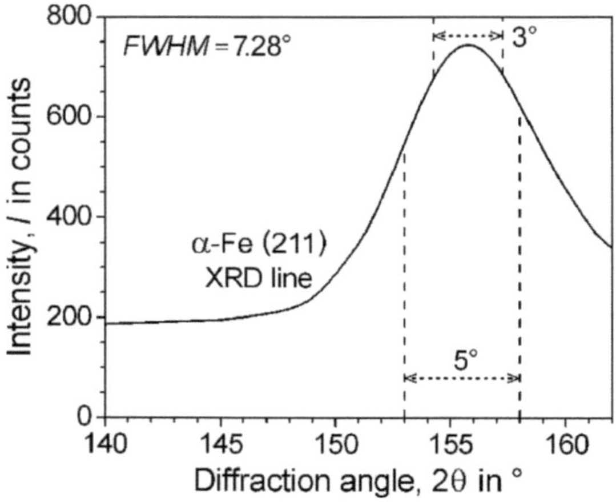

Fig. 2. Position of the angular ranges for the pre- and subsequent main measurement on the diffraction line

by a fitting polynomial regression. For each sample orientation-dependent diffraction line, the required symmetric arrangement of the base points around the peak maximum is obtained by means of a pre-measurement across an angular range of 5°. Here, a much lower accuracy with respect to the counting statistics is demanded. The rapid pre-analysis provides the identification of the peak maximum with an error of $\pm 0.2°$. The described iterative technique yields the sought evaluation points of the subsequent highly accurate main measurement for the sample orientation-dependent line shift. Figure 2 schematically illustrates the position of the two measuring ranges on an α-Fe (211) diffraction line of exemplary line width (full width at half maximum, FWHM) of 7.28°.

A large number of investigations applying this automated measuring technique have shown that seven different sample orientations suffice for an individual residual stress measurement, which contributes, together with other values, to a depth profile from the edge to the core of the specimen. The additional time saving amounts to 60 %. With these measures, an XRD residual stress measurement on an irradiated sample area of $2 \times 3\,\text{mm}^2$ can be performed within 5 min.

2.3 Measurement of the XRD Line Width and the Retained Austenite Volume Fraction

For the determination of the line width, the complete diffraction line must be considered and adequately registered, which is usually done at only one sample orientation angle. Also for this task, pertinent evaluation algorithms are developed. For a diffraction line of approximately 7°, the variance equals 0.06 to 0.09°, depending on the method.

The XRD phase analysis for the determination of the retained austenite content is performed applying Mo Kα radiation. The use of the square focal spot of the X-ray tube provides for the corresponding gain in intensity.

3 Application Fields of XRD Residual Stress Analysis

Over the last three decades, the described measuring time optimizations have allowed for the industrial application of XRD residual stress measurements and retained austenite determinations to diverse work areas in sufficient number to adequately extend the experience:

- residual stresses and material behavior
- material response analysis of run parts, such as rolling bearings or camshafts [5]
- failure analysis [6]
- heat treatment residual stresses, e.g., for martensitic and bainitic structures, case hardening, thermochemical processes, decarburization
- distortion problems related to hardening and grinding
- machining residual stresses, e.g., for grinding and honing, hard turning, shot peening

4 Residual Stress Generation by a Cold Working Treatment

Numerous series of measurements on deburred hardened steel balls for process optimization have been performed in the 1970s, with high demands on sample throughput of the improved XRD analysis technique. Figures 3 and 4 respectively show the residual stress and line width profile from the edge to the core of a deburred and subsequently final ground and lapped ball of martensitic rolling bearing steel 100Cr6 (SAE 52100) with 14.3 mm in diameter. The distance curves are explained as follows:

The balls bounce on a steel plate, which can be interpreted as a reverse shot peening process. As a result of the ball-plate fall contact, compressive residual stresses are built up in the edge zone of the balls, where the v. Mises equivalent stress exceeds the yield stress $R_{p0.2}$ and thus display a similar distribution below the surface. The line width decreases despite the increasing hardness because of the rearrangement of the dislocation structure, which is, for instance, also the case for grinding of high-strength steels. The object of this edge zone strengthening is the lowering of the indentation sensitivity (flattening) of the balls during handling.

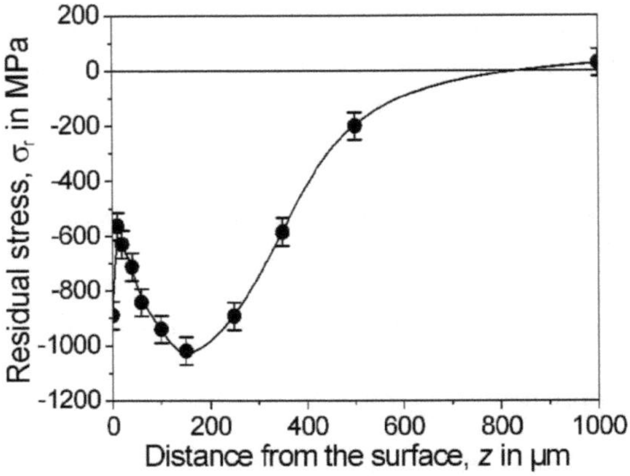

Fig. 3. Residual stress depth profile below the surface of a steel ball of 14.3 mm diameter after drum deburring and finishing

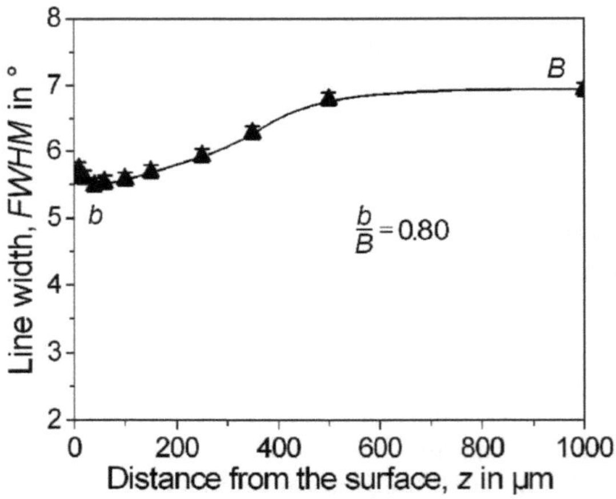

Fig. 4. α-Fe (211) diffraction line width depth profile related to the residual stress distribution of Fig. 3

5 Material Response Analysis on Run Rolling Bearings

From residual stress and line width profiles below the bearing raceways, conclusions can be drawn to location and intensity of the material response to stressing.

5.1 Surface- and Sub-Surface Mode of Rolling Contact Fatigue

Numerous measurements on run non-failed bearings from the field application and on bearings from rig tests make clear the differentiation between

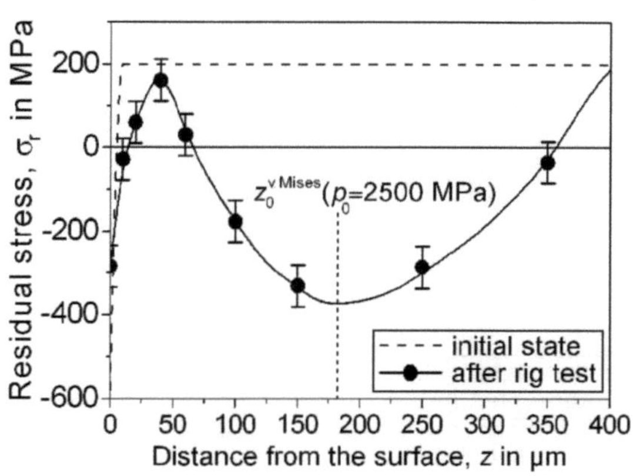

Fig. 5. Residual stress depth profile below the inner ring raceway of a taper roller bearing after a rig test at elevated temperature of 160°C; the *broken line* represents the initial state after martensite hardening and surface finishing

near-surface positioned material aging and the classical sub-surface failure mode. Figures 5 and 6 present the change of residual stresses and line width below the inner ring raceway of a taper roller bearing run in a rig test at a temperature of 160°C. Compressive residual stresses, with the peak in the depth z0 of maximum v. Mises equivalent stress corresponding to a Hertzian pressure of 2500 MPa, are generated. Decrease of line width can be observed in two zones: near the surface and in a depth close to the maximum orthogonal shear stress.

The course of the equivalent stresses can cause the alteration of the residual stress and line broadening, measured by XRD, in the edge zone or in a defined depth below the raceway: the established nomenclature for these cases is surface and sub-surface mode of fatigue. To quantify the change of line width, its minimum b (at the surface or in the depth of the material) is divided by the initial value B. In the case of the taper roller bearing from the rig test in Fig. 6, these normalized line broadenings of $b/B = 0.6$ in the depth and $b/B \geq 0.79$ at the surface represent a certain degree of material aging that occurred at the denominated positions.

Near-surface material changes are mostly the consequence of dense indentations caused by contaminated lubrication, which form a statistical waviness at the contact areas. Increased stress acts in Hertzian micro contacts. Another cause for a shift of the maximum stressing towards the surface arises from the superposition of tangential and radial stresses in the case of sliding friction that are often linked to engine vibration.

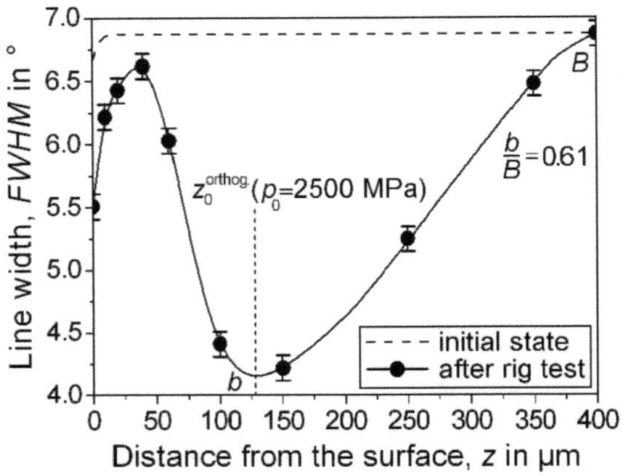

Fig. 6. Line width depth profile related to the residual stress distribution of Fig. 5

5.2 Time-Dependent Alteration of the XRD Material Characteristics and Calibration of the Material Response Analysis

The parameters of two rig tests are purposefully set in order to produce surface distress at taper roller bearings and classical rolling contact fatigue (i.e., sub-surface failure mode) at ball bearings. Compressive residual stress creation and decrease of line width below the raceways of the ball bearings result from a Hertzian pressure of 3300 MPa at the inner ring and ideal elastohydrodynamic lubrication. Controlled lubricant contamination and a contact pressure of 2000 MPa lead to the surface failure mode in the case of the taper roller bearings that failed after a running time similar to car gearbox bearings.

Figures 7 and 8 show the decrease of the minimum normalized line width b/B as a function of the number of revolutions of the rotating inner ring (IR), measured at the IR raceway surface of the taper roller bearings and in the depth of its maximum change below the IR raceway of the ball bearings [7, 8]. The L_{10} lifetime parameter, i.e., the running time related to the 10 % failure accumulation, is indicated in both diagrams. With respect to the failure modes, these calibration curves enable an estimation of the so-called residual life. The usage of this key task in the case of bearings from the field application, however, should be considered very carefully.

5.3 Characterization of Short-Cycle Fatigue and Failure in the Middle or Lower Fatigue Endurance Limit on the Basis of Known Rig Tests

In the described calibrating rig tests, dark etching areas (DEA) are formed in the microstructure below the raceway of the martensitically hardened bearings after a certain running period, which is consistently related to a reduction

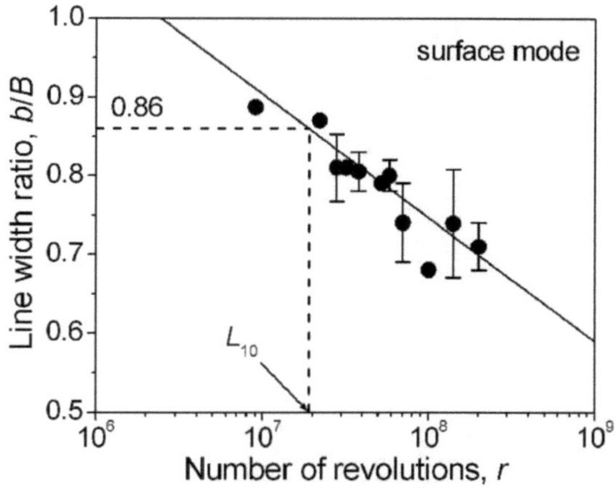

Fig. 7. Calibration diagram for the surface failure mode of rolling contact fatigue

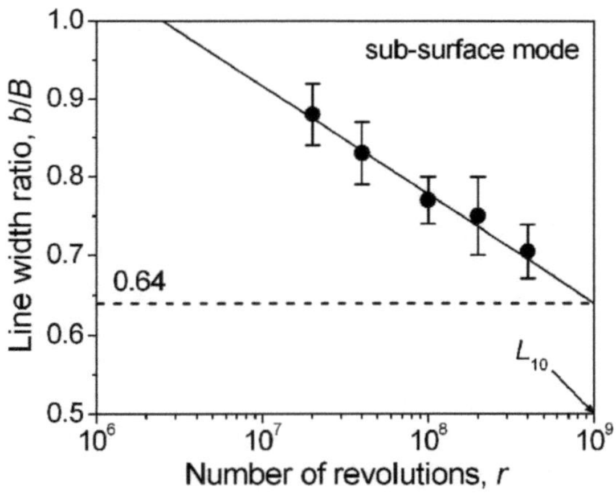

Fig. 8. Calibration diagram for the sub-surface failure mode of rolling contact fatigue

of the minimum normalized line broadening b/B to about 0.83. Consequently, it is additionally indicated by metallographic investigation that the material alteration occurs in the life cycle stage of instability. This carbon diffusion-induced microstructure transformation extends to a depth of 50 µm below the raceway in the case of the taper roller bearing. The center of the DEA of the ball bearing inner ring is located about 200 µm below the raceway.

DEA's adjacent to the surface, which indicate the surface mode of fatigue, occur locally irregularly distributed. They disappear gradually and partly extend only to a depth of 5 µm. In Fig. 9, a mild grain boundary etching reveals sliding bands in such a near-surface DEA.

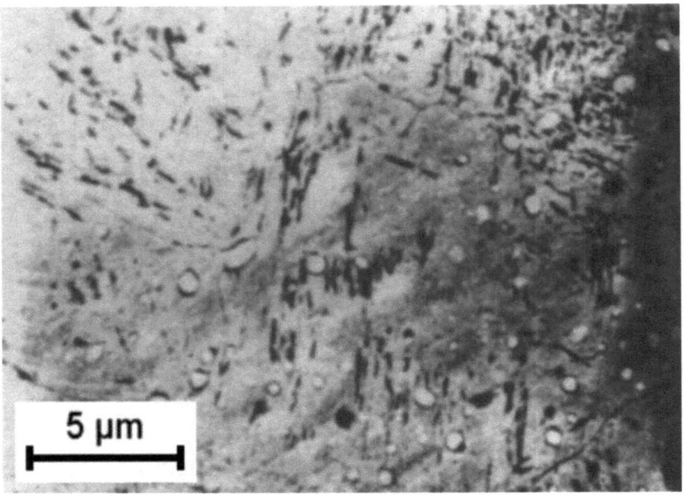

Fig. 9. Light-optical cross-sectional micrograph of a dark etching area positioned immediately below the inner ring raceway surface of a taper roller bearing from the calibrating rig test of Fig. 7

In the calibration tests, the decrease of the minimum normalized line broadening to $b/B = 0.83$ requires a running period of 3 × 107 to 4 × 107 revolutions. Therefore, the tests meet the range of middle or lower fatigue endurance limit. Material softening as a consequence of carbon diffusion-controlled martensite decay induces a further build-up of compressive residual stresses to −800 MPa after a running period of 1×109 revolutions in the depth of maximum v. Mises equivalent stress below the inner ring raceway of the ball bearings. Related to this high compressive residual stress, a decrease of the minimum normalized line broadening to $b/B = 0.64$ occurs. The test mentioned in Sect. 5.1. (cf., Figs. 5 and 6) can also be classified as a test in the range of middle or lower fatigue endurance limit. The compressive residual stress of −370 MPa and the decrease of the minimum normalized line width to $b/B = 0.61$ belongs to a lower Hertzian pressure of 2500 MPa but to a very high running temperature of 160°C. A marked homogenous DEA around a depth of 150 µm is found in the metallographic cross-section.

According to Figs. 3 and 4, the formation of compressive residual stress up to more than −900 MPa in the course of the cold working process of hardened balls is associated with a reduction of the minimum normalized line width to $b/B = 0.80$. However, no DEA occurs in the affected microstructure. In this case, the material alteration is restricted to the higher fatigue endurance limit and based on the change of the dislocation structure but not on carbon diffusion-controlled martensite decay.

Figure 10 shows the inner ring raceway of a ball bearing used in a car gear box with heavily contaminated lubrication. Despite a Hertzian pressure

Fig. 10. Scanning electron microscope image (secondary electron mode) of the indentation-covered inner ring raceway of a ball bearing after a rig test under heavily contaminated lubrication

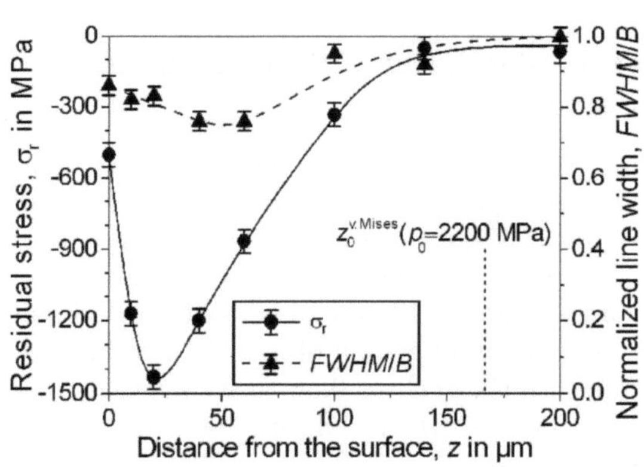

Fig. 11. Residual stress and normalized line width depth profile below the inner ring raceway of the ball bearing of Fig. 10

of 2200 MPa, the bearing failed after 1.8 × 106 IR revolutions by surface distress. The bearing life corresponds to only 1/60 of the nominal L_{10} value. As indicated in Fig. 11, the compressive residual stresses formed near the inner ring raceway surface with a maximum of −1450 MPa are connected with a b/B decrease to about 0.77 to 0.80. The bearing suffers from material ratcheting in this low cycle fatigue test.

By comparing the absolute values of formed compressive residual stresses with the associated decrease of the minimum normalized line broadening it should be possible to draw similar conclusions for bearings from the field application.

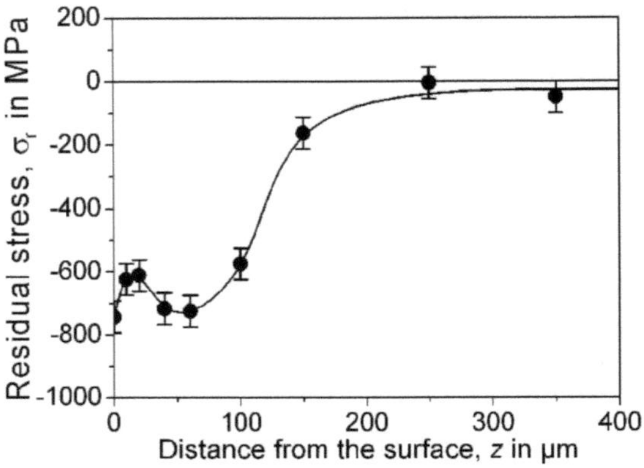

Fig. 12. Residual stress depth profile in the load zone below the outer ring raceway of a cylindrical roller bearing exposed to axial vibrations in a rig test

5.4 Characterization of the Stressing Level of Rolling Bearings in Case of Additional Loading by Vibrations

Engine vibrations generate sliding friction and are thus the cause of tangential forces in rolling contacts. The superposition of tangential to the radial loads shifts the equivalent stress towards the raceway surface. A rig test of a cylindrical roller bearing that is exposed to axial vibrations and lubricated with aged grease serves as an example. Figures 12 and 13 present the residual stress and line broadening profiles at a running time of 5×10^6 revolutions. The decrease of the normalized line width, b/B, to 0.58 to 0.66 reveals material aging similar to an annealing process, which relieves the generation of residual stresses by material softening.

The material response analysis on a cylindrical roller bearing from a field application points to three dimensional vibrations as damaging cause. The residual stress and line width profiles below the raceway in the outer ring load zone, given in Figs. 14 and 15, show a build-up of compressive residual stresses to −750 MPa and a related decrease of the normalized line broadening to 0.84. According to the considerations in Sect. 5.3. this combination of residual stress formation and line width reduction can be explained by short-cycle fatigue. However, with respect to the running period of 2×10^8 revolutions, severe vibrations are acting intermittently.

6 Summary

For the successful application of residual stress analyses to rolling bearings, it is necessary to gain experience based on a large quantity of measurements. A high throughput of samples is achieved by both a specific modification of the beam path of conventional diffractometers in order to increase the

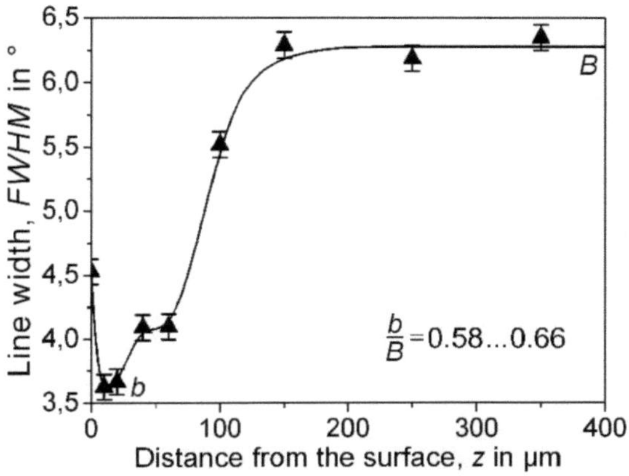

Fig. 13. Line width depth profile related to the residual stress distribution of Fig. 12

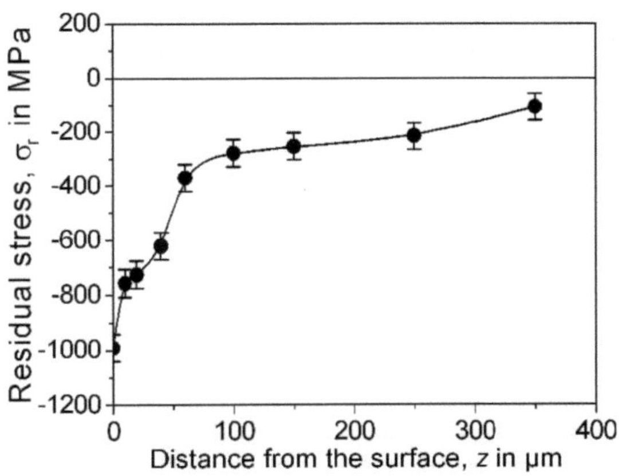

Fig. 14. Residual stress depth profile in the load zone below the outer ring raceway of a cylindrical roller bearing after 2×10^8 revolutions showing the impact of vibrations

recorded X-ray intensity and a rationalization of the automated measuring process control. Selected examples of application cover the generation of compressive residual stresses by production processes and the material response analysis on run rolling bearings, including the characterization of different failure modes and material aging.

The causes and mechanisms of failure in rolling contact can be divided into two distinct types: the classical sub-surface and the more complex (near-) surface mode of fatigue, the latter of which predominates in field applications. The failure causes of lubrication contamination and machine vibrations serve

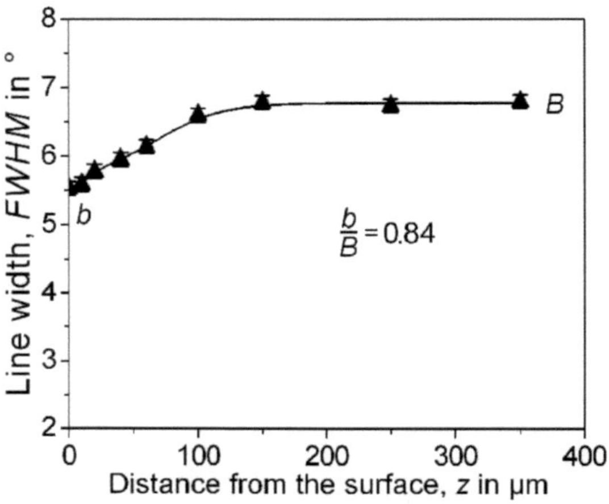

Fig. 15. Line width depth profile related to the residual stress distribution of Fig. 14

as examples of the surface mode of fatigue. It is also shown how to distinguish between material ratcheting and the range of lower fatigue endurance limit.

References

[1] E. Macherauch, P.Müller: Das $\sin^2 \psi$-Verfahren der röntgenographischen Spannungsmessung, Z. angew. Physik **13**, 305–312 (1961)
[2] M. Moore, W. Evans: Mathematical correction for stress in removed layers in x-ray diffraction residual stress analysis, SAE Trans. **66** (1958)
[3] U. Wolfstieg: Personal communication, unpublished
[4] W. Nierlich, J. Gegner: Material response analysis of rolling bearings using x-ray diffraction measurements, Materials Week (October 2001) proc. 4th Int. Congress Werkstoffwoche-Partnerschaft Frankfurt 2002 (CD-ROM ISBN 3-88355-302-6)
[5] W. Nierlich, J. Gegner, M. Brückner: X-ray diffraction in failure analysis of rolling bearings, Mater. Sci. Forum **524–525**, 147–152 (2006)
[6] W. Nierlich, J. Volkmuth: Schäden und Schadensverhütung bei Wälzlagern Teil ii: Beanspruchungsanalyse und Schadensverhütung, Antriebstechnik **40**, 49–53 (2001)
[7] W. Nierlich, U. Brockmüller, F. Hengerer: Vergleich von Prüfstand- und Praxisergebnissen an Wälzlagern mit Hilfe von Werkstoffbeanspruchungsanalysen, HTM **47**, 209–215 (1992)
[8] A. Voskamp: Ermüdung und Werkstoffverhalten im Wälzkontakt, HTM **53**, 25–30 (1998)

Piezoelectric Graded Materials – Preparation and Characterization

Ralf Steinhausen

Institute of Physics, Martin-Luther-University Halle,
Friedemann-Bach-Platz. 6, 06108 Halle, Germany
ralf.steinhausen@physik.uni-halle.de

Abstract. Monolithic piezoelectric ceramics with an inhomogeneous chemical composition are suitable for applications where internal mechanical stresses should be reduced. A chemical gradient can be prepared using layered structures with different materials and different thicknesses of the layers. Several combinations of piezoelectric high and low active, piezoelectric hard and soft as well as electroconductive and electrostrictive ceramics are discussed in this work. In all cases the chemical gradient have to be transformed to the piezoelectric gradient by a poling process. An inhomogeneous electric field is induced in the sample during the poling process due to the different ferroelectric properties. This yields to a gradient of polarization. The poling process can be described by a simple multilayer model. The single layers were simulated using the Preisach-model. The role of conductivity was discussed. The gradient of polarization was investigated using a thermal wave method (LIMM).

1 Introduction

Functional graded materials (FGM) have a one or more dimensional gradient of at least one material coefficient. Usually, several material properties changes their values continuously in the direction of the gradient. Thus, a gradient of the piezoelectric coefficients in electroceramics is often combined with a gradient of the dielectric or elastic coefficients. Piezoelectric functional graded materials are suitable in applications where materials with different properties should be mechanically connected. They can reduce the mechanical stresses at this interfaces. Piezoelectric graded materials with a gradient of the electromechanical properties can be used in micromechanical systems (MEMS) or actuators to connect piezoelectric active parts with substrates or other inactive parts. In this way reliability and lifetime can be improved.

A special application are bending actuators. The bending effect strongly depends on the difference of the lateral strain at the top and the bottom surface of the actuator. Usually, two layers with different piezoelectric properties or with an electrode between them were bonded together. It was shown that piezoelectric FGM can reduce the mechanical stresses at the interface [1].

A good performance of the bending actuator depends on the quality of the piezoelectric gradient. In monolithic samples the gradient have to be induced by a poling process. The characterization of this piezoelectric gradient in

R. Haug (Ed.): Advances in Solid State Physics,
Adv. in Solid State Phys. **47**, 315–326 (2008)

monolithic ceramics is difficult. However, the local piezoelectric coefficient is correlated with the polarization degree. For the characterization of the polarization profile a thermal wave method, for instance LIMM is a helpful tool [2].

The first version of a monolithic ceramic with various piezoelectric properties along the thickness direction is the RAINBOW (Reduced And INternally Biased Oxide Wafer) actuator described by *Haertling* et al. [3]. A piezoelectric ceramic disk (PZT or PZT) with a high content of lead is thermally reduced at one side. This lead reduce layer becomes piezoelectric inactive. Furthermore, piezoelectric ceramics can be combined with electroconductive or electrostrictive materials. The combination of soft and hard piezoelectric materials allows to polarize the monolithic sample in different directions. In this work, these different types of piezoelectric FGM are discussed. Experimental data of lead free actuators based on pure and doped $BaTiO_3$ ceramics were presented.

2 Preparation of a Piezoelectric Graded Material

The common way to prepare a piezoelectric gradient in monolithic ceramics starts with a chemical gradient in one or more dimensions of the sample. By adding dopants or changing the volume ratio in multicomponent systems a series of parameters can be influenced. In ferroelectrics the microstructure (grain sizes, porosity, and domain structure) depends not only on the sintering conditions but as well on the chemical composition. Slight shifting of the the volume ratio in the near of a morphotropic phase boundary in piezoelectrics yields to huge differences in the ferroelectric parameters (spontaneous and remnant polarization, coercive field strength) as well as the linear material coefficients of the material. The maximum of the piezoelectric, dielectric, and elastic properties of PZT at the morphotropic phase boundary is a well-known example [4].

The preparation of the chemical gradient is based on a multilayer structure in the green body. In dependence on the final thickness of the sample and the used materials different technologies are suitable. All kinds of thin and thick film techniques are usable for thinner samples at the nano- and micrometer scale. In the micrometer range the technology of pressing green foils can be used. The samples investigated in this work are prepared with the powder pressing method advantageously used for graded bulk ceramics.

In all methods the chemical composition of each layer slightly differs from that one of the neighboring layer. The gradient itself can be varied by the number of different layers and the value of the difference of the chemical composition between the layers. *Jeon* et al. used up to 21 powder layers to prepare a graded barium-strontium titanat ceramic [5]. In our work we pressed only 2, 3 or 4 powder layers together.

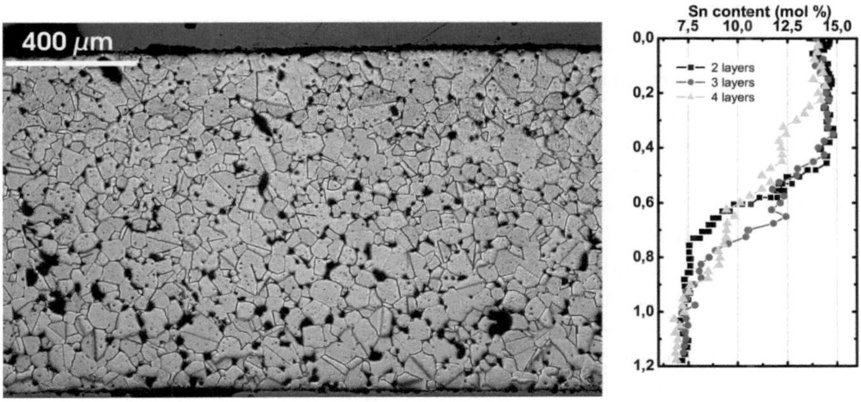

Fig. 1. Micrograph of a Ba(Ti,Sn)O_3 ceramic with the chemical composition BTS7.5-BTS10-BTS12.5-BTS15 (*left*) and the Sn-distribution along the thickness in BTS samples prepared with different number of powder layers

We investigated different graded materials based on $BaTiO_3$ ceramics. By adding different dopants the ferroelectric and electromechanical properties can be changed in a wide range. Thus, combinations of piezoelectric ceramics with electrostrictive, electroconductive and hard piezoelectric ceramics are investigated. The ceramic powders needed for the different layers were prepared by the conventional mixed-oxide technique. The green body was formed by successive pressing of the granulated powder with different chemical composition. The samples were sintered one hour at 1400°C under a uniaxial pressure of about 1 kPa to prevent their deformation due to the different shrinkage properties of the different ceramics. The final sample's thickness amounted to about 1.2 mm.

Figure 1 shows the microstructure of a $BaTiO_3$-$BaSnO_3$ ceramic with a gradient of the Sn content. The sample was prepared from four powder layers with Sn content of 7.5, 10, 12.5, and 15 mol%, respectively. In the following the description BTSX is used, where X is the amount of tin (mol%). The local Sn content along the thickness, i.e., along the direction of the chemical gradient, was characterized by electron probe micro analysis (WDX-EPMA). The results of samples with 2, 3, and 4 powders layers were compared in the right graph of Fig. 1. Due to the diffusion processes during sintering the Sn concentration changes gradually at the interfaces between the different layers.

Finally, the samples are metallized with aluminum and polarized applying a DC-voltage to transform the chemical gradient into a gradient of the piezoelectric properties.

3 Poling Model

The last step of preparation of piezoelectric graded ceramics is the poling process. The ferroelectric properties like spontaneous polarization and coercive field strength as well as the dielectric permittivity depend of course on the chemical composition of the ceramics. Therefore, the electric field distribution will be very inhomogeneous inside the sample when the poling voltage is applied. Thus, it is not self-evident that all parts of the monolithic ceramics are completely polarized.

The poling process of a piezoelectric functional graded material can be described by a multilayer model. The gradient of the ferroelectric properties along the thickness have to be replaced by a stepwise function. Each layer of the model structure consists of a homogeneous material with constant parameters. The layers are connected only electrically. Mechanical effects like clamping are neglected. The layers can be regarded as serial connected RC-elements [6]. The electric conductivity σ and the dielectric polarization P are taken into account for each layer. The conductivity is considered as constant and independent on the electric field. Using the constitutive equations for each layer and basic electric laws, a System of Ordinary Differential Equations (ODE) is resulting. For the simplest case, a two layer system follows

$$\sigma_1 E_1 + \epsilon_0 \dot{E}_1 + P_1' \dot{E}_1 = \sigma_2 E_2 + \epsilon_0 \dot{E}_2 + P_2' \dot{E}_2 \tag{1}$$

$$\dot{E}_{appl} = \frac{t_1}{t}\dot{E}_1 + \frac{t_2}{t}\dot{E}_2 \,, \tag{2}$$

where E_1 and E_2 are the local electric field strength in the layers, and t_1 and t_2 the thickness of the layers. The applied electric field E_{appl} in (2) is the applied voltage per overall thickness t.

In order to solve the ODEs the knowledge of the ferroelectric behavior is necessary. That means the function $P(E)$ and, in particular the derivation P′(E) is needed. This quantity is estimated by means of the Preisach model. The advantage of this model bases on the feasibility to simulate incomplete hysteresis loops and partial depolarization or switching processes. The theoretical background of the poling model is described in detail in [7]. The results of the modelling will be discussed for different combinations of materials in the next section.

4 Different Concepts of Piezoelectric Graded Materials

The aim of application of piezoelectric graded materials is the reduction of mechanical stresses at the interface between materials with different electromechanical properties. The preparation of stress-reduced bending actuators is another field of interest [1]. Therefore, the piezoelectric coefficients changes from a high value to a zero or low value. If the piezoelectric activity disappears, the material is usually still electrostrictive. In the case of bending actuators it should be helpful to shift the sign of the piezoelectric coefficient to improve the bending deformation. Moreover, the performance of an actuator can be enhanced matching the piezoelectric gradient with a dielectric one. In the following different concepts of monolithic graded materials will be discussed

4.1 Combination of Piezoelectric and Electroconductive Ceramics

If we combine a piezoelectric ceramic with an electroconductive, i.e., non-piezoelectric active ceramic, the system is called unimorph or monomorph [8]. A well-known representative of this group is Haertling's RAINBOW actuator.

The conductivity of $BaTiO_3$ ceramics can be dramatically increased by doping with Lanthanum. Already at low concentrations of 0.2 mol% La the conductivity reaches about $10(\Omega\text{m})^{-1}$. Thus, the La-doped layer can be used as a ceramic electrode. The voltage drop at this layer can be neglected in comparison with the $BaTiO_3$ layer. Hence, the electric field strength in the piezoelectric part is approximately equal to the applied electric field.

The thickness of such a ceramic electrode influences the bending performance of the monolithic actuator. Samples were prepared with different ratios of the thickness of the piezoelectric layer (t_p) to the overall sample thickness t. The bending deflection of an one-side fixed bending actuator was measured with a capacitive displacement sensor [1]. If the applied electric field is constant, the maximum value of the bending deflection was found at the thickness ratio $r = t_p/t = 0.5$. This result can be confirmed for low electric field strength by a linear Finite-Element-Modelling (FEM) (Fig. 2). The commercial FEM program code ANSYS was used for the simulation. Here, the electromechanical response is calculated by an linear coupling of the strain and the electric field.

The voltage applied to samples with a thinner piezoelectric layer is lower than in samples with a higher ratio r to induce the same electric field strength. Due to the fact that the applied voltage is one of the significant parameters in applications Fig. 3 shows the bending deflection normalized to the applied voltage. If the applied voltage increases nonlinear behavior is observed and the bending deflection increases.

4.2 Combination of Piezoelectric and Electrostrictive Ceramics

The electromechanical strain is the product of the piezoelectric coefficient and the electric field strength. Therefore, the performance of a piezoelectric graded actuator can be improved by an additional gradient of the local electric field strength, i.e., the dielectric permittivity. In the region of low permittivity the bigger part of the voltage is applied. In the layer with higher permittivity the induced electric field strength is consequently lower. Hence, the gradient of the permittivity should be contrary to the piezoelectric gradient.

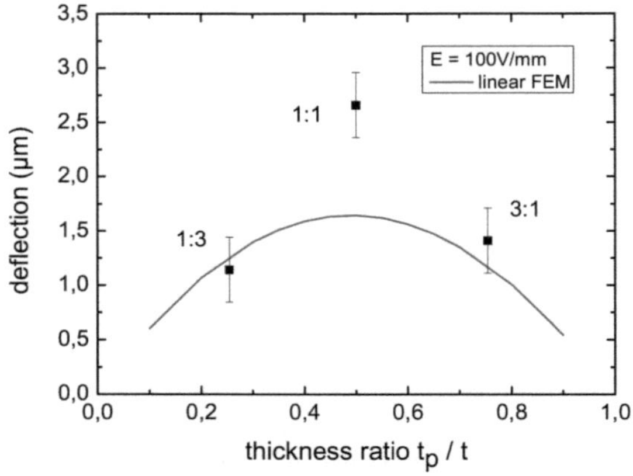

Fig. 2. Dependence of the bending deflection of $BaTiO_3$/$BaTiO_3$+La monomorphs on the thickness ratio at constant applied electric field

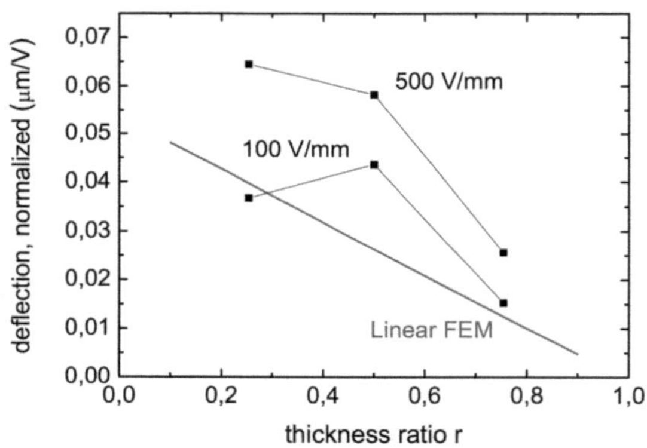

Fig. 3. Bending deflection of $BaTiO_3$/$BaTiO_3$+La monomorphs with different thickness ratios, normalized to the applied voltage

In the ceramic system Ba(Ti,Sn)O_3 the dielectric and piezoelectric coefficients strongly depend on the Sn content (Fig. 4, left). In the region of interest with a Sn content between 7.5 and 15 mol% the coefficients vary in the manner described.

Let's keep in mind that the piezoelectric gradient should be induced by the poling process. Hence, it is necessary to consider the virgin loops of polarization. The ceramics $BaTi_{0.925}Sn_{0.075}O_3$ (BTS7.5) has a rhombohedral structure. The ceramics with 15 mol% Sn (BTS15) is near the phase boundary to the cubic phase [9]. Correspondingly, the dielectric properties change from ferroelectric to paraelectric. The virgin loops of the polarization of BTS ceramics with different Sn content are shown in Fig. 4, right. The ferroelectric BTS7.5 reaches a higher maximum polarization. The P(E) curve has a hys-

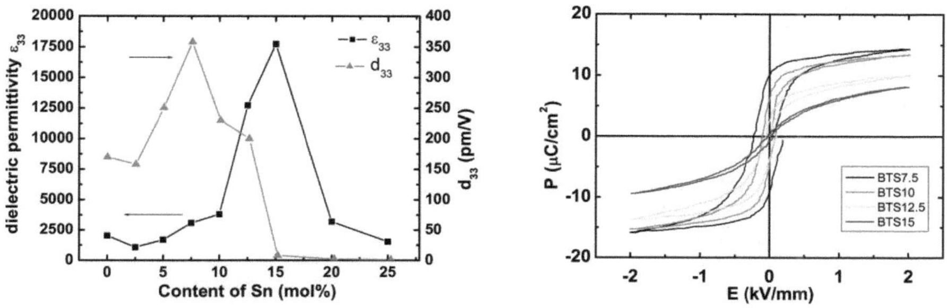

Fig. 4. *Left*: permittivity and piezoelectric coefficient d_{33} of $Ba(Ti,Sn)O_3$ ceramics; *right*: virgin loops in dependence on the Sn content

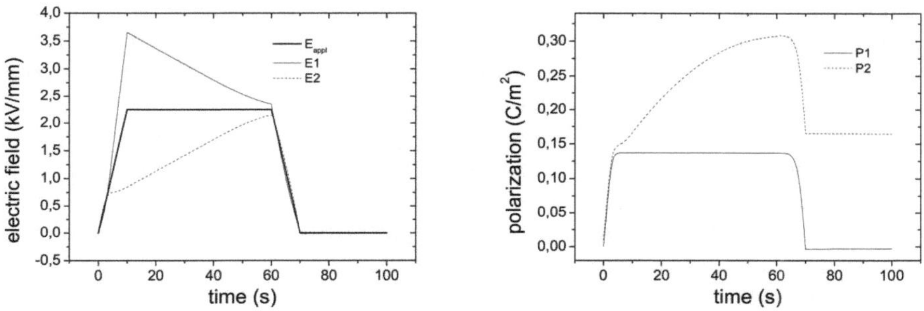

Fig. 5. Time dependence of the poling field and the local electric field strength (*left*) as well as the local polarization (*right*) in a 2-layer system of ferroelectric and paraelectric materials

teretic shape. The polarization of BTS15 depends also nonlinear, but nearly non-hysteretic on the applied electric field.

In the following the simplest case of the poling process of a 2-layer system will be discussed. A DC electric field of 2.5 kV/mm is applied to the system for 50 seconds to polarize it. The voltage is switched on and off by a ramp of 10 seconds, respectively (Fig. 5). The local electric field strength in the layer with the lower maximum polarization rises very fast to a maximum value and then decreases slowly. In the other layer the local electric field increases slowly to a saturation value. When saturation is reached the electric field strength in each layer doesn't depend on time. From (1) follows that

$$E_1\sigma_1 = E_2\sigma_2 \,. \tag{3}$$

The ratio of the saturated values of the electric field in both layers depends on the electric conductivity of the materials. In a first approximation the conductivity was chosen as equal in each layer.

A relaxation of the local fields and the polarization was observed after switching off the poling field. The polarization of the paraelectric material

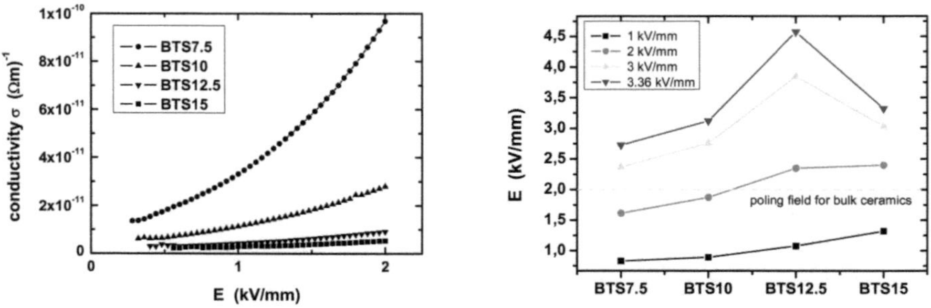

Fig. 6. Electrical conductivity of Ba(Ti,Sn)O_3 ceramics (*left*); electric field distribution in a 4-layer model structure consisting on BTS7.5-10-12.5-15 (*right*)

tends to zero. In the ferroelectric layer a remnant polarization remains (Fig. 5, right). In this way, the polarization in a chemical graded material should be changed from the remnant polarization of the ferroelectric material to the nearly zero polarization of the paraelectric part.

The monolithic BTS ceramics with a gradient of the Sn content were polarized with an poling field strength of about 2 kV/mm for 4 minutes. The time should be enough to reach the saturation of polarization. In the real ceramics the conductivity slightly depends on the Sn content. The data shown in Fig. 6 (left) were measured with an high resistance meter Keithley 6517A. The conductivity increases with increasing voltage for all samples. Thus, the approximation of constant conductivities can provide only qualitative results. The distribution of the electric field strength was measured at glued model structures with the same chemical composition and inner electrodes [10]. The electric field distribution of a 4 layer sample (Fig. 6, right) corresponds for field strengths up to 2 kV/mm qualitatively with the measured data of conductivity, considering (3).

The polarization distribution of a monolithic graded ceramic along the thickness was characterized by the thermal wave method [2]. The measured pyroelectric coefficient profile is shown in Fig. 7. The sample is prepared from four powder layers with BTS7.5, BTS10, BTS12.5, and BTS15, respectively. The induced gradient of the polarization is proportional to the pyroelectric signal. For comparison the remnant polarizations of the corresponding ceramics with the same chemical composition are plotted in arbitrary units (red lines).

4.3 Combination of Piezoelectric Hard and Soft Ceramics

In the first two concepts there is on one side of the actuator a piezoelectric non- or low-active layer. Higher bending deflections of actuators can be achieved if one side of the actuator contract and the other side expand in the same time. That problem can be solved if the layers can be polarized in opposite directions. The idea of a gradient of polarization with changing sign can

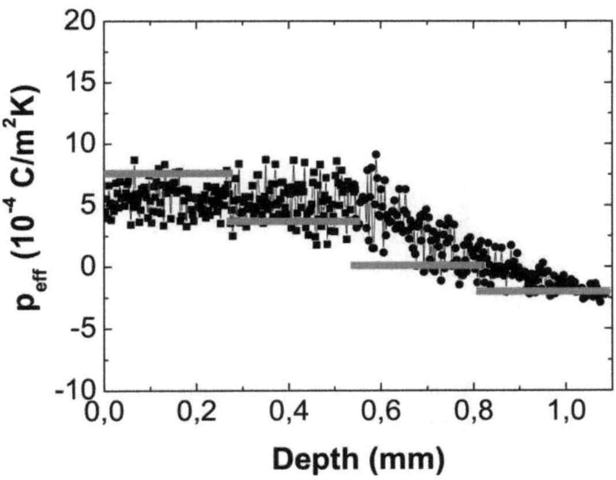

Fig. 7. Thickness profile of the pyroelectric coefficient of a monolithic BTS sample with a Sn gradient

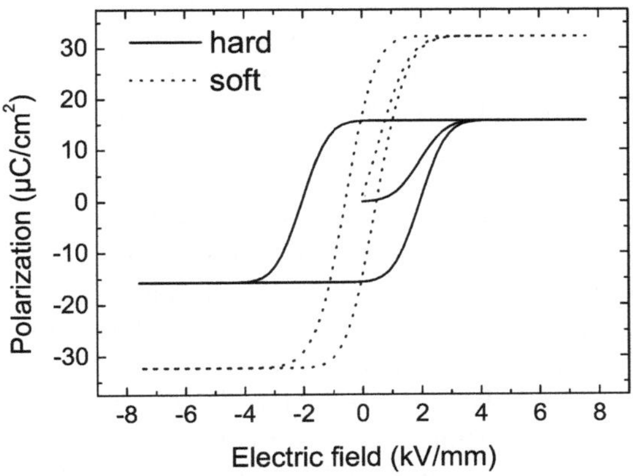

Fig. 8. Hysteresis loops of ferroelectric soft and hard materials modelled by the Preisach Model

be realized by using ferroelectric soft and hard materials. Typical hysteresis loops using in the modelling are shown in Fig. 8. The soft material is characterized by low coercive field strength and high spontaneous polarization. The hard material exhibits an higher coercive field and lower polarization.

The poling regime for such a system is more complicated. In Fig. 9 the applied voltage and the local electric field in the soft and hard material (left), as well as the polarization of the materials (right) are shown. In a first step, a positive voltage is applied to the bilayer system. We used a ramp in the calculation to avoid numerical problems. In the experiment a DC voltage for a short time can be more useful. The poling time and maximum value of the poling field should be high enough to polarize the hard material completely.

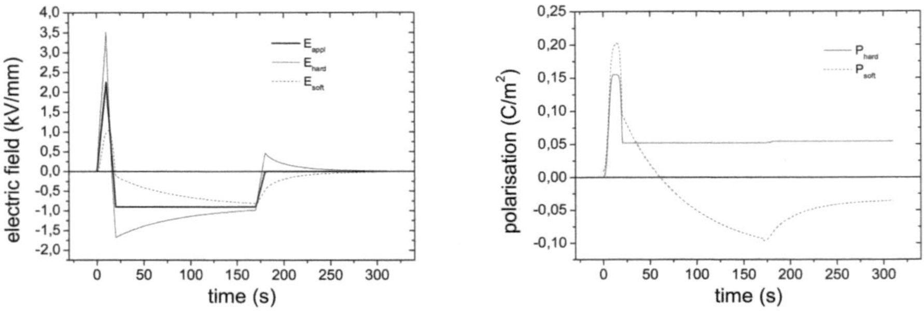

Fig. 9. Time-dependence of the local electric field (*left*) and the polarization (*right*) in hard and soft layers during the poling process

The poling degree of the soft material is not so important. In contrast, the lower the polarization of the soft material the better the polarization can be switched.

Then the applied voltage is slowly reduced. The speed of this reduction depends on the conductivity of the materials. The higher the conductivity the faster the voltage can be reduced. If the conductivity is lower, more time is required to collect space charges at the interface compensating the difference in the local polarization.

The electrical field strength is then reduced to a negativ value. This value should be higher than the coercive field strength of the soft material to shift its polarization. But it must be lower than the coercive field strength of the hard material to protect the positive polarization in this layer. The negative DC voltage is applied for a longer time to permit the switching of polarization. The length of time depends primarily on the conductivity of the materials. When the saturation is achieved the voltage is switched off. The remnant polarizations show different signs after a short relaxation time.

Ferroelectric soft and hard materials based on $BaTiO_3$ are needed for the preparation of a gradient from positive to negative polarization. $BaTiO_3$ doped with 0.5 mol% Mn was used instead of the hard material. The non-poled ceramics show antiferroelectric behavior of the polarization (Fig. 10, left). The strain is constant below a threshold value of 0.5 kV/mm. The sample was polarized with 2 kV/mm for 30 min at 80°C. After the poling the strain has a small hysteresis with a shift to the negative voltage (Fig. 10, right). Thus, the remnant piezoelectric response $d_{33}=dS_3/dE_3$ at small voltages remains positive after applying a high field strength of 2 kV/mm for a short-time. The strain loop was measured at 10 Hz by a capacitive displacement sensor.

We used a model structure to investigate the poling process and to characterize the local piezoelectric coefficients. Two single layers of $BaTiO_3$ and $BaTiO_3$ + 0,5 mol% Mn were connected by a wire and poled together. This allows measuring the piezoelectric coefficients d_{33} of each single layer at sev-

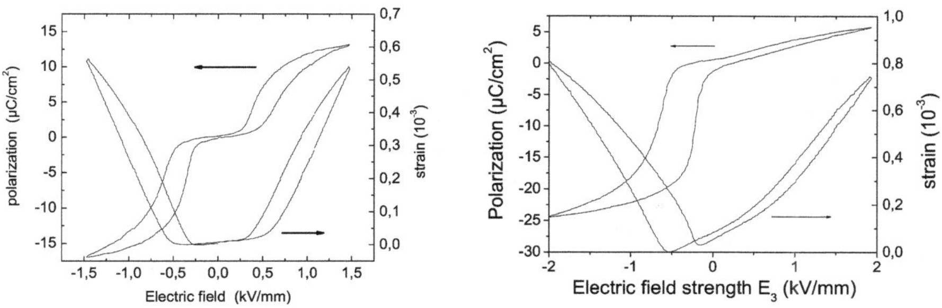

Fig. 10. Polarization and strain of unpoled (left) and poled (right) $BaTiO_3$+Mn in dependence on the electric field

eral times during the poling process. After 30 min poling at 80°C both layers were completely poled. The piezoelectric coefficients of $d_{33,BT}$=198 pm/V and $d_{33,BTMn}$=175 pm/V were measured. After cooling down to room temperature a negative electric field strength of 2 kV/mm was applied for few seconds. After 20 seconds, the $BaTiO_3$ layer was nearly completely depolarized. That means that the absolute value $d_{33,BT}$ was of about zero. The piezoelectric coefficient of BTMn decreases to the half of the original value. After another 30 seconds poling with −2 kV/mm the polarization switched back in the $BaTiO_3$ layer and was still remain in BT+Mn. The piezoelectric coefficients $d_{33,BT} = -200$ pm/V and $d_{33,BTMn} = 80$ pm/V were measured. The negative sign means that the piezoelectric response is in the opposite direction to the first poling field direction. Thus, we could demonstrate that it is possible to prepare a piezoelectric gradient where the sign of d_{33} switched from positive to negative.

5 Summary

Monolithic samples with a chemical gradient were prepared by the powder pressing technique. The poling process transforming the chemical into a piezoelectric gradient can be described by a multilayer model. Several lead free actuators based on $BaTiO_3$ ceramics were investigated. The combination of piezoelectric and electroconductive materials is easily to polarize. The bending deflection can be increased be decreasing the thickness ratio of the layers. Samples with a gradient of the ferroelectric properties should be poled for a longer time to induce a polarization gradient. The good bending performance of such actuators was shown elsewhere [11]. A poling regime was proposed to polarize monolithic actuators in different directions. The principle was demonstrated using hard and soft piezoelectric materials.

References

[1] T. Hauke, A. Z. Kouvatov, R. Steinhausen, W. Seifert, H. Beige, H. T. Langhammer, H. Abicht, Ferroelectrics **238**, 195 (2000)

[2] A. Movchikova, O. Malyshkina, G. Suchaneck, G. Gerlach, R. Steinhausen, H. T. Langhammer, C. Pientschke, H. Beige, accepted to be published in J. Electroceramics (2007)

[3] G. H. Haertling: Am. Ceram. Soc. Bull. **73** (1), 93 (1994)

[4] B. Jaffe, W. R. Cook, H. Jaffe: *Piezoelectric Ceramics* (Academic Press Ltd, London, 1971).

[5] J.-H. Jeon, Y.-D. Hahn, H.-D. Kim: J. Europ. Cer. Soc. **21**, 1653 (2001)

[6] Y. T. Or, C. K. Wong, B. Ploss, F. G. Shin: J. Appl. Phys. **93**, 4112 (2003)

[7] C. Pientschke, R. Steinhausen, A. Kouvatov, H. T. Langhammer, H. Beige: Ferroelectrics **319**, 181, (2005)

[8] K. Uchino: *Piezoelectric actuators and ultrasonic motors* (Kluwer Academic Publishers, Boston/Dortrecht/London, 1997)

[9] N. Yasuda, H. Ohwa, S. Asano: Jpn. J. Appl. Phys. **35**, 5099 (1996)

[10] R. Steinhausen, A. Z. Kouvatov, C. Pientschke, H. T. Langhammer, H. Beige: Proc. IEEE Int. UFFC Joint 50th Anniversary Conference, Montreal 2004, 118 (2004)

[11] R. Steinhausen, H. Th. Langhammer, A. Z. Kouvatov, C. Pientschke, H. Beige, H.-P. Abicht: Materials Science Forum **494**, 167 (2005)

Part VII

Materials

^{59}Co, ^{23}Na, and ^{1}H NMR Studies of Double-Layer Hydrated Superconductors $Na_xCoO_2 \cdot yH_2O$

Y. Itoh[1], H. Ohta[1], C. Michioka[1], M. Kato[2], and K. Yoshimura[1]

[1] Department of Chemistry, Graduate School of Science, Kyoto University, Oiwake-cho, Kitashirakawa, Sakyo-ku, 606-8502 Kyoto, Japan
kyhv@kuchem.kyoto-u.ac.jp

[2] Department of Molecular Science and Technology, Faculty of Engineering, Doshisha University,
Tatara-Miyakodani, 610-0394 Kyotanabe, Japan

Abstract. We present our NMR studies of double-layer hydrated cobalt oxides $Na_xCoO_2 \cdot yH_2O$ ($x \approx 0.35$, $y \approx 1.3$) with various $T_c = 0$–$4.8\,K$ and magnetic transition temperatures. High-resolution ^{1}H NMR spectrum served as an evidence for the existence of H_3O^+ oxonium ions. ^{23}Na nuclear spin-lattice relaxation rates served to detect local field fluctuations sensitive to T_c. ^{59}Co nuclear quadrupole resonance (NQR) spectra served to classify the various T_c samples. From the classification by ^{59}Co NQR frequency, the double-layer hydrated compounds were found to have two superconducting phases closely located to a magnetic phase. In the normal state and at a magnetic field in the *ab*-plane, two ^{59}Co NMR signals with different Knight shifts and different ^{59}Co nuclear spin-lattice relaxation times $^{59}T_1$ were observed. The two ^{59}Co NMR signals suggest magnetic disproportionation of two Co sites or XY-anisotropy of a single Co site. Non Korringa behavior and power law behavior in zero-field NQR $1/^{59}T_1$ above and below T_c suggest non-Fermi liquid and unconventional superconductivity.

1 Introduction

$Na_{0.7}CoO_2$ is a high-performance thermoelectric power material [1]. The crystal structure is constructed from the stacking of a Co triangular lattice in the layered edge-sharing CoO_2 octahedrons and of a Na layer. The discovery of superconductivity through soft chemical treatment of sodium de-intercalation and hydration has renewed our interests in spin and charge frustration effects on the CoO_2 triangular lattice [2]. Figures 1a and 1b illustrate the crystal structure and the soft chemical treatment, respectively. The double-layer hydrated cobalt oxides $Na_xCoO_2 \cdot yH_2O$ ($x \approx 0.35$, $y \approx 1.3$) are triangular-lattice superconductors with the optimal $T_c \approx 5\,K$ [2]. Except the charge ordered insulator Na_xCoO_2 with $x \approx 1/2$, Na_xCoO_2 and the hydrated compounds are itinerant electronic systems. For $x \approx 0.7$, *A*-type spin fluctuations (in-plane ferromagnetic and out-of-plane antiferromagnetic spin

R. Haug (Ed.): Advances in Solid State Physics,
Adv. in Solid State Phys. **47**, 329–341 (2008)

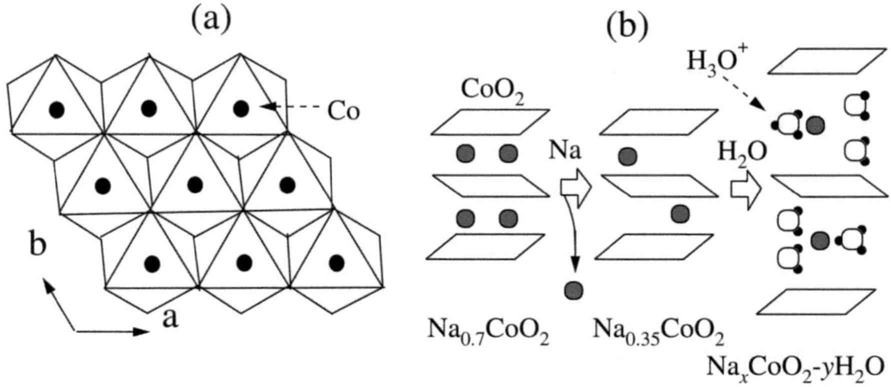

Fig. 1. (**a**) The top view of a CoO_2 plane. (**b**) The side views of the crystal structures of a parent $Na_{0.7}CoO_2$, a Na-deintercalated $Na_{0.35}CoO_2$ and a *bi*layer hydrated $Na_{0.35}CoO_2 \cdot yH_2O$. *Open arrows* indicate soft chemical treatment

fluctuations) [3, 4], the motion of Na ions [5, 6] and Co charge disproportionation [6–9] have been reported.

For anhydrated Na_xCoO_2, the Na concentration changes the carrier doping level of a CoO_2 plane, which develops the electronic state [10]. For double-layer hydrated $Na_xCoO_2 \cdot yH_2O$, the Na concentration is not a unique parameter of T_c [11–13]. Since no one has ever observed superconductivity of the anhydrated compounds, the carrier doping level is not a unique parameter of T_c. The chemical diversity of the site ordering of Na ions [14] and the charge compensation by oxonium ions H_3O^+ [11] may realize a delicate electronic system on the triangular lattice. The effect of the intercalated water molecules on the electronic state of the CoO_2 plane is poorly understood.

The followings have been debated so far, whether the normal-state spin fluctuations are ferromagnetic, antiferromagnetic or the other type, whether or not the superconductivity occurs in the vicinity of the magnetic instability, whether the charge fluctuations play a key role in the superconductivity, how different the triangular-lattice superconductivity is from the square-lattice one, what the pairing symmetry is.

In this paper, we present the highlights of our NMR studies and findings for the double-layer hydrated compounds; the various T_c samples synthesized in controllable way, ^{1}H NMR evidence for the existence of the H_3O^+ oxonium ions, T_c-sensitive ^{23}Na nuclear spin-lattice relaxation, two superconducting phases classified by ^{59}Co nuclear quadrupole resonance (NQR) frequency, a magnetic phase located between the two superconducting phases, in-plane two components in ^{59}Co NMR Knight shift and nuclear spin-lattice relaxation, non-Korringa behavior above T_c, and power law behaviors in nuclear spin-lattice relaxation rates.

2 Synthesis and Experiments

Polycrystalline samples of the starting compound $Na_{0.7}CoO_2$ were synthesized by conventional solid-state reaction methods. The powdered samples of $Na_{0.7}CoO_2$ were immersed in Br_2/CH_3CN solution to deintercalate Na^+ ions and then in distilled water to intercalate H_2O molecules [15–17] or in Br_2/H_2O solution to make ion-exhange reaction of Na and H_2O [18, 19]. After the powders of $Na_{0.35}CoO_2 \cdot yH_2O$ were separated from the solution by filtration, they were exposed in various humidity air. For the synthesized samples, we observed various duration (keeping time in the humidity-controlled chamber in a daily basis) effects on T_c [15–17, 19–21].

^{59}Co (spin $I = 7/2$), ^{23}Na (spin $I = 3/2$), and ^{1}H (spin $I = 1/2$) NMR experiments were performed by a pulsed NMR spectrometer for the powder samples. The nuclear spin-lattice relaxation times were measured by inversion recovery techniques.

3 Experimental Results

3.1 Magnetic Phase Diagram Classified by ^{59}Co NQR

Zero-field NQR is sensitive to local crystal structure and charge distribution around the nuclear site. Although no distinct changes in X-ray diffraction patterns have been observed for the various double-layer hydrated compounds, systematic changes in the peak values of ^{59}Co NQR spectra were observed [15, 21–23]. Thus, we present the results of ^{59}Co NQR spectrum measurments to characterize the samples and to classify the electronic states.

For an electric field gradient tensor $V_{\alpha\beta}$ ($\alpha\beta$ = xx, yy, and zz, principal axis directions) with an asymmetry parameter $\eta[\equiv (V_{xx} - V_{yy})/V_{zz}] < 1$, three transition lines of ^{59}Co (spin $I = 7/2$) NQR should be observed as ν_{Q1} ($I_z = \pm 3/2 \leftrightarrow \pm 1/2$), ν_{Q2} ($I_z = \pm 5/2 \leftrightarrow \pm 3/2$), and ν_{Q3} ($I_z = \pm 7/2 \leftrightarrow \pm 5/2$). For $\eta \ll 1$, one easily finds $\nu_{Q3} \approx 1.5\nu_{Q2} \approx 3\nu_{Q1}$.

In Fig. 2, superconducting transition temperatures T_c and magnetic transition temperatures T_M are plotted against ^{59}Co NQR frequency ν_{Q3} at 10 K [15, 19, 21, 22, 24]. The existence of two superconducting phases were confirmed. We call the two superconducting phases at $\nu_{Q3} < 12.5$ MHz and at $\nu_{Q3} > 12.65$ MHz *α-phase* and *β-phase*, respectively. In the magnetic phase diagram of Fig. 2, the optimal samples of T_c = 4.5–4.8 K in the α-phase are located in 12.1 MHz $< \nu_{Q3} \leq$ 12.4 MHz. The magnetic ordering phase is located between two superconducting phases. Thus, the magnetic instability and the superconductivity occur closely to each other.

It should be emphasized that for the samples with 12.5 MHz $< \nu_{Q3} <$ 12.65 MHz, a possibility of charge-density-wave (CDW) ordering is excluded by the temperature dependence of both ^{59}Co NQR spectra of ν_{Q2} and ν_{Q3} [21, 22]. The CDW ordering at a transition temperature T_M must result in the

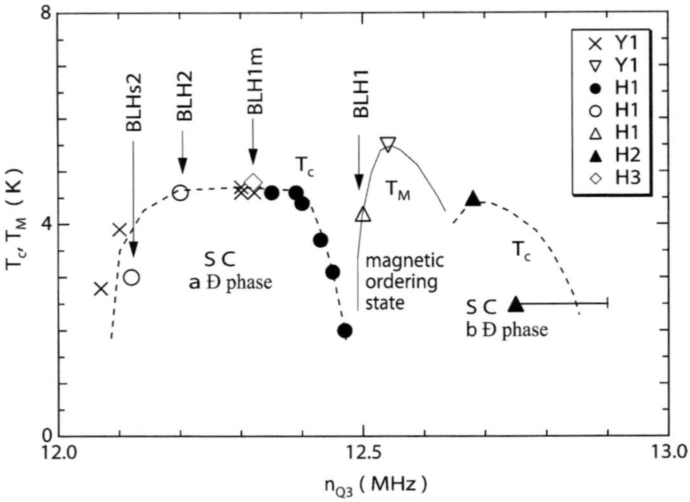

Fig. 2. Magnetic phase diagram of T_c and T_M plotted against ^{59}Co NQR frequency ν_{Q3} at 10 K. We reproduced the data of Y1 from [22], H1 from [15,21], H2 from [19], and H3 from [24]. The *solid* and *dashed curves* are guides for eyes. We call two superconducting phases characterized by $\nu_{Q3} < 12.5$ MHz and by $\nu_{Q3} > 12.6$ MHz *α-phase* and *β-phase*, respectively. The samples denoted by BLH (*bi*layer hydrates) are referred to the NMR experiments

broad NQR line widths of $\Delta\nu_{Q3} \approx 1.5\Delta\nu_{Q2} \approx 3\Delta\nu_{Q1}$. However, $\Delta\nu_{Q2} \gg \Delta\nu_{Q3}$ was observed below T_M [21, 22]. Thus, an internal magnetic field due to magnetic ordering is concluded. Here, since no divergence behavior at T_M was observed in nuclear spin-lattice relaxation rates, this magnetic ordering is unconventional [20, 21]. No wipeout effect below T_M excludes the existence of slow modes due to spin glass transition [21, 22].

It should be noted that the experimental magnetic phase diagram against ν_{Q3} in Fig. 2 is similar to the theoretical one against the thickness of a CoO_2 layer [25, 26]. In the theoretical phase diagram, two superconducting phases are separated through a magnetic ordering phase and result from two competing Fermi surfaces of a_{1g} cylinders and $\acute{e}_g$ hole pockets. Two superconducting phases should have different pairing symmetry [25, 26]. We present the NMR studies of the α-phase below.

3.2 ^{1}H NMR of Oxonium Ion

In the double-layer hydrated superconductors, not only the concentration of Na^+ ions [10] but also the H_3O^+ ions [11–13] play key roles in the occurrence of superconductivity. The existence of the oxonium ions H_3O^+ was evidenced by the observation of a bending mode and stretching modes of H_3O^+ ions

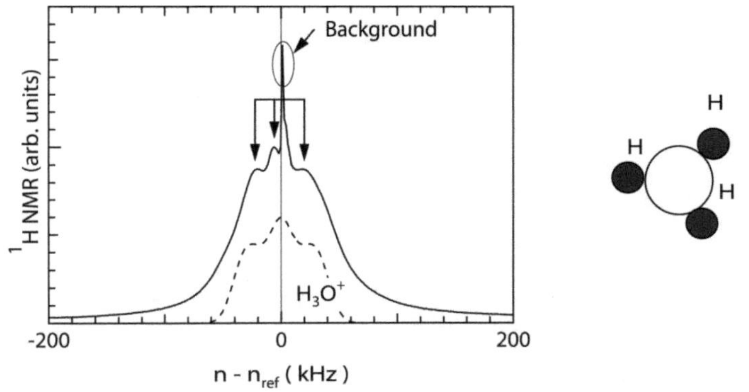

Fig. 3. ^{1}H proton NMR spectrum at 200 MHz for a double-layered hydrated sample at room temperature. The *dashed curve* is a numerical simulation of ^{1}H NMR powder pattern for H_3O^+ illustrated in the *right* figure

using Ramann spectroscopy [11]. NMR is also a powerful technique to detect such a local structure.

Figure 3 shows a high resolution Fourier-transformed ^{1}H NMR spectrum at 200 MHz for a double-layered hydrated sample in the α-phase at room temperature. We observed one peak and two shoulders in the ^{1}H NMR spectrum. For a frozen water molecule, a Pake doublet of two peaks should be observed [27]. In Fig. 3, a numerical simulation for an oxonium ion H_3O^+ with triangular coordination of three protons is illustrated by a dashed curve [27]. Similarity between the multiple structure in the ^{1}H NMR spectrum and the numerical simulation in Fig. 3 indicates the existence of the oxonium ion H_3O^+ in the double-layer hydrated compound. It is a future work to estimate how much H_3O^+ ions and H_2O molecules are involved in each sample.

3.3 *A*-Type Spin Fluctuations via ^{23}Na NMR

Since Na ions are located between CoO_2 layers, it is expected that Na nuclear spins can probe interlayer correlations.

Figure 4a shows the central transition lines ($I_z = \pm 1/2 \leftrightarrow \mp 1/2$) of the Fourier transformed ^{23}Na (nuclear spin $I = 3/2$) NMR spectra for $Na_{0.7}CoO_2$, a double-layer hydrated non-superconducting BLH1 ($T_c <$ 1.8 K) and a superconducting BLH2 ($T_c \approx 4.5$ K), both of which are denoted in Fig. 2, and a dehydrated $Na_{0.35}CoO_2$ [16, 20]. The dehydrated $Na_{0.35}CoO_2$ was obtained by heating a double-layer hydrate at about 250°C. In $Na_{0.7}CoO_2$, the Na NMR lines are affected by large Knight shift and electric quadrupole shift. For the double-layer hydrates BLH1 and BLH2, however, the ^{23}Na NMR lines show small Knight shifts. For the $Na_{0.35}CoO_2$ dehydrated from $Na_{0.35}CoO_2 \cdot yH_2O$, the ^{23}Na NMR lines show large Knight shift once again. The recovery of Knight shift by dehydration indicates that

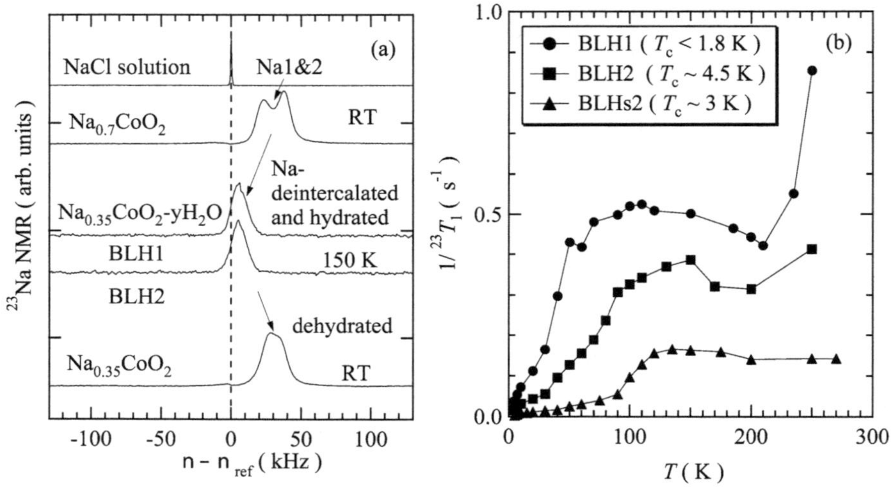

Fig. 4. (**a**) Fourier-transformed ^{23}Na NMR frequency spectra of $Na_{0.7}CoO_2$, double-layer hydrated $Na_{0.35}CoO_2 \cdot yH_2O$ (BLH1 and BLH2 denoted in Fig. 2), and a dehydrated $Na_{0.35}CoO_2$. The *dashed line* is $\nu_{\rm ref} = 84.2875\,{\rm MHz}$ ($\nu_{\rm ref} = {}^{23}\gamma_{\rm n} H$, ${}^{23}\gamma_{\rm n} = 11.262\,{\rm MHz/T}$ and $H \approx 7.484\,{\rm T}$). (**b**) Temperature dependence of ^{23}Na nuclear spin-lattice relaxation rates $1/^{23}T_1$ for double-layer hydrated $Na_{0.35}CoO_2 \cdot yH_2O$ (BLH1, BLH2, and BLHs2 denoted in Fig. 2)

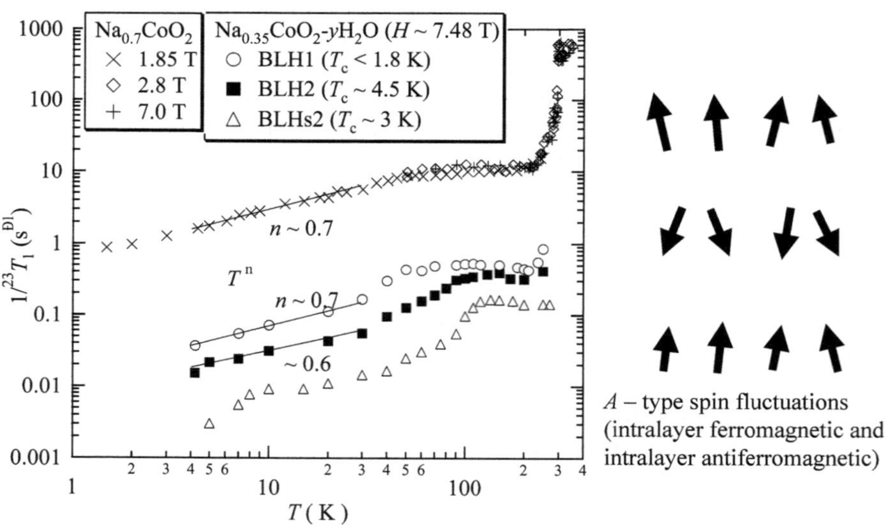

Fig. 5. Log-log plots of ^{23}Na nuclear spin-lattice rate $1/^{23}T_1$ against temperature for $Na_{0.7}CoO_2$ [5,31], double-layer hydrated $Na_{0.35}CoO_2 \cdot yH_2O$ (BLH1, BLH2, and BLHs2 denoted in Fig. 2). The *solid lines* are fits by power laws of T^n. The *right* figure illustrates a snapshot of A-type spin fluctuations

not the Na deficiency but the intercalated water molecule blocks the Co-to-Na hyperfine field.

Figure 4b shows the temperature dependence of the ^{23}Na nuclear spin-lattice relaxation rate $1/^{23}T_1$ for BLH1, BLH2, and BLHs2 ($T_c \approx 3$ K) [16, 20, 28]. $1/^{23}T_1$ largely decreases with duration from BLH1 to BLH2 and then BLHs2. The local field fluctuations probed by $1/^{23}T_1$ [29, 30] are sensitive to the duration effect.

Figure 5 shows the log-log plots of $1/^{23}T_1$ for the parent $Na_{0.7}CoO_2$ which are reproduced from [5, 31] and for the double-layer hydrated BLH1, BLH2, and BLHs2 [16, 20]. At low temperatures in the normal states above T_c, $1/^{23}T_1$'s show the power-law behaviors (solid lines) of T^n with $n \approx 0.7$ and 0.6 for BLH1 and BLH2, respectively. These non-Korringa behaviors indicate that the samples are not conventional Fermi liquid systems. The magnitude of $1/^{23}T_1$ significantly decreases from the non-hydrated to the double-layer hydrated samples.

The temperature dependence of $1/^{23}T_1$ in BLH1 is nearly the same as that in the parent $Na_{0.7}CoO_2$ [20]. Both systems show the power law of $T^{0.7}$ in $1/^{23}T_1$ at $T < 20$ K and the rapid increase at $T > 200$ K. The electron spin dynamics probed at the Na site in the double-layer hydrates is similar to that in $Na_{0.7}CoO_2$. The A-type spin fluctuations, i.e., intra-plane ferromagnetic and inter-plane antiferromagnetic fluctuations, are observed in $Na_{0.7}CoO_2$ by inelastic neutron scattering experiments [3, 4]. The intercalated water molecules may block the inter-plane antiferromagnetic couplings. Nearly ferromagnetic intra-plane spin fluctuations may persist in the double-layer hydrates.

4 ^{59}Co NMR for α-Phase

4.1 Normal State

From ^{59}Co NQR and NMR experiments for an optimal sample of $T_c = 4.8$ K (BLH1m denoted in Fig. 2), we obtained the significant results that the electric field gradient tensor $V_{\alpha\beta}$ at the ^{59}Co nuclear site is asymmetric ($\eta \approx 0.2$), that the in-plane ^{59}Co Knight shift has two components (XY-anisotropic $K_x \neq K_y$ or magnetic disproportionation), and that the ^{59}Co nuclear spin-lattice relaxation rate $1/^{59}T_1$ is different at the two signals [24].

Figure 6a shows the temperature dependence of ^{59}Co nuclear quadrupole resonance frequency ν_Q and the asymmetry parameter η [24]. The temperature dependence of ν_Q in Fig. 6a is widely observed in many materials. The finite asymmetry parameter $\eta \approx 0.2$ suggests an unique underlying electronic wave funtion and/or charge distribution at the Co site.

Figure 6b shows the temperature dependence of the central transition ($I_z = \pm 1/2 \leftrightarrow \mp 1/2$) of ^{59}Co NMR spectra at H(=7.4847 T) //ab-plane [24]. For the ^{59}Co NMR experiments, we prepared the powder sample $Na_{0.35}CoO_2 \cdot yH_2O$ oriented by a magnetic field of $H \approx 7.5$ T in Fluorinert

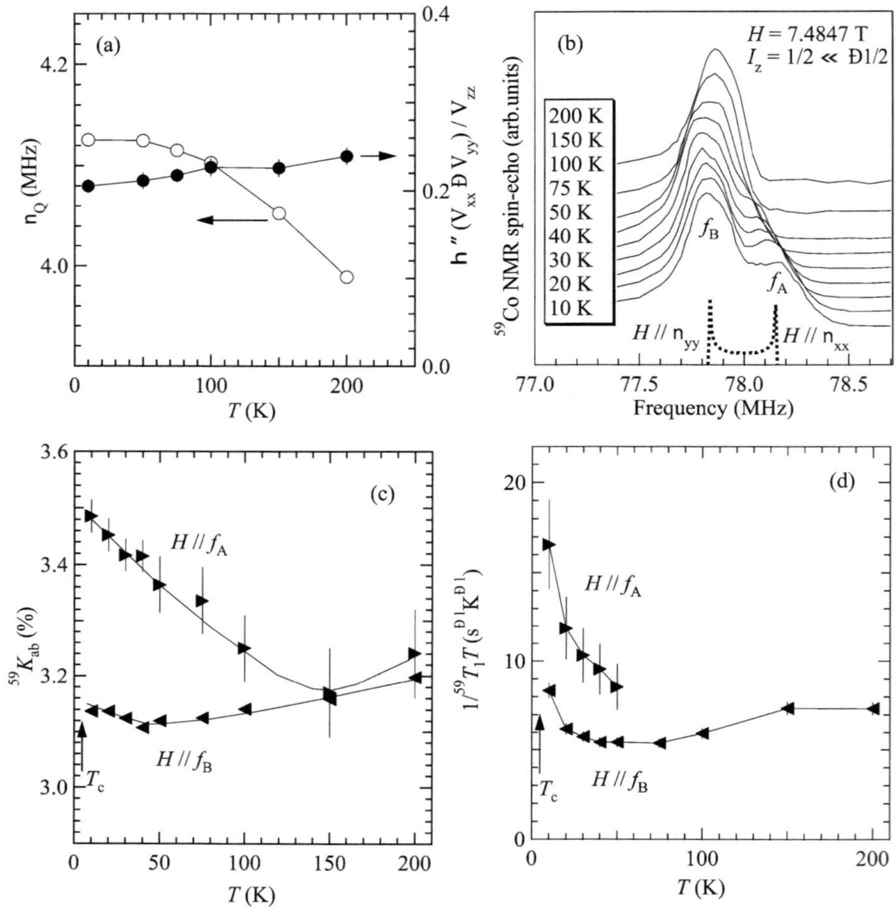

Fig. 6. ^{59}Co NQR and NMR for an optimal sample (BLH1m in Fig. 2) of $T_c = 4.8\,K$. (**a**) ^{59}Co nuclear quadrupole resonance frequency ν_Q and the asymmetry parameter η. (**b**) The central transition lines ($I_z = 1/2 \leftrightarrow -1/2$) of ^{59}Co NMR spectra at $H//ab$ plane and a numerical simulation (*dotted curve*) of two dimensional powder pattern of the central transition including anisotropic Knight shift of $K_x > K_y$. (**c**) ^{59}Co Knight shifts K_A and K_B at f_A and f_B. (**d**) ^{59}Co nuclear spin-lattice relaxation rates divided by temperature $1/T_1T$ at f_A and f_B

FC70 (melting point of 248 K). Stycast 1266 was the worst for the hydrated samples. Hexane was much better than Stycast 1266 and actually applied to NMR experiments [32] but hard to handle. Fluorinert was the best. The *ab*-planes of powder grains are aligned to the external magnetic field. Thus, the observed NMR spectrum should be a two dimensional powder pattern.

The dotted curve in Fig. 6b is a numerical simulation of two dimensional powder pattern including anisotropic Knight shift of $K_x > K_y$. Here we define the Knight shift K_x for the line $H//\nu_{xx}$ and K_y for the line $H//\nu_{yy}$ of the central transition ($I_z = \pm 1/2 \leftrightarrow \mp 1/2$). The full-swept ^{59}Co NMR spectra and their field dependences were well reproduced from two dimensional powder pattern due to quadrupole shifts with second order perturbation and anisotropic Knight shifts [32]. In principle, the asymmetric quadrupole shift with $\eta \approx 0.2$ yields two peaks in the central transition but could not reproduce the actual width between the two resonance peaks (f_{A} and f_{B}) in Fig. 6b. From the magnetic field dependence of the two peaks, two signals have nearly the same quadruple shifts [32]. Since twice magnetic field yields twice split width between the two resonance peaks in Fig. 6b, we should introduce at least two components of Knight shift [32]. Since imperfect orientation of powder grains easily suppresses the singularity at $H//\nu_{xx}$, then the disagreement of the intensity ratio of f_{A} to f_{B} in the NMR spectrum may not be significant. However, if the intensity difference of f_{A} and f_{B} is intrinsic, one must introduce two magnetic Co sites with nearly the same quadruple shifts. The charge disproportionation was observed in the parent $Na_{0.7}CoO_2$ [6–9]. Then, magnetic disproportionation may occur in the double-layer hydrates in that case.

After the second order correction of quadrupole shifts based on ν_Q and η in Fig. 6a, we estimated the respective Knight shifts K_{A} and K_{B} at the lines f_{A} and f_{B} in Fig. 6b [24]. Figure 6c shows the temperature dependences of the Knight shifts K_{A} and K_{B} [24]. The Knight shift K_{A} shows a Curie–Weiss-type upturn below 50 K, being similar to a low temperature bulk magnetic susceptibility [32]. Thus, a Curie–Weiss behavior in the magnetic susceptibility below 50 K is intrinsic.

Figure 6d shows the temperature dependence of $1/^{59}T_1T$ at f_{A} and f_{B} [24]. The difference in $1/^{59}T_1T$ indicates XY-anisotropy or disproportionation in the Co electron spin fluctuations.

4.2 Superconducting State

Figure 7a shows zero-field ^{59}Co nuclear spin-lattice relaxation rates $1/(^{59}T_1)_{\mathrm{NQR}}$ for BLH1, the sample of $T_{\mathrm{c}} = 4.6$ K [21] and the sample of $T_{\mathrm{c}} = 4.7$ K [33]. No Korringa behavior in $1/(^{59}T_1)_{\mathrm{NQR}}$ just above T_{c} and no coherence peak just below T_{c} suggest non-Fermi liquid and unconventional superconductivity. The power law behavior of $1/(^{59}T_1)_{\mathrm{NQR}}$ below T_{c} indicates the existence of line nodes on a superconducting gap parameter. In a triangu-

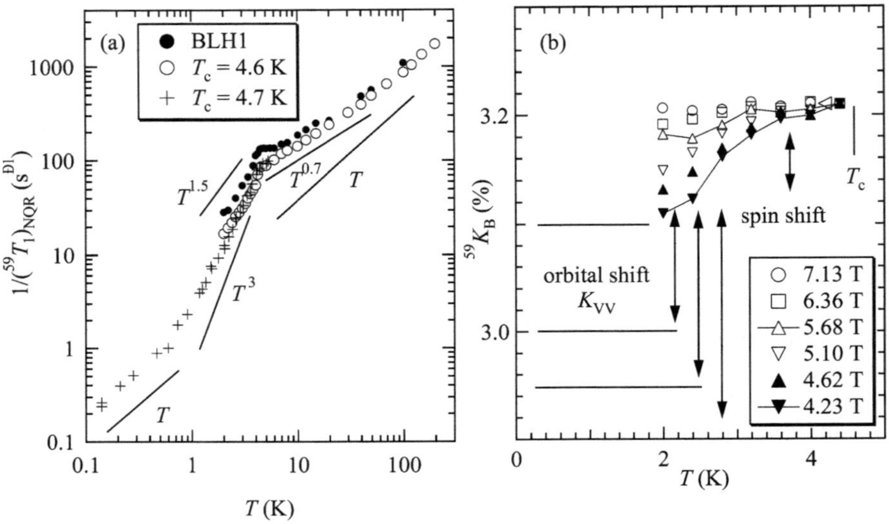

Fig. 7. (**a**) Zero-field ^{59}Co nuclear spin-lattice relaxation rates $1/(^{59}T_1)_{NQR}$ for BLH1 and BLH sample of $T_c = 4.6$ K [21] and the sample of $T_c = 4.7$ K [33]. (**b**) Magnetic field dependence of the superconducting ^{59}Co Knight shift K_B for the optimal sample of $T_c = 4.7$ K reproduced from [32]. The horizontal lines indicate the orbital shifts $K_{VV} = 2.9$, 2.95, 3.0, and 3.1 % reported so far. The *arrows* indicate the respective spin shifts K_{spin}

lar lattice, such a line-nodal order parameter is expected for d_{xy}-wave, p-wave and f-wave pairing [34–37] and extended s-wave inter-band pairing [38].

Figure 7b shows the magnetic field dependence of the superconducting ^{59}Co Knight shift K_B for the optimal sample with $T_c = 4.7$ K [32]. The horizontal bars indicate the reported orbital shifts $K_{VV} = 2.7$ (out of the figure scale) [39], 2.9 [9], 2.95 [40], 3.0 [24], and 3.1 % [32]. The Knight shift K is expressed by the sum of the spin shift K_{spin} and the Van Vleck orbital shift K_{VV}. Then, the arrows indicate the respective spin shifts K_{spin}. If $K_{VV} = 2.9$ %, the finite K_{spin} indicates a spin-triplet pairing. If $K_{VV} = 3.1$ %, the diminished K_{spin} indicates a spin-singlet pairing. Thus, the choice of K_{VV} changes the conclusion for the issue whether the spin singlet or triplet is realized.

The existing data on the impurity Ga- and Ir-substituion effects [41] support non-s-wave pairing. Although any impurity substitution effects on T_c as Anderson localization effect [42] could not serve as the tests of identification of any pairing symmetry, weak impurity potential scattering for the line-nodal pairing could account for the observed weak suppression of T_c.

5 Conclusion

Double-layer hydrated cobalt oxides $Na_xCoO_2 \cdot yH_2O$ were found to have two superconducting phases labeled by ^{59}Co NQR frequency ν_{Q3}. Non-superconducting magnetic phase is located between two phases in the ν_{Q3} classification. The peculiar electronic states, magnetic disproportionation or XY anisotropy of ^{59}Co local spin susceptibility, was observed for the *α-phase*. Although the superconducting gap parameter with line nodes seems to be established, there does not seem to be a robust answer to a question which is realized, a spin singlet or triplet state.

Acknowledgements

We thank Y. Yanase, M. Mochizuki, M. Ogata, J. L. Gavilano, J. Haase, T. Imai, and M. Takigawa for their fruitful discussions. This study was supported by a Grant-in-Aid for Science Research on Priority Area, "Invention of anomalous quantum materials" from the Ministry of Education, Science, Sports and Culture of Japan (Grant No. 16076210).

References

[1] I. Terasaki, Y. Sasago, K. Uchinokura, Phys. Rev. B **56**, 12685(R) (1997)
[2] K. Takada, H. Sakurai, E. Takayama-Muromachi, F. Izumi, R. A. Dilanian, T. Sasaki: Nature **422**, 53 (2003)
[3] A. T. Boothroyd, R. Coldea, D. A. Tennant, D. Prabhakaran, L. M. Helme, C. D. Frost: Phys. Rev. Lett. **92**, 197201 (2004)
[4] L. M. Helme, A. T. Boothroyd, R. Coldea, D. Prabhakaran, D. A. Tennant, A. Hiess, J. Kulda: Phys. Rev. Lett. **94**, 157206 (2005)
[5] J. L. Gavilano, D. Rau, B. Pedrini, J. Hinderer, H. R. Ott, S. M. Kazakov, J. Karpinski: Phys. Rev. B **69**, 100404(R) (2004)
[6] J. L. Gavilano, B. Pedrini, K. Magishi, J. Hinderer, M. Weller, H. R. Ott, S. M. Kazakov, J. Karpinski: Phys. Rev. B **74**, 064410 (2006)
[7] F. L. Ning, T. Imai, B. W. Statt, F. C. Chou: Phys. Rev. Lett. **93**, 237201 (2004)
[8] I. R. Mukhamedshin, H. Alloul, G. Collin, N. Blanchard: Phys. Rev. Lett. **93**, 167601 (2004)
[9] I. R. Mukhamedshin, H. Alloul, G. Collin, N. Blanchard: Phys. Rev. Lett. **94**, 247602 (2005)
[10] M. L. Foo, Y. Wang, S. Watauchi, H. W. Zandbergen, T. He, J. Cava, N. P. Ong: Phys. Rev. Lett. **92**, 247001 (2004)
[11] K. Takada, K. Fukuda, M. Osada, I. Nakai, F. Izumi, R. A. Dilanian, K. Kato, M. Takata, H. Sakurai, E. Takayama-Muromachi, T. Sasaki: J. Mater. Chem. **14**, 1448 (2004)
[12] H. Sakurai, K. Takada, T. Sasaki, E. Takayama-Muromachi: J. Phys. Soc. Jpn. **74**, 2909 (2005)

[13] C. J. Milne, D. N. Argyriou, A. Chemseddine, N. Aliouane, J. Veira, S. Landsgesell, D. Alber: Phys. Rev. Lett. **93**, 247007 (2004)
[14] H. W. Zandbergen, M. Foo, Q. Xu, V. Kumar, R. J. Cava: Phys. Rev. B **70**, 024101 (2004)
[15] H. Ohta, C. Michioka, Y. Itoh, K. Yoshimura: J. Phys. Soc. Jpn. **74**, 3150 (2005)
[16] H. Ohta, C. Michioka, Y. Itoh, K. Yoshimura, H. Sakurai, E. Takayama-Muromachi, K. Takada, T. Sasaki: Physica B **378**, 859 (2006)
[17] H. Ohta, C. Michioka, Y. Itoh, K. Yoshimura: J. Mag. Mag. Mater. **310**, e141 (2007)
[18] P. W. Barnes, M. Avdeev, J. D. Jorgensen, D. G. Hinks, H. Claus, S. Short: Phys. Rev. B **72**, 134515 (2005)
[19] H. Ohta, C. Michioka, Y. Itoh, K. Yoshimura: unpublished works
[20] H. Ohta, Y. Itoh, C. Michioka, K. Yoshimura: Physica C **445–448**, 69 (2006)
[21] C. Michioka, H. Ohta, Y. Itoh, M. Kato, K. Yoshimura: J. Phys. Soc. Jpn. **75**, 063701 (2006)
[22] Y. Ihara, K. Ishida, C. Michioka, M. Kato, K. Yoshimura, K. Takada, T. Sasaki, H. Sakurai, E. Takayama-Muromachi: J. Phys. Soc. Jpn. **74**, 867 (2005)
[23] Y. Ihara, H. Takeya, K. Ishida, H. Ikeda, C. Michioka, K. Yoshimura, K. Takada, T. Sasaki, H. Sakurai, E. Takayama-Muromachi: J. Phys. Soc. Jpn. **75**, 124714 (2006)
[24] C. Michioka, Y. Itoh, H. Ohta, M. Kato, K. Yoshimura: cond-mat/0607368
[25] M. Mochizuki, M. Ogata: J. Phys. Soc. Jpn. **75**, 113703 (2007)
[26] M. Mochizuki, M. Ogata: J. Phys. Soc. Jpn. **76**, 013704 (2007)
[27] A. Abragam: *The Principles of Nuclear Magnetism* (Oxford University Press, Oxford 1986)
[28] K. Yoshimura, H. Ohta, C. Michioka, Y. Itoh: J. Mag. Mag. Mater. **310**, 693 (2007).
[29] T. Moriya: Prog. Theor. Phys. **16**, 23 (1956)
[30] T. Moriya: J. Phys. Soc. Jpn. **18**, 516 (1963)
[31] Y. Ihara, K. Ishida, C. Michioka, M. Kato, K. Yoshimura, H. Sakurai, E. Takayama-Muromachi: J. Phys. Soc. Jpn. **73**, 2963 (2004)
[32] M. Kato, C. Michioka, T. Waki, Y. Itoh, K. Yoshimura, K. Ishida, K. Takada, H. Sakurai, E. Takayama-Muromachi, T. Sasaki: J. Phys.: Condens. Matter **18**, 669 (2006)
[33] K. Ishida, Y. Ihara, Y. Maeno, C. Michioka, M. Kato, K. Yoshimura, K. Takada, T. Sasaki, H. Sakurai, E. Takayama-Muromachi: J. Phys. Soc. Jpn. **72**, 3041 (2003)
[34] M. Vojta, E. Dagotto: Phy. Rev. B **59**, 713 (R) (1999)
[35] Y. Yanase, M. Mochizuki, M. Ogata: J. Phys. Soc. Jpn. **74**, 430 (2005)
[36] Y. Yanase, M. Mochizuki, M. Ogata: J. Phys. Soc. Jpn. **74**, 2568 (2005)
[37] Y. Yanase, M. Mochizuki, M. Ogata: J. Phys. Soc. Jpn. **74**, 3351 (2005)
[38] K. Kuroki, S. Onari, Y. Tanaka, R. Arita, T. Nojima: Phys. Rev. B **73**, 184503 (2006)
[39] Y. Kobayashi, H. Watanabe, M. Yokoi, T. Moyoshi, Y. Mori, M. Sato: J. Phys. Soc. Jpn. **74**, 1800 (2005)
[40] G-q. Zheng, K. Matano, D. P. Chen, C. T. Lin: Phys. Rev. B **73**, 180503(R) (2006)

[41] M. Yokoi, H. Watanabe, Y. Mori, T. Moyoshi, Y. Kobayashi, M. Sato: J. Phys. Soc. Jpn. **73**, 1297 (2004)
[42] Z. P. Guo, Y. G. Zhao, W. Y. Zhang, L. Cui, S. M. Guo, L. B. Luo: J. Phys.: Condens. Matter **18**, 4381 (2006)

Index

MIX
Papier aus verantwortungsvollen Quellen
Paper from responsible sources
FSC® C105338

If you have any concerns about our products,
you can contact us on
ProductSafety@springernature.com

In case Publisher is established outside the EU,
the EU authorized representative is:
Springer Nature Customer Service Center GmbH
Europaplatz 3, 69115 Heidelberg, Germany

Printed by Libri Plureos GmbH
in Hamburg, Germany